S0-CVL-637

PROCEEDINGS OF THE INTERNATIONAL SYMPOSIUM ON FRONTIERS IN SCIENCE

ON THE OCCASION OF THE 65th BIRTHDAY OF PROFESSOR HANS FRAUENFELDER

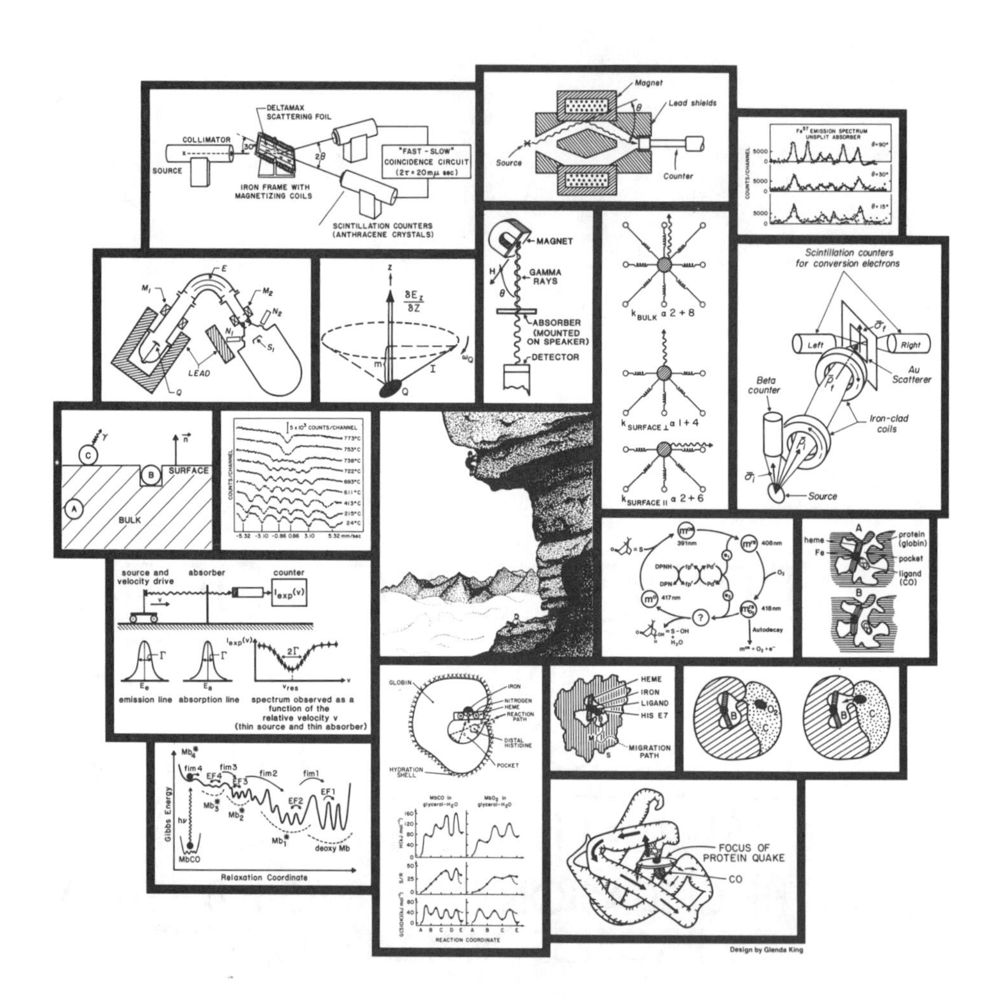

Design by Glenda King

AIP CONFERENCE PROCEEDINGS 180

RITA G. LERNER
SERIES EDITOR

PROCEEDINGS OF THE INTERNATIONAL SYMPOSIUM ON FRONTIERS IN SCIENCE

ON THE OCCASION OF THE 65th BIRTHDAY OF PROFESSOR HANS FRAUENFELDER

URBANA, IL 1987

EDITORS:

SHIRLEY S. CHAN
RUTGERS UNIVERSITY

PETER G. DEBRUNNER
UNIVERSITY OF ILLINOIS

AMERICAN INSTITUTE OF PHYSICS NEW YORK 1988

L.C. Catalog Card No. 88-83526
ISBN 0-88318-380-3
DOE CONF 8705337

Printed in the United States of America.

Contents

Portrait of Professor Hans Frauenfelder

Preface

This volume contains the Proceedings of the International Symposium on Frontiers in Science that took place at the University of Illinois in Urbana on May 1–3, 1987. Organized in honor of Professor Hans Frauenfelder on the occasion of his 65th birthday, the symposium featured eminent speakers from many different fields of science; it was attended by over 300 visitors, friends, and students who gathered for the talks and discussions at Loomis Laboratory of Physics.

The wide range of topics covered by the speakers reflects the remarkably broad spectrum of Hans Frauenfelder's scientific contributions, which began with nuclear physics, touched on surface and condensed matter physics, moved on to weak interactions, parity and time reversal, then the Mössbauer effect, to focus eventually on biomolecular physics, in particular protein dynamics. The collage of figures on the face page lifted from Frauenfelder publications, highlights the breadth of his interests; it was the centerpiece of a Symposium poster, and it is left to the reader to trace the pieces of the puzzle.

Hans Frauenfelder began his career in physics at the Federal Institute of Technology in Zurich, Switzerland, where he finished his Ph.D. under Paul Scherrer in 1950. The Zurich Institute, though small, was an exciting and widely known center, as both Pauli and Scherrer, chairing theoretical and experimental physics, respectively, attracted many brilliant visitors and taught generations of promising young scientists. Life-long friendships developed with Pauli, Scherrer, and many other, now prominent physicists.

In Zurich Hans set up a laboratory for angular correlation of nuclear radiation. Experimental and theoretical advances followed in rapid succession, and the first pioneering applications of the technique to problems of nuclear, condensed matter, and surface physics were made. In 1952 he moved to the University of Illinois at Urbana-Champaign and continued along the same lines. Later, when word was out that weak interactions might not conserve parity, Hans was among the first to demonstrate and measure the effect in beta decay. Tests of time reversal symmetry followed thereafter, done both at low and high energies.

In 1957 Rudolf Mössbauer's discovery of nuclear gamma resonance opened up a new field drawing on nuclear and condensed matter physics. It caught Han's interest immediately, as Harry Lipkin describes in his account of the early history of the Mössbauer Effect at the University of Illinois. Collaborations ensued with Darragh Nagle at Los Alamos, later with Harry Drickamer in Urbana and I.C. Gunsalus. The latter contact was to be particularly fateful, since it exposed Hans to biomolecular problems, which captured his imagination for the last fifteen years. His stimulating contributions in this field are referred to in several articles of the Proceedings; they also formed the basis for the Discussions held on the last day of the Symposium, which appears in transcription towards the end of this book.

It is a pleasure to acknowledge the numerous contributions, financial and otherwise, that made this Symposium possible. Thanks go to the University of Illinois Department of Physics and its staff, the School of Chemical Sciences, the Institute for Advanced Study; to the American Institute of Physics for publication of the Proceedings; and to all the speakers. Edgar Lüscher conceived the idea of the symposium and laid the ground work; Ralph Simmons, Head of the Department, embraced the concept and invited the speakers officially, while an Organizing Committee consisting of Peter Debrunner, I.C. Gunsalus, David Pines and Peter Wolynes arranged for the details.

Two other people deserve special mention. Mrs. Verena Frauenfelder, the spouse and faithful supporter of Hans for 38 years, handled most questions of taste, and her personal touch was apparent in all the events. Mrs. Mary Ostendorf, who has handled Hans' correspondence for the last 15 years, did double duty for the Symposium, taking care of all the paperwork and many organizational matters in a truly outstanding fashion.

Finally, this volume could not have been completed in the present uniform typesetting without the help of Ms. Lorraine Nelson, Physics Department, Princeton University, with her unusual mastery of the TEX word processor and of scientific symbols, equations and tables.

Shirley S. Chan, Editor

Peter G. Debrunner, Co-editor

October 1988

Hans Frauenfelder and Rudolf Mössbauer

David Pines, I. C. Gunsalus and Ernest Henley

Ernest Henley and Peter Debrunner; Last Minute Arrangements

Edgar Lüscher and Ernest Henley

Manfred Eigen, John Bardeen and Harry Lustig

Manfred Eigen and Harry Drickamer in deep thoughts

Relaxing Moments: Charlie Slichter, Hans and Verena Frauenfelder

Edgar Lüscher and Verena Frauenfelder

Two Old Friends: Vitalii Goldanskii and Hans Frauenfelder

Hans Frauenfelder Chairing the Discussions

Hans Frauenfelder and Peter Wolynes working together

Hans' Other Passion

Frauenfelder Family: Anne, Katterli, Hans, Vreneli, Uli, Adriano;
Loredana taking picture

Three Generations of Frauenfelder: Uli, Adriano, Hans

LEARNING WITH HANS FROM IRON AND OXYGEN

I. C. Gunsalus
School of Chemical Sciences
University of Illinois at Urbana-Champaign
Urbana, IL 61801
and
International Centre for Genetic Engineering and
Biotechnology, Padriciano 99, 34012 Trieste, Italy

ABSTRACT

To present a paper in tribute to Hans Frauenfelder on the occasion of his 65th birthday is a rare privilege. His work on iron protein reaction centers has changed our perception of protein dynamics and our view of the essential role of oxygen in biological systems. Most importantly, his continued efforts to increase the experimental precision and to develop new models have inspired a generation of multidisciplinary investigators and have thus enriched the chemistry and genetics of modern biology. I beg your indulgence for the following account of the gradual unravelling of an enzyme system in our labs, with strong participation from Physics and friends elsewhere, an account of peaks, valleys and a few crevasses encountered in a quest that is still in progress.

INTRODUCTION

The system of interest consists of three proteins that act in concert to catalyze the transformation of hydrocarbons, a rather inert material. The three proteins, found in the microorganism *Pseudomonas putida*, represent the most thoroughly understood example of the large and diverse family of the so-called P450 enzymes. Some of the materials formed by these enzymes are essential to human health, others to the recycling of organic matter to useful intermediates and eventually to carbon dioxide.

Hydrocarbon transformation requires an enzyme center in a highly reactive state. Nature solved the problem by juxtaposition, in a protein, of binding sites for a molecule of hydrocarbon and for activated oxygen. The oxygen derives from atmospheric O_2, which binds to the iron of a heme group embedded in the protein; the heme group is the familiar red pigment of muscle and blood in the O_2 carrier molecules myoglobin and hemoglobin. The highly activated state of the enzyme is necessarily unstable and short-lived. With a specific hydrocarbon as substrate, the enzyme-directed process leads predominantly to a single product. Other decay routes, however, may lead to alternate products or may inactivate the enzyme or nearby proteins in processes related to ageing. Free-radical reactions, in particular, are fast and non-specific. Like fire supported by oxygen, the decay process lacks control.

Of the three proteins mentioned above only one binds substrates, i.e., hydrocarbon and O_2. This protein was first noticed on the basis of a characteristic absorption band of the heme group at 450 nm and is since called P450. Only later was it realized that a bond of the heme iron to the cysteine sulfur of the amino acid chain was responsible for the unique electronic properties of P450.

The second protein, abbreviated Rd for Redoxin, functions in the transfer of energy or reducing power and as a P450 modulator. Rd is a small iron protein consisting of roughly 100 amino acid residues; it contains a unique reaction center made up of two iron atoms, two bridging sulfide ions and four cysteinyl sulfurs from the polypeptide chain.

The third protein is a vitamin B2 flavoprotein reductase; it initiates energy transfer from the reducing power of cellular intermediates via Rd to P450.

Selected references will be cited to document the interdisciplinary nature of the work done here and the central role of two specific systems as models for the larger family of P450's. Papers and reviews with Hans Frauenfelder, Peter Debrunner, Steve Sligar and collaborators, others with Pierre Douzou and colleagues, have given a basic standard to the P450 field. The biological importance of the reactions catalyzed by P450's and the large, rapidly increasing literature on this topic lie outside the scope of this contribution; they are conveniently reached through monographs and journals.

RESULTS

In the enzymatic reaction each of the three proteins introduced above moves through a series of states with defined dynamics to return to the initial or ground state. The proteins recognize one another and bind together reversibly. The binding affinities or equilibrium coefficients vary with the protein state. The changes act as switches to allow the flow of energy from the cell supply to the activated complex in several single steps.

The cycle of protein states is controlled by the presence of substrates, a hydrocarbon and atmospheric oxygen. The substrates diffuse to their binding sites in the P450 protein and undergo a series of transitions to form new products. The overall process, termed oxygenation, depends on the supply of energy from a cellular reducing agent. The oxygen and the reducing agent are common to all P450 systems, whereas the hydrocarbon and the product are specific. In each enzymatic cycle only half of the dioxygen molecule, i.e., one atom of oxygen, reaches the product, while the second atom emerges in a molecule of water. As a class of enzymes the P450 systems are therefore termed monoxygenases.

Three figures will serve to introduce the characteristics of the P450 monoxygenases, the mosaic of multidisciplinary growth and two decades of learning. Figure 1 expresses the features common to the organization of the active centers in the two known classes of P450 systems; at left, the promiscuous liver microsomal type; at right, the more selective mitochondrial/microbial type. We shall examine the latter with two iron protein centers in most detail. The redoxin, Rd, a small protein with an iron-sulfur active center, reacts with P450 in a redox/effector role. Rd, in turn, is coupled to the redox energy supply of the cell through the reductase, a flavine-containing protein. The flavoprotein, although specific to the system, is less unique than the two iron centers and will therefore be discussed in less detail. Recently, a student in the Sligar's

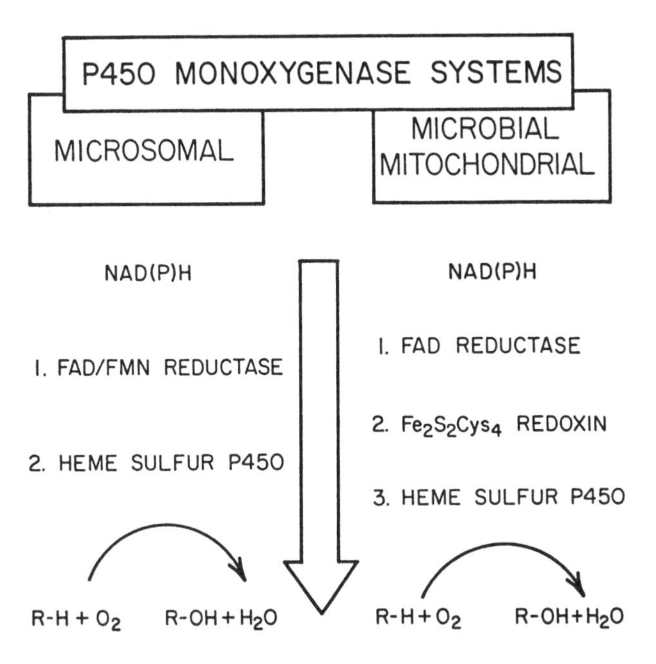

Figure 1. Organization, composition and enzymatic reaction of the two families of P450 monoxygenases.

group produced this natively scarce protein in quantity by gene modification, thus more data may soon become available.

The mitochondria are energy-generating cell inclusion bodies that contain a variety of essential P450 synthetic systems of the three-protein type. They function, for example, in prostaglandin and regulatory steroid synthetic pathways. In microorganisms similar systems catalyze the first steps in the oxidation of numerous natural and synthetic hydrocarbon structures. Both the mitochondrial and microbial P450 enzymes are highly stereo- and carbon-group-specific. The microsomal proteins, in contrast, are much less selective and remove a broad array of foreign substances from the circulation. Other, less well studied P450's of similar function are found in the respiratory tract, etc. The reaction general to all P450 systems is indicated in the lower part of Figure 1 as the incorporation of an oxygen into a methylene group of a hydrocarbon to form an alcohol.

Figures 2 and 3 suggest graphically the interactions of the groups in Physics and Biochemistry. Students and faculty worked together, and informal or more structured, but still open, interactions developed. Problems and concepts were discussed in seminars and microcolloquia, and people as well as equipment moved freely among laboratories.

Figure 2 depicts the iron reporter group, Fe, which stands out prominently in FrauenFelder, with Debrunner evident at all levels in the web of collaborators. My role was to ask questions, to formulate genetic systems, select and isolate enzymes, native or labelled. The students and postdocs in Loomis Lab of Physics and ours in Roger Adams had different primary disciplines and laboratory skills, yet they collaborated most effectively as they helped, inspired and checked each other. The montage, prepared by perusing bound reprint volumes, includes, I hope, all the players. The montage attemps to trace actual scenarios and interactions; to the participants each name means more than science, recalling memories, small and large tales. Greetings to all, who learned together, especially to Hans for many interesting ideas, conversations, lunches, problems, students and colleagues.

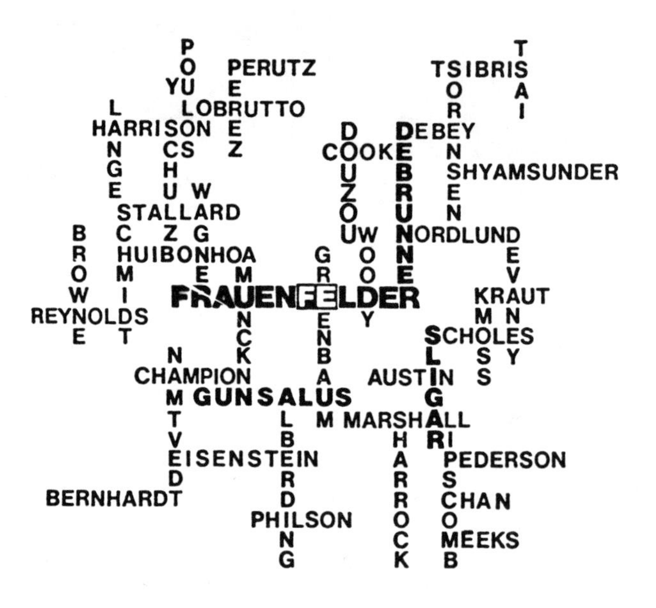

Figure 2. Participants in the monooxygenase collaboration.

Figure 3 provides a rough chronology in a ladder of interactions between biochemists on one hand and physicists on the other. In the early studies microorganisms were exposed to different substrates in search of new processes and enzyme centers. Bill Bradshaw enriched and isolated such an organism, *Pseudomonas putida*, Ed Conrad isolated reaction intermediates in the fermentation broth, and E. J. Corey determined the structures of these intermediates[1]. Cell-free P450 was first prepared by Hedegaard[2], while Dave Cushman isolated and purified redoxin[3], opening the way for EPR studies with Helmut Beinert and Bill Orme-Johnson[4], ^{57}Fe replacement by John Tsibris and Randy Tsai[5], and Mössbauer measurements by Roger Cooke[6]. The iron-sulfur center of the redoxin was recognized to have a unique and novel structure[7], and the diffuse concept of "non-heme iron" was replaced. The mixture of a red and yellow protein described in Bimal Ganguli's thesis was resolved into pure P450$_{CAM}$ and flavoprotein by Dr. Katagiri[8], a truely rare professional on a six-month sabbatical from Kanazawa to Urbana.

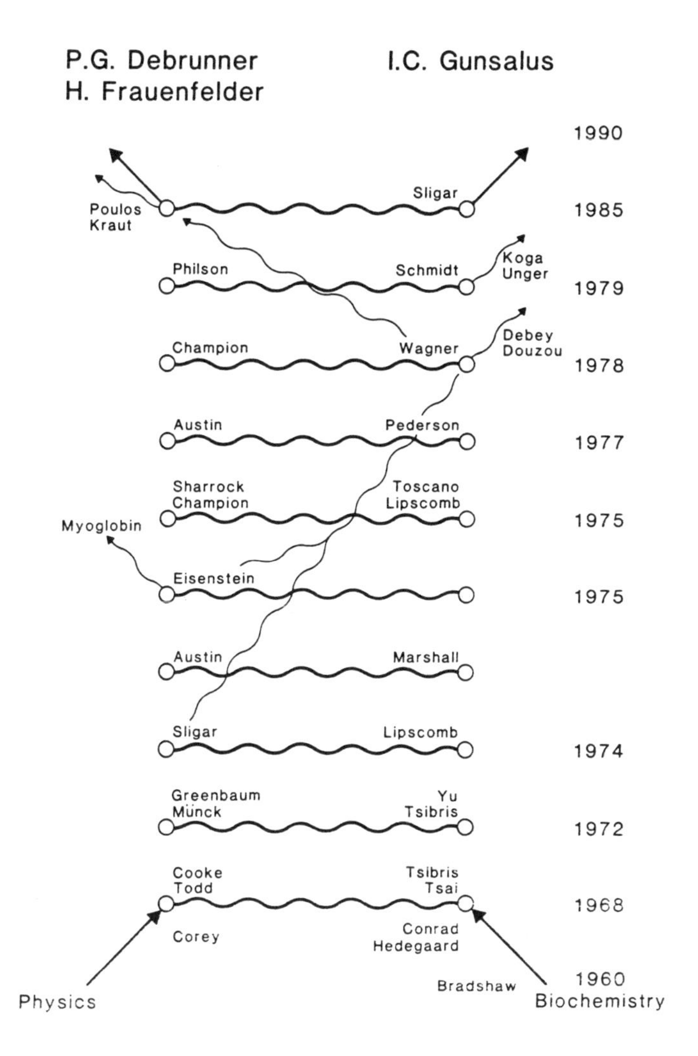

Figure 3. Evolution of the Biochemistry-Physics collaboration on P450 monozygenase.

Sligar and Lipscomb developed the physics, chemistry and enzymology of the system, and much was learned about the structure, the dynamics, the equilibria and the energetics of Rd and P450[9]. Flash photolysis of carbonmonoxy-P450 at helium temperature started with Marshall and was polished by Austin and Laura Eisenstein[10]. Oxy-P450 was first studied optically by Laura with Pierre Douzou in Paris[11] and by Sharrock using the Mössbauer effect[12]. G. C. Wagner succeeded in crystallizing two new forms of P450[13] in addition to the form obtained by Yu ten years earlier[14]. The new crystals were suitable for x-ray diffraction, and Poulos determined a structural model of the substrate

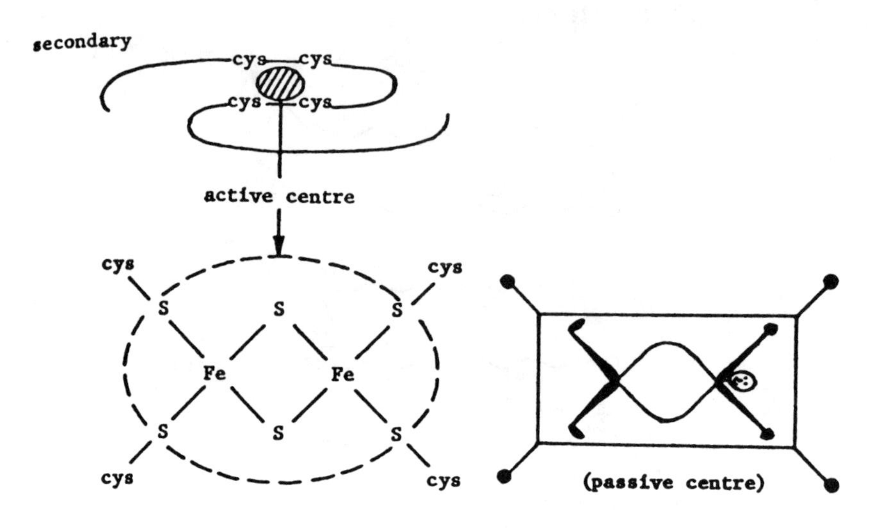

Figure 4A. Schematic diagram of the redoxin active site[17].

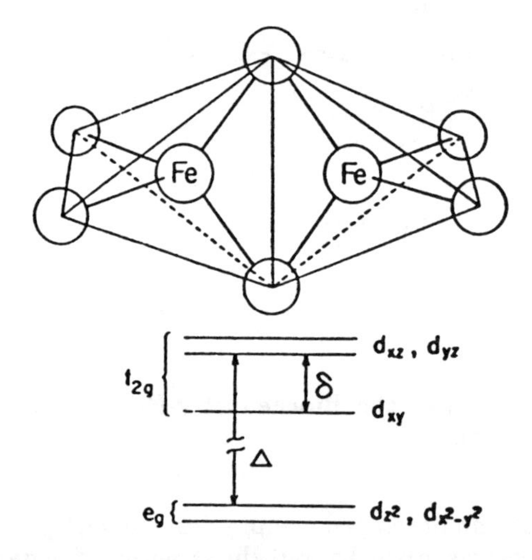

Figure 4B. The Quasi-tetrahedral coordination of the two iron atoms in the active center and the corresponding energy levels of each iron[6].

complex of $P450_{CAM}$ at 2.6 Å resolution[15] and later also of the substrate-free enzyme[16].

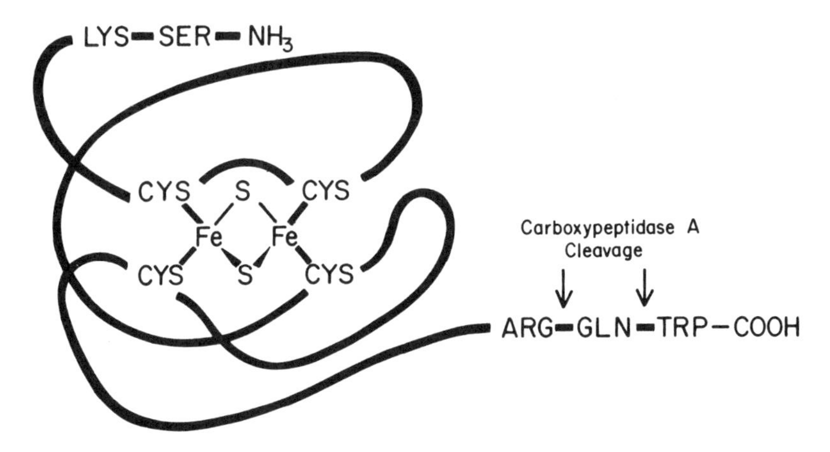

Figure 4C. Carboxypeptidase-A cleavage sites at the C-terminus of Rd[9].

I. REDOXIN

Figures 4A-4C illustrate the salient properties of Rd. The term used originally, putidaredoxin, was changed to Rd_{CAM}, Rd_{LIN}, etc., following the recognition of several similar, but distinct iron-sulfur redox proteins among the P450 systems of *P. putida* that are specific for different substrates.

Figure 4A[17] indicates the active site of Rd_{CAM} with specifically placed cysteine residues, abbreviated as Cys, and the folded primary sequence to form the redox center, $Fe_2S_2^*Cys_4$, where S^* represents an inorganic sulfide ion. The two iron atoms were found to be non-equivalent as only one participated in the Fe(III)/Fe(II) redox reaction, transferring one electron at a time from the flavoprotein to P450. Thus, two separate single-electron reactions occur as will be discussed later.

The reversible unfolding of Rd with release/re-incorporation of $Fe_2S_2^*$ permitted the substitution of ^{33}S and $^{77}Se/^{80}Se$ for ^{32}S and of ^{57}Fe for ^{56}Fe. EPR[4,5] and Mössbauer spectra[7] of the substituted proteins unequivocally proved the composition, $Fe_2S_2^*$, of the active center, revealed spin coupling among the irons and many structural details.

No matter how simply we express new data and our presumed understanding, Hans always finds an added depth, or a simpler analogy as here in Figure 4A for the resting, passive state of Rd[17].

Figure 4B illustrates the distorted tetrahedral iron-sulfur coordination in Rd and the corresponding energy level diagram as suggested in the first Cooke paper[6]. This model is compatible with the EPR spectra taken at Wisconsin[4,5] and with other data.

Figure 4C[9] is a cartoon of the N- and C-termini of Rd and a folding pattern that allows the formation of the active center with a pair of Cys residues spaced by two intervening residues and two others from distant parts of the primary structure. The C-terminal tryptophan residue (Trp) was found to be important both to the Rd-P450 affinity and the activity of the complex. Trp and glycine (Gln) can be removed by carboxypeptidase, which is unable to cleave the following arginine (Arg) linkage. The decrease in affinity was shown by Sligar in measurements of fluorescence quenching, the decrease in activity by dynamic measurements with Lipscomb[9].

This brief sketch of the early developments illustrates the power, in the hands of highly qualified and motivated students, of the physical concepts and methods introduced by Hans and Peter Debrunner, and how much could be learned from an in-depth study of the iron reporter group.

II. HEME SULFUR PROTEIN POCKETS

The character of P450, the main actor in this play, is indicated in three figures, each with several views. These concern: (1) the pockets for the heme with the iron axial ligand, for the oxygen and the substrate; (2) the redox and binding equilibria; and (3), the organization and regulation of the gene, which controls the formation and action of the enzyme system.

Figure 5A[18] is a cartoon of the two axial heme ligands in the resting state that impart the P450 character. A cysteinyl sulfur remains an intrinsic iron ligand throughout the reaction cycle while the opposite, extrinsic iron ligand changes. The iron sulfur linkage is crucial for the reaction and is common to all P450 systems. The sixth ligand in the resting state was suggested to be water by Philson[19]; the three-dimensional model of Poulos later showed six water molecules to be bound in the heme pocket[16]. More details of the heme environment will emerge later.

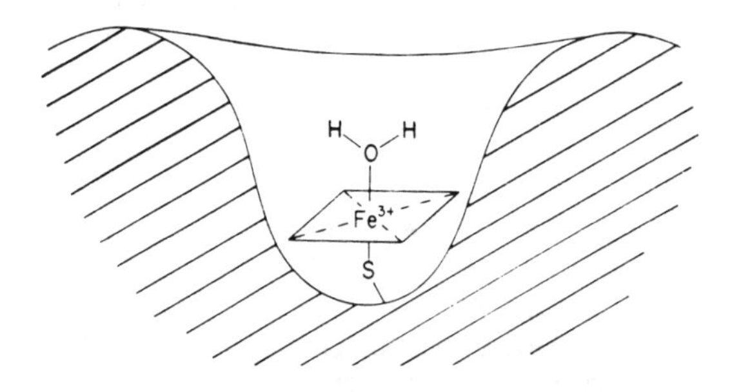

Figure 5A. Schematic diagram of the heme pocket in $P450_{CAM}$ with cysteine sulfur as axial iron ligand and a bound water molecule[18].

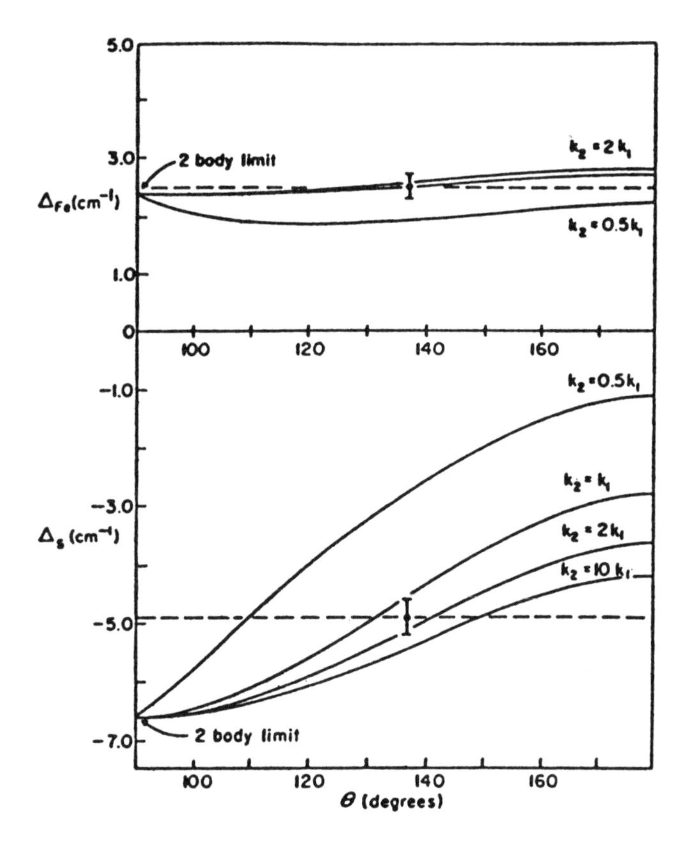

Figure 5B. Shifts in the resonance Raman spectra of the Fe-S stretching frequencies in $P450_{CAM}$ on substitution of ^{54}Fe for ^{56}Fe and of ^{34}S for ^{32}S[20].

Following an early suggestion of Howard Mason much circumstantial evidence had been accumulated for sulfur as a potential iron ligand. The first conclusive proof of the sulfur ligation, however, came from Paul Champion's resonance Raman spectra on samples differentially enriched in $^{54}Fe/^{56}Fe$ and $^{34}S/^{32}S$ as illustrated in Figure 5B[20]. Gerry Wagner prepared the samples from cells grown on enriched isotopes. Other metal replacements were made possible by the preparation of the heme-free apo-P450 and reconstitution with variously substituted metalloporphyrins, also by Gerry Wagner, as illustrated in Figure 5C[21].

The first structural model of P450 by Tom Poulos provided a wealth of new information about the nature of the heme pocket and the surrounding residues, which are essential to the orientation of the substrates[15,16]. The right side of the molecule, as presented in Figure 5D, is largely helical, while the left side is mainly random coil and pleated sheet. The protein is roughly triangular, about 60 Å on edge and about 35 Å deep. The cysteine that binds

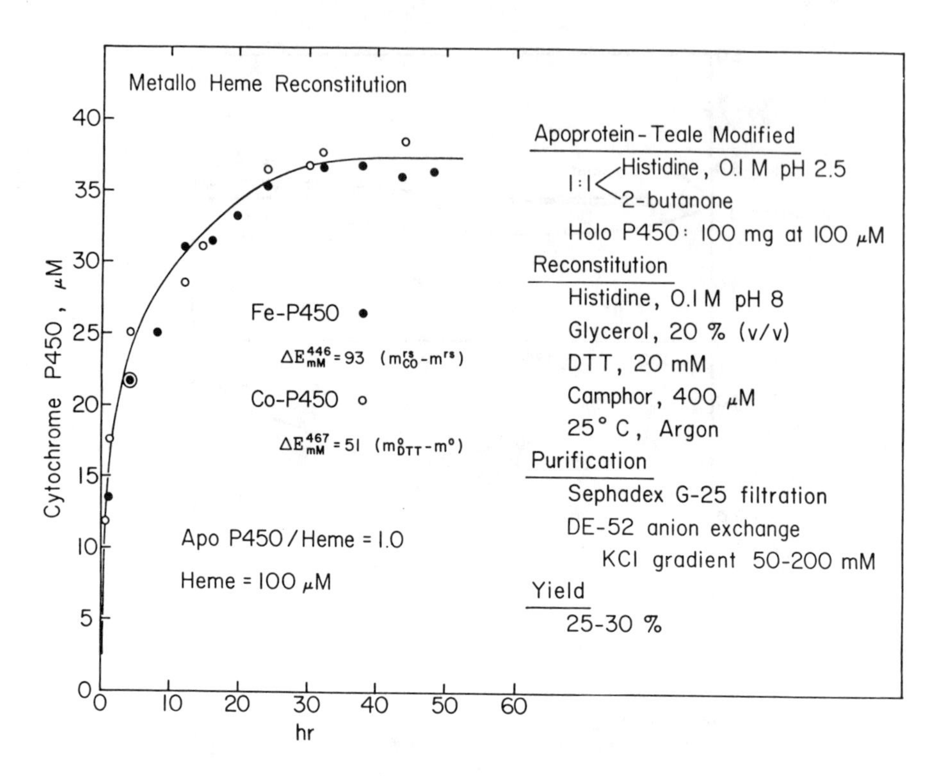

Figure 5C. Reconstitution of $P450_{CAM}$ with iron and cobalt protoporphyrin IX[21].

axially to the heme iron is part of a helix in the right center. An opening for the entry of hydrocarbon and oxygen is evident, but the dynamics of substrate and protein motion remains to be analyzed. Two close-ups of the active site are shown in Figures 5E and 5F. In the resting enzyme, Figure 5E, six water molecules are seen, which must be sufficiently stabilized to be identified in the electron density map. In the substrate complex, Figure 5F, the bound hydrocarbon, camphor, is held in place and oriented appropriately by a hydrogen bond from tyrosine 96 to the 2-keto group of camphor and by van der Waals interactions of valine 295 with the geminal dimethyl groups of camphor. Steve Sligar's group has shown a lack of hydroxylation selectivity with a substrate lacking the C8 geminal dimethyl groups. Much in this structure still remains to be interpreted, both directly and by site specific replacement of residues considered to be essential.

III. EQUILIBRIA

The collaboration between Biochemistry and Physics has greatly advanced our understanding of the equilibria and dynamics of the enzymatic process, of the nature of the stable intermediates in the reaction cycle, and of the reaction

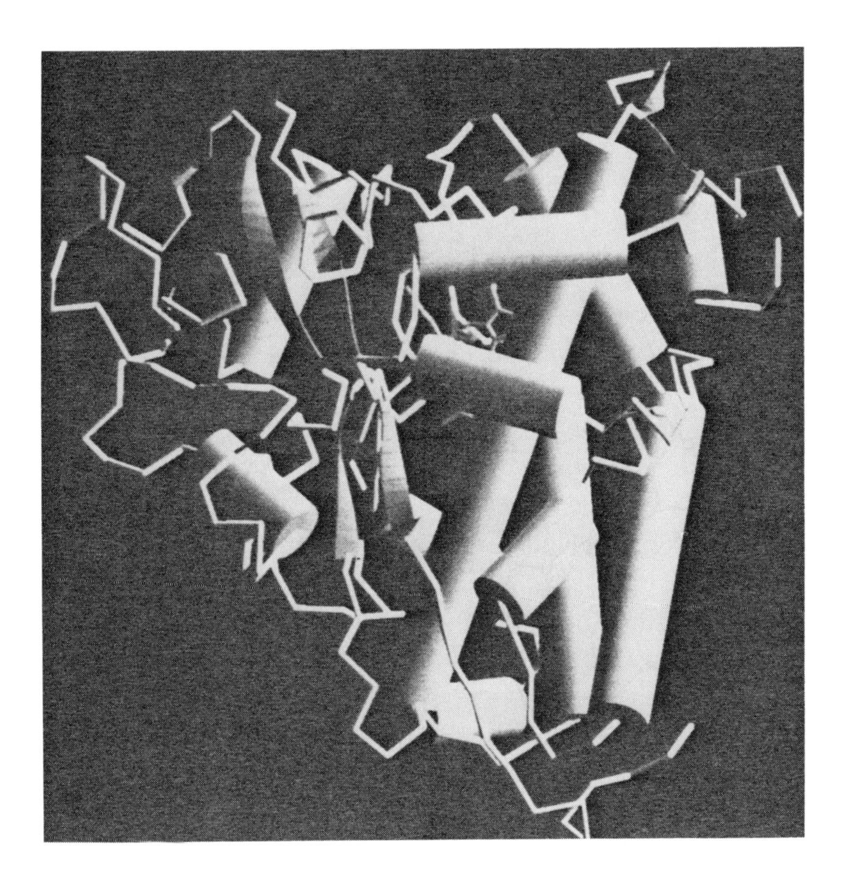

Figure 5D. Model of the 3-dimensional structure of the $P450_{CAM}$ substrate complex at 2 Å resolution. (T.L. Poulos, Personal communication.)

mechanism. The availability of clean protein in quantity was a prerequisite as it permitted precise measurements otherwise not possible, but Hans's probing questions regarding the physical concepts were always a major driving force, challenging the skills of the collaborators. There is no room here for a critical assessment of everybody's contributions, and the following discussion will be sketchy.

The reaction cycle, Figure 6A[22], was deduced fairly early from the processes observed following single-step addition of hydrocarbon, a single reducing equivalent and dioxygen. The stable intermediates were subsequently characterized in detail, and the dynamics of the transitions was measured. As indicated in Figure 6B[22], the intense Soret absorbance band of the heme group with a millimolar extinction coefficient of ca. 10^3 $mM^{-1}cm^{-1}$ near 400 nm is

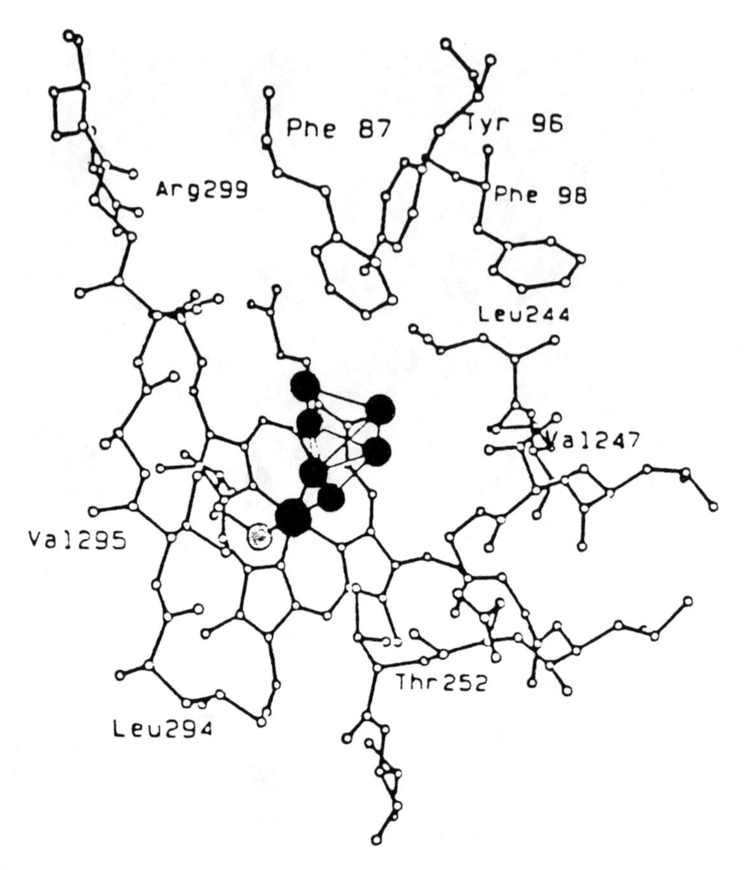

Figure 5E. Heme environment in substrate-free $P450_{CAM}$. The large black dot represents the heme iron bound to an axial cysteine sulfur shown in grey. The six smaller black dots represent water molecules. (T.L. Poulos, Personal communication.)

modulated by the oxidation and spin state of the heme iron and by the axial iron ligand(s). It thus serves as a spectral fingerprint of P450 in the stable states of the reaction cycle that not only has descriptive value, but also allows one to draw structural conclusions. As Figures 6A and 6B indicate, the Soret band of the ferric P450 ground state is at 417 nm; substrate addition shifts the peak absorbance to 391 nm; reduction shifts the maximum to 408 nm, and oxygenation brings the maximum to 418 nm with a decrease in absorbance. Figure 6C shows the distinctive Mössbauer spectra of three ferrous forms of $P450_{CAM}$, (a) the ground state, (b) the oxygenated state [with an admixture of (a)], and (c) the CO-adduct[12].

Figure 6D presents an expanded scheme of the reaction intermediates of $P450_{CAM}$, which allows each of the previously discussed four states (ferric, ferrous with and without substrate) to have high spin or low spin. This scheme

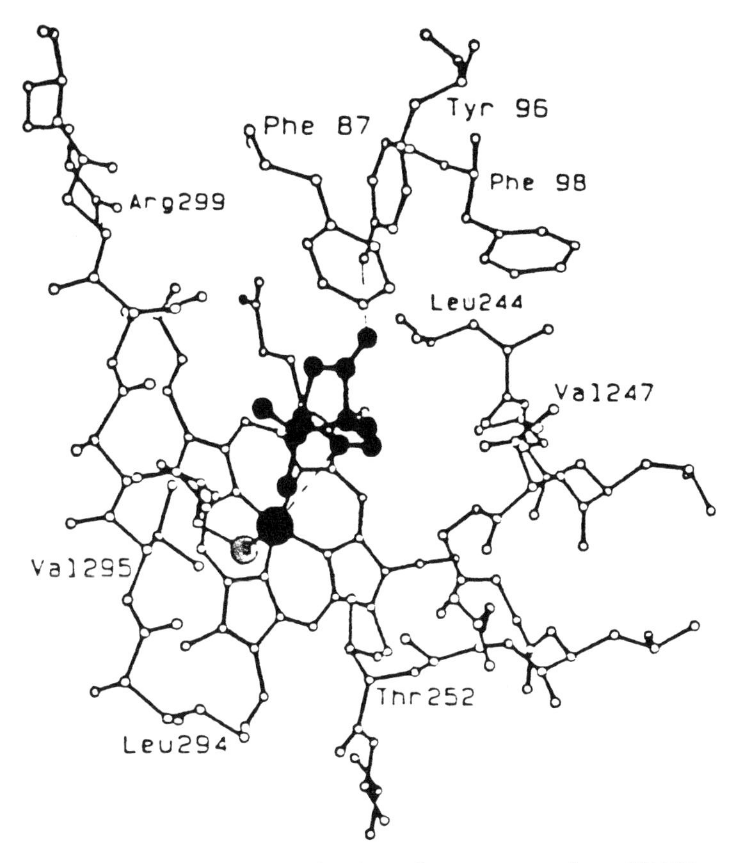

Figure 5F. Heme environment in the substrate complex of $P450_{CAM}$. The hydrocarbon (camphor) occupies the space on the proximal side of the heme that was occupied by water in Figure 4E. (T.L. Poulos, Personal communication.)

was developed by Steve Sligar[23] to relate the measured equlibrium coefficients, redox potentials and transition rates among the various states. All the equilibrium coefficients and therefore the Gibbs free energy differences have been determined either directly or indirectly.

IV. GENETICS: CAM GENES AND THE HYDROXYLASE OPERON

In order to understand the organization and control of biological systems an understanding of the underlying genetic basis is necessary. One approach toward this goal is to screen genetically modified bacteria for altered protein catalytic activity and enzyme formation. The interpretation of these data

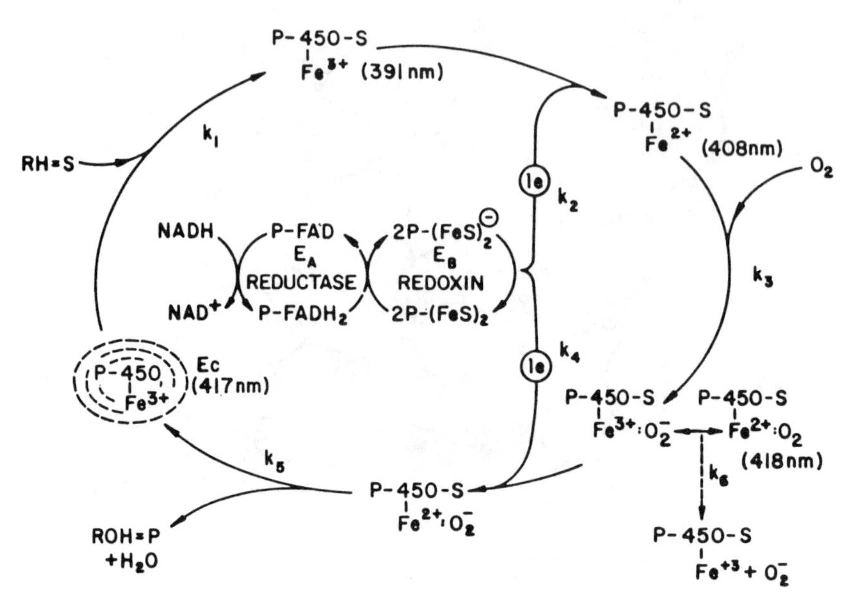

Figure 6A. Reaction cycle and stable intermediates of P450$_{CAM}$. Wavelengths of Soret peaks are given in parentheses[22].

depends critically on an understanding of the reaction sequence and of the composition of the enzyme system.

In the course of our studies of hydrocarbon oxygenation new enzyme systems were found. Early genetic studies of the organization of the CAM system failed to show linkage of the CAM genes to chromosomal auxotrophic markers, suggesting that the CAM genes resided on extrachromosomal elements. It was later shown that the genes encoding the early and middle enzymes of camphor metabolism resided on a large DNA plasmid (230 kb). In addition, it was shown that this plasmid was capable of initiating the genetic transfer of plasmid-encoded genes as well as of chromosomal genes.

It thus came to light that the CAM plasmid encodes at least two monoxygenases: (1) P450$_{CAM}$ that starts the process, and (2) ketolactonases that cleave each carbocyclic ring of the monoterpene.

Genetic mapping studies using point mutations in genes encoding the early and middle segment of camphor metabolism indicated that the CAM genes were clustered in two groups, with less than 50% linkage between the cluster encoding the CAM hydroxylase and the ketolactonases. The shared linkage groups among the genes encoding camphor oxidation and the pathway are shown in Figure 7A[24].

Figure 7B indicates the organization of the CAM hydroxylase operon and the clones that were used to establish the regulation and gene order of this operon. This work was carried out by Professor Koga of Kushu University, who visited our laboratory for two years and subsequently continued in his own department[25].

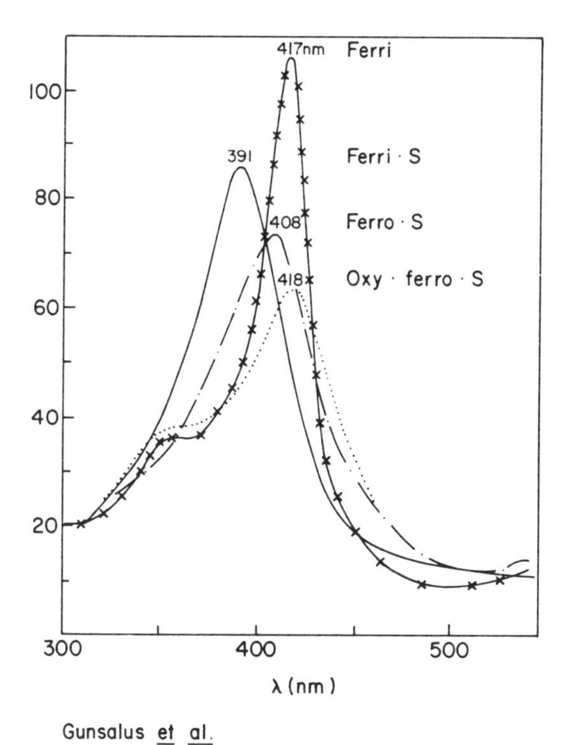

Figure 6B. Soret bands in the near UV of various states of $P450_{CAM}$. Millimolar extinction coefficients of native $P450_{CAM}$ (Ferri), its substrate complex (Ferri S), the reduced substrate complex (Ferro S), and its O_2-adduct (Oxyferro S)[22].

The plasmid pKG201 was isolated by its ability to complement a *camC* ($P450_{CAM}$) mutation. However, transcription of the *camC* gene was initiated from a promoter on the vector. The plasmid pJP1 contains the *camC* and *camD* (5-hydroxycamphor dehydrogenase) genes. When this plasmid is transformed into a CAM-plasmid-free host, expression of the *camC* and *camD* genes is induced by the presence of camphor, suggesting the existence of regulatory elements on this plasmid. The regulatory elements were identified as the *trans* acting repressor protein encoded by the *camR* gene, which interacts with the promoter-operator region to depress expression of the CAM hydroxylase operon in the absence of camphor. The plasmid pJP32 contains the *camA* and *camB* genes, encoding the reductase and the redoxin, Rd, respectively. These genes are not transcribed unless a promoter 5' to the *camA* gene is provided. The plasmid pJP45 was constructed by joining the *Hind* III inserts of pJP1 and pJP32 to reconstruct the full CAM hydroxylase operon in the gene order

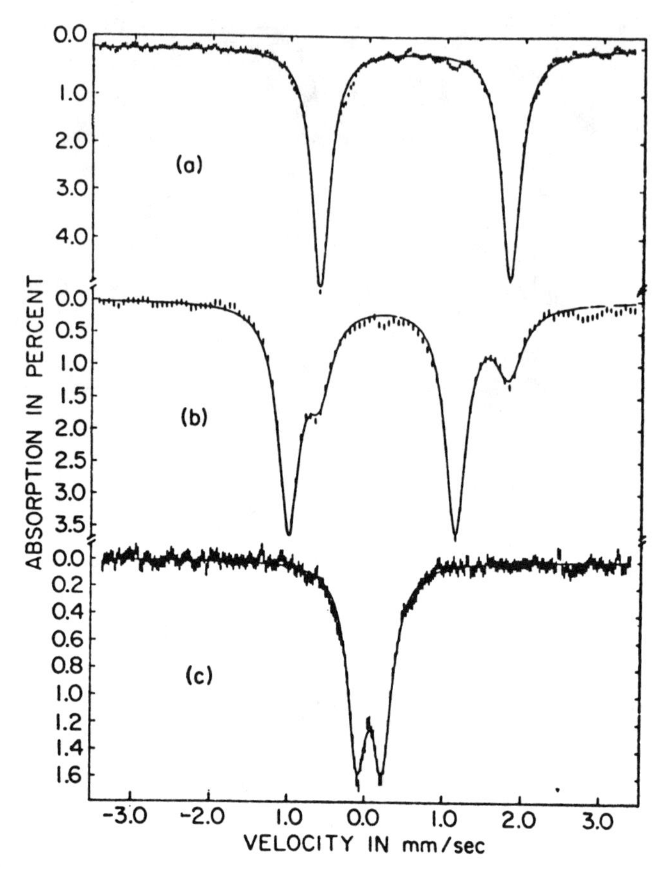

Figure 6C. Mössbauer spectra of (a) the reduced substrate complex, (b) the O_2-adduct [with admixture of (a)], and (c) the CO-adduct[12].

camRDCAB. In pJP45 the expression of the four structural genes is indeed regulated by camphor.

The nucleotide sequences of these genes are known and thus the complete protein primary structures of the $P450_{CAM}$ hydroxylase components have been deduced. The gene assignments have been confirmed by N-terminal amino acid sequences of hydroxylase components[26].

V. RELATED P450 SYSTEMS AND PROPERTIES

Figures 8 and 9 look beyond the properties of the model, $P450_{CAM}$, and ask more general questions. To what extent does the vast array of microsomal heme-thiolate monoxygenases differ from $P450_{CAM}$? What controls the substrate specificity or lack thereof? Is the enrichment for microorganisms or exposure of various compounds to the organs of mammals a good way of

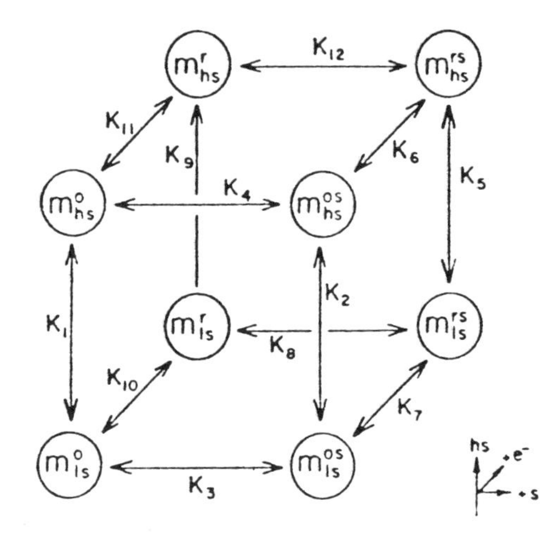

Figure 6D. Redox, spin and ligand states of $P450_{CAM}$, here designated m for monoxygenase. Superscripts *o*, *r*, and *s* stand for oxidized, reduced and substrate complex, subscripts *ls* and *hs* stand for low-spin and high-spin states of the heme iron. The single-step reactions are labeled with equilibrium coefficients K_i. Conservation of Gibbs free energy limits the number of independent parameters to seven[23].

learning the functions in nature? In several instances the isolation and even the structure determination of colored proteins preceded the discovery of their function. An example is alcohol dehydrogenase of hepatic tissue, which shows poor affinity for ethanol and turned out later to be a dehydrogenase for a long, branched-chain or polycyclic alcohol.

Figures 8A and 8B illustrate the hydrocarbon selectivity and the tolerance for substrate modification as realized from analogues. Figure 8A compares the camphor (CAM) and linalool (LIN) systems at the level of substrate affinity and activity[27]. Each of these systems has three components of similar size, which can be separated by the protocol developed earlier for the CAM system. The substrates are monoterpenes of different structures, and the two P450's are distinctly different. The affinity of the CAM enzyme for the substrate is ca. 1 μM, for the alcohol product 200 μM, and for the ketone even less. Thus, in the presence of camphor as substrate, only the alcohol appears. If only the alcohol, but no camphor is present, the ketone is formed. With the LIN system, in contrast, the affinities for linalool and 8-hydroxy linalool are similar, thus the observed reaction sequence is the appearance and subsequent decrease of the alcohol. The second reaction cycle leads to aldehyde accumulation. A third system, p-cymene, also yields a P450 similar to $P450_{CAM}$, so far without revealing the nature of the second and third components.

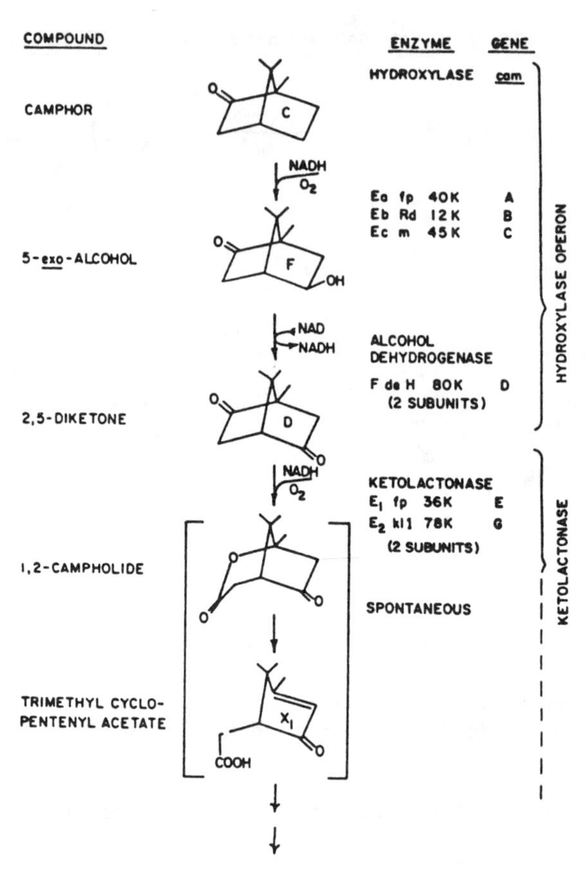

Figure 7A. D-camphor oxidation pathway of *Pseudomonas putida* PpG1 and the enzymes involved. Enzymes: E_a fp, flavoprotein; E_b Rd, redoxin; E_c m, monoxygenase $P450_{CAM}$. E_1 fp, NAD-FMN ketolactonase-1 reductase; E_2 kl1, D-camphor or 2,5-diketocamphane ketolactonase[24].

Permissive modifications of the linalool substrate are extensive as illustrated in Figure 8B. Aromatic substitution at the 5-carbon does not alter the affinity, although the saturation level of the transition from low spin to high spin decreases. Aromatic substitution at the 3-carbon does not modify the extent of the spin-state saturation but decreases the affinity; similar changes are found on lengthening the carbon-3 substitution with two rather than one carbon alkyl on the bridgehead, e.g., saturation of the 2-3 double bond may be critical. As indicated by earlier data, substitution of a methyl group by H abolishes the transition to high spin. Some 70 analogues have been synthesized and studied.

Figure 8C adds a third P450 that reacts with the aromatized monocyclic monoterpene-like *p*-cymene. The tolerance for saturation is broad, but much

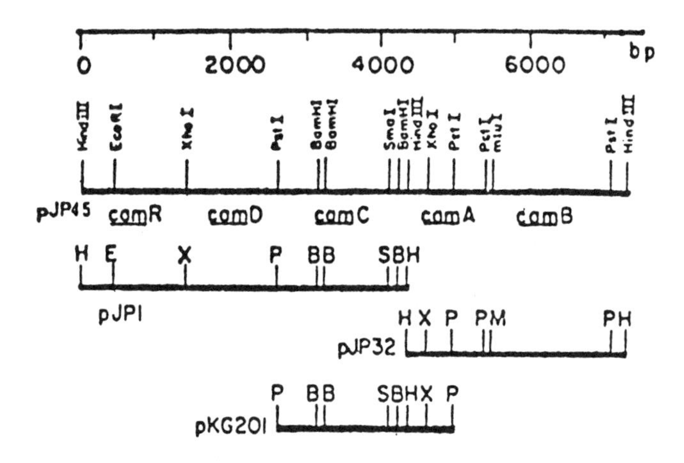

Figure 7B. Physical map of the *cam RDCAB* operon as discussed in the text[25].

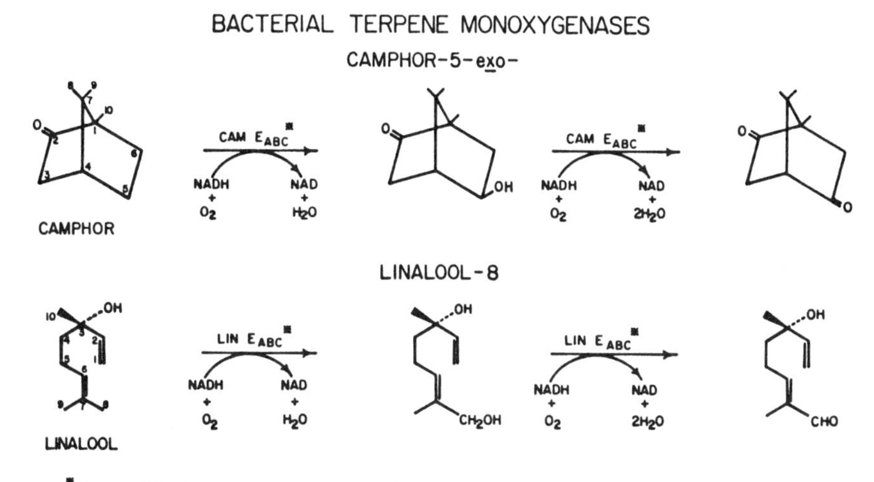

Figure 8A. Comparison of the camphor and linalool monoxygenase systems[27].

less for hydroxyl or ketone. The latter are cytotoxic and may bind without inducing a spin transition, possibly because the oxygen function resembles the bound water in the low-spin ferric resting state. These observations leave room for further questions as they may be relevant for the model system $P450_{CAM}$.

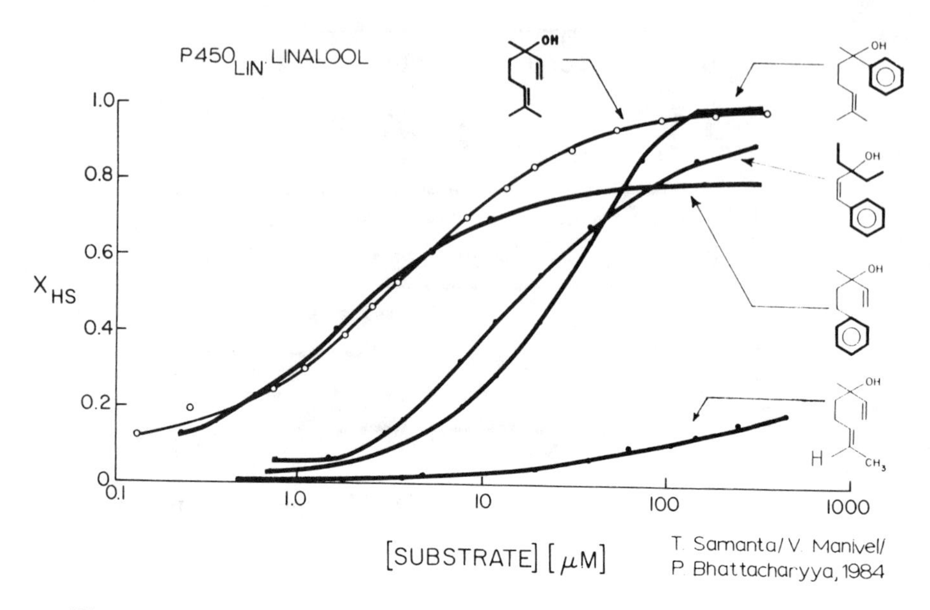

Figure 8B. High-spin fraction, x_{HS}, of $P450_{LIN}$ for various substrates as a function of their concentration[27].

Figures 9A and 9B compare some physical properties of $P450_{CAM}$ and $P450_{LIN}$, in particular their stabilization by substrate. Figure 9A[28] illustrates the unfolding with increasing temperature. Similar data were obtained from pressure studies in the laboratories of Pierre Douzou and Steve Sligar. These results show a substantial stabilization by substrate for $P450_{CAM}$, but none for $P450_{LIN}$ nor the hepatic, microsomal LM2 induced by phenobarbitol. We are pursuing the $P450_{LIN}$ system further for comparison with the $P450_{CAM}$ substrate and redoxin interactions. Figure 9B[27] compares spectral changes on binding of inhibitors. $P450_{LIN}$ reacts with KCN as strongly as $P450_{CAM}$, but much less with molecules of larger size, again suggesting a more rigid, tighter structure for the former. Site specific mutations combined with binding studies of substrate analogues should shed further light on the range of substrates the P450 family of enzymes can accommodate.

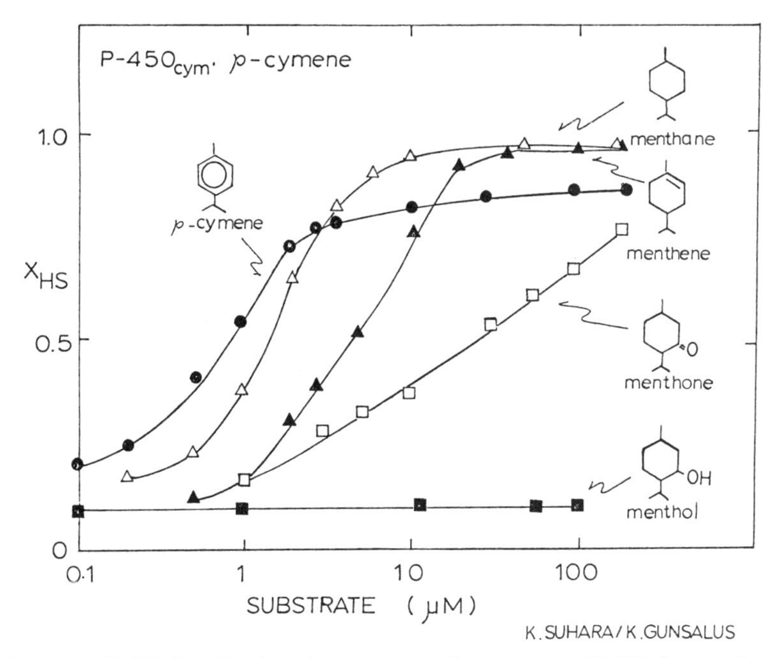

Figure 8C. High-spin fraction, x_{HS}, of *p*-cymene P450 for various substrates as a function of their concentration[27].

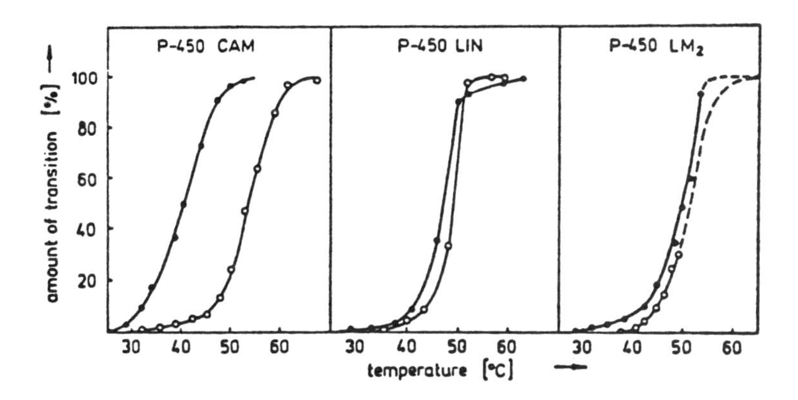

Figure 9A. Thermal unfolding as monitored by the decrease of the Soret peak absorption of (•) substrate-free low-spin (418 nm), and (o) substrate-bound high-spin (387 nm) P450's. CAM and LIN refer to the camphor and linolool monoxytenase of *P. putida*, respectively, while LM_2 refers to phenobarbitol-induced hepatic P450[28].

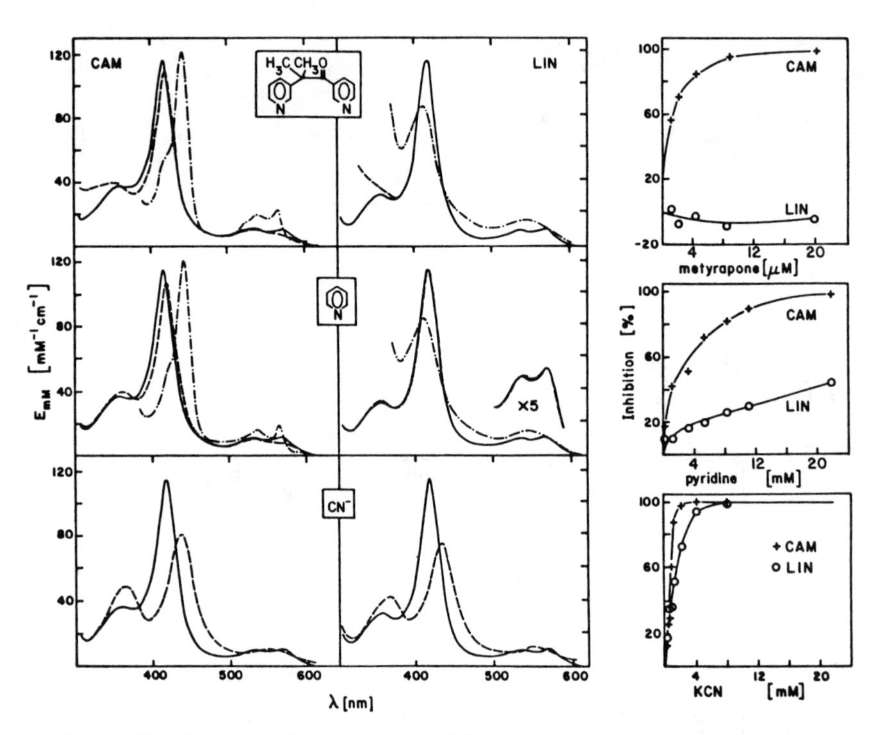

Figure 9B. Spectral changes and inhibition of $P450_{CAM}$ and $P450_{LIN}$ due to metyrapone (top), pyridine (middle) and KCN (bottom)[27].

REFERENCES

1. W. H. Bradshaw, H. E. Conrad, E. J. Corey, I. C. Gunsalus and D. Lednicer, J. Am. Chem. Soc. 81, 5507 (1959).
2. J. Hedegaard and I. C. Gunsalus, J. Biol. Chem. 240, 4038 (1965).
3. D. W. Cushman, R. L. Tsai and I. C. Gunsalus, Biochem. Biophys. Res. Commun. 26, 577 (1967).
4. D. V. DerVartanian, W. H. Orme-Johnson, R. E. Hansen and H. Beinert, Biochem. Biophys. Res. Commun. 26, 569 (1967).
5. J. C. M. Tsibris, R. L. Tsai, I. C. Gunsalus, W. H. Orme-Johnson, R. E. Hansen and H. Beinert, Proc. Natl. Acad. Sci. USA 59, 959 (1968).
6. R. Cooke, J. C. M. Tsibris, P. G. Debrunner, R. Tsai, I. C. Gunsalus and H. Frauenfelder, Proc. Natl. Acad. Sci. USA 59, 1045 (1968).
7. E. Münck, P. G. Debrunner, J. C. M. Tsibris and I. C. Gunsalus, Biochemistry 11, 853 (1972).
8. M. Katagiri, B. N. Ganguli and I. C. Gunsalus, J. Biol. Chem. 243, 3543 (1968).
9. S. G. Sligar, P. G. Debrunner, J. D. Lipscomb, M. J. Namtvedt and I. C. Gunsalus, Proc. Natl. Acad. Sci. USA 71, 3906 (1974).
10. R. H. Austin, K. Beeson, L. Eisenstein, H. Frauenfelder, I. C. Gunsalus and V. P. Marshall, Science 181, 541 (1973).
11. L. Eisenstein, P. Debey and P. Douzou, Biochem. Biophys. Res. Commun. 77, 1377 (1977).
12. M. Sharrock, P. G. Debrunner, C. Schulz, J. D. Lipscomb, V. Marshall and I. C. Gunsalus, Biochim. Biophys. Acta 420, 8 (1976).
13. T. L. Poulos, M. Perez and G. C. Wagner, J. Biol. Chem. 257, 10427 (1982).
14. I. C. Gunsalus, R. L. Tsai, C. A. Tyson, M.-C. Hsu and C.-A. Yu, Abstracts, Fourth International Conference on Magnetic Resonance in Biological Systems, (Oxford, 1970), p. 26.
15. T. L. Poulos, B. C. Finzel, I. C. Gunsalus , G. C. Wagner and J. Kraut, J. Biol. Chem. 260, 16122 (1985).
16. T. L. Poulos, B. C. Finzel and H. A. Howard, Biochemistry 25, 5413 (1986).
17. H. Frauenfelder, I. C. Gunsalus and E. Münck, *Mössbauer Spectroscopy and its Applications* (International Atomic Energy Agency, Vienna, 1972) p. 231.
18. R. White and M. J. Coon, Ann. Rev. Biochem. 49, 315 (1980).
19. S. B. Philson, P. G. Debrunner, P. G. Schmidt and I. C. Gunsalus, J. Biol. Chem. 254, 10173 (1979).
20. P. M. Champion, B. R Stallard, G. C. Wagner and I. C. Gunsalus, J. Am. Chem. Soc. 104, 5469 (1982).
21. G. C. Wagner, M. Perez, W. Toscano and I. C. Gunsalus, J. Biol. Chem. 256, 6266 (1981).
22. C. A. Tyson, R.L.Tsai, J. D. Lipscomb and I. C. Gunsalus, J. Biol. Chem. 247, 5777 (1972).
23. S. Sligar, Biochemistry 15, 5399 (1976).
24. I. C. Gunsalus and G. C. Wagner, Methods Enzymol. 52, 166 (1978).

25. (a) H. Koga, B. Rauchfuss and I. C. Gunsalus, Biochem. Biophys. Res. Commun. 130, 412 (1985). (b) H. Koga, H. Aramaki, E. Yamaguchi, K. Takeuchi, T. Horiuchi and I. C. Gunsalus, J. Bacteriol. 166, 1089 (1986).
26. B. P. Unger, I. C. Gunsalus and S. G. Sligar, J. Biol. Chem. 261, 1158 (1986).
27. I. C. Gunsalus, P. K. Bhattacharyya and K. Suhara, Current Topics in Cellular Regulation 26, 295 (1985).
28. C. Jung, P. Bendzko, O. Ristau and I. C. Gunsalus, *Cytochrome P450 Biochemistry, Biophysics and Induction* (L. Vereczkey and K. Magyar, eds., Akadémiai Kiadó, Budapest, Hungary, 1985), p. 19.

NUCLEAR MAGNETIC RESONANCE STUDIES OF SIMPLE MOLECULES ON METAL SURFACES

Charles P. Slichter
University of Illinois at Urbana-Champaign

ABSTRACT

The author describes the use of nuclear magnetic resonance to study simple molecules (CO, C_2H_2, C_2H_4) adsorbed on small particles composed of 6 of the Group VIII metals (Ru, Rh, Pd, Os, Ir, Pt). The data yield information about bonding of the molecules to the surface, the structure of the molecules, rupture of the C-C bonds on heating, the observation of isolated C atoms on the metal surface, and measurement of diffusion rates of C atoms or CO molecules on the metal surface.

I. INTRODUCTION

My topic, the use of nuclear magnetic resonance to study simple molecules on metal surfaces, has a special significance on this happy occasion, the celebration of the birthdate of Hans Frauenfelder, because the entry of Hans into science in 1950 was his use of radioactivity to study surface processes of solids. Here is the title of his thesis and the abstract of its contents[1]:

Die Untersuchung von Oberflachenprozessen mit
Radioaktivitat von Hans Frauenfelder

Summary. A short survey on surface processes on solids is given. New methods for measuring desorption probabilities, sticking coefficients and surface diffusion coefficients by means of radioactive substances are described. Recoil atoms from K capture are recorded; this allows the measurement of the adsorption of thin gas layers at low pressures and of extremely small diffusion coefficients. Experimental results are given and compared to former investigations with other methods. Applications to nuclear physics are the rapid separation of radioactive elements and the preparation of very thin sources.

What a prophetic document. Here in capsule form is displayed the essence of Hans as a scientist, a preview of his entire career. What an innovative concept it was to use radioactivity to study surfaces. Here is a concept which opens a whole new field. Techniques of one field of physics, nuclear physics, are brought to bear on important problems in another field, study of surface processes. Indeed, study of surfaces, so clearly recognized as important by Hans in 1950, is today widely recognized as being one of the most important areas of physics, chemistry, and even biology.

Today, surfaces are studied by a variety of methods, made possible by development of techniques of first measuring and then obtaining ultrahigh vacuums. There is a veritable alphabet soup of experimental methods (for example LEED, UPS, HREELS). My students, post doctoral research associates[2], and I have been working with Dr. John Sinfelt of the Exxon Research and Engineering Laboratories, to apply nuclear magnetic resonance (NMR) to the study of metal surfaces. Dr. Sinfelt, whom I am proud to say received his Ph.D. from the University of Illinois, working with Professor Drickamer, is widely known and has been frequently honored for his pioneering studies of catalysis. We have been studying six of the nine group VIII metals (Ru, Rh, Pd, Os, Ir, Pt) which are so important as catalysts. We have not studied the other three (Fe, Co, and Ni) which are ferromagnets, since we expect the NMR lines would be too broad to observe[3].

Nuclear magnetic resonance has been such a powerful technique for study of problems in condensed matter physics, in chemistry, and in biology, that one may ask why it has not previously been widely used to study surfaces. The difficulty is of course one of weak NMR signal strength. One square centimeter of surface has only about 10^{15} atoms, much too small a number to be seen by NMR. The solution to this problem is to go to samples of large surface area, as with real catalysts. Our samples consist of many small metal particles, each one 10-100 Å in diameter, supported on alumina (Al_2O_3). The metal constitutes 5-10% of the sample weight. The total sample volume is about 1 cm^3. We characterize the particle size by quoting the dispersion of the sample. The dispersion, usually measured by hydrogen chemisorption, is the percentage of the metal atoms which are on the surface of the metal particles. One can also calculate the dispersion if one measures the particle size distribution by means of electron microscope photos. We use a notation such as Pt-76-CO to denote the characteristics of a sample (Pt particles whose dispersion is 76% onto which CO has been adsorbed, usually to full coverage unless otherwise specified). Such samples have several square meters of surface area.

Although the increased area provided by such samples makes an enormous difference, we still could not do what we do without the fruits of the revolution in magnetic resonance which has taken place in recent years: superconducting solenoids, (which greatly increase the strength of the static field H_o), digital signal averaging which permits one to average for many hours (until the signal to noise ratio reaches the desired level), laboratory computers to control the application of complex pulse sequences, to simulate experimental spectra and to manipulate data (as in performing Fourier transforms). Lastly, the understanding of magnetic resonance itself has grown continuously deeper, enabling one to conceive of experiments today not previously thought to be possible.

In thinking about our experiments, it is useful to consider that the individual metal particles are regular polyhedra, probably cubooctohedra, so that the exposed metal surfaces are largely low index crystal faces (For a cubooctahedron 80% of the surface is (111), 20% (100)).

We have studied both the metal and adsorbed molecules. Our first studies were of the ^{195}Pt nucleus in Pt samples, and led to the discovery of a resonance which was clearly associated with the surface layer of Pt atoms[4]. We proved this conclusion[5] by use of a double resonance method called Spin Echo Double Resonance (SEDOR). We adsorbed ^{13}CO onto the surface. Then we observed the ^{195}Pt signal alternately with and without flipping the ^{13}C spins. Since the surface Pt nuclei experience a magnetic field from the ^{13}C nucleus, the signal from the surface layer of ^{195}Pt's is changed by flipping the ^{13}C nuclei. The signal from ^{195}Pt in the interior of the particles is unaffected since the nuclear dipolar coupling to the ^{13}C falls off strongly with distance (inversely proportional to the cube of the internuclear distance). Subtracting the ^{195}Pt signal with the ^{13}C flip from that without it cancels the signal from all the ^{195}Pt nuclei except those on the surface. This example illustrates the type of experiment NMR can bring to bear on the study of surfaces.

In this paper, I will describe some of our studies of adsorbed molecules. I will take up the bonding of CO to the surface, determination of the structure of simple hydrocarbons, C-C bond rupture under heating, the observation of isolated C atoms on surfaces, and studies of diffusion on surfaces.

II. THE BONDING OF CO TO METALS

CO adsorbed on metal surfaces has been the subject of many studies. In part, its importance arises because it is a simple molecule whose properties one might hope could be understood in precise theoretical terms. It is also important, however, in actual reactions. Thus, the reaction

$$3H_2 + CO \rightarrow CH_4 + H_2O \tag{1}$$

which is catalyzed by Ni, is used to form the useful fuel, methane (CH_4). Indeed, CH_4 is an example of one of the simplest hydrocarbons formed by reacting H_2 with CO. The term Fischer-Tropsch reaction is used to cover the class of such catalyzed reactions which form hydrocarbons (including CH_2CH_2, CH_3CH_3, and so on).

Our studies give us information about the carbon-metal bond. In particular, they demonstrate and provide quantitative information about the mixing of the electron wave functions of the molecule with the wave functions of the conduction electrons, the phenomenon which provides the bonding.

Figure 1 shows the ^{13}C NMR line (measured at a temperature of 77 K) of CO adsorbed on Ru, Rh, Pd, Os, Ir, and Pt. The CO has been enriched to 90 with ^{13}C. The data were obtained by Susan Shore and Zhiyue Wang using spin echoes. In a spin echo experiment one applies two radio frequency pulses (a $\pi/2$ and a π pulse) separated in time by τ. At a time τ after the second pulse, the NMR echo signal forms. We record the echo digitally (for averaging and signal processing). The NMR intensity plotted in the figure is the area of the echo as a function of the frequency of the radio frequency pulses. The frequency is measured relative to the ^{13}C resonance in the conventional reference compound tetramethyl silane (TMS). (In addition to measuring the echo area we on occasion measure the echo peak height, thereby measuring the total spin magnetization. We also process data by taking the Fourier transform of the second half of the echo.)

It is immediately apparent from Fig. 1 that the ^{13}CO resonance is strongly affected by the metal to which the CO is bonded, hence that the spectra contain information about the carbon-metal bond.

In molecules such as Pt carbonyls, CO may bond to either one or two metal atoms ("linear" and "bridge" bonding respectively). The ^{13}CO resonance for linearly bonded Pt-CO occurs at about 185 ppm (parts per million) shift, and bridge bonded Pt-CO occurs at about 240 ppm. ^{13}CO on the Pt metal surface occurs at about 340 ppm, far above either linear or bridge bonded positions

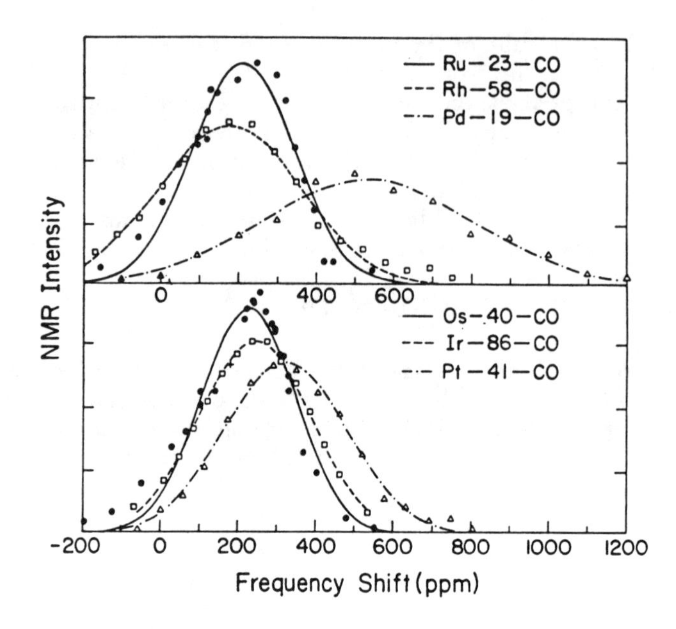

Figure 1. ^{13}CO NMR lineshapes at 77 K showing the fact that the NMR absorption spectrum depends on the metal on which the CO is adsorbed: NMR absorption versus shift δ (ppm downfield) relative to TMS: Upper, • Ru, □Rh, Δ Pd; lower, • Os, □Ir, Δ Pt. The various lines are the gaussians that best fit the data, centered at the average shift of the data. The shift for CO on Pd is the largest shift known for ^{13}C. The notation Pt-41-CO means 41% dispersion Pt coated with CO.

of molecules. The shift of CO on Pd is even more dramatic (540 ppm). It is to my knowledge the largest known ^{13}C shift.

The normal ^{13}C chemical shift is well-known to arise from the magnetism induced in the electronic orbital motion by the applied magnetic field (the Van Vleck temperature independent paramagnetism[6]). It seems unlikely that this effect should differ so greatly between a metal-C bond in a molecule and the same bond at a surface. Rather, we looked for a mechanism which could arise with a metal but not in a diamagnetic molecule. Such a mechanism is the polarization of electron spins, an effect well known in metals, where it is called the Knight shift after its discoverer. The Knight shift arises from the so-called Fermi contact interaction between the conduction electron spins and the nucleus. It requires that the electron wave function be non-zero at the position of the nucleus. This could arise, for example, if the otherwise fully occupied CO 5σ orbital (the famous lone pair bond of the isolated CO molecule) were partially emptied by bonding to the metal[7,8].

One test of the hypothesis that the shift comes from such a mixing of metal with molecular wave functions is to measure the temperature dependence of the ^{13}C spin-lattice relaxation time, T_1. In metals, the relaxation rate $(1/T_1)$ is proportional to temperature[9]. Figure 2 shows data taken by Rudaz and by Ansermet of the spin-lattice relaxation rate of ^{13}CO on Pt[10]. It indeed has a temperature dependence which gives the metallic signature. If there is a Knight shift, it must be accompanied by a relaxation rate which is at least as fast as that predicted by the famous equation due to Korringa[9,11]:

$$T_1 T K^2 = \frac{\hbar}{4\pi k_B} \tag{2}$$

where K is the fractional shift in resonance frequency, k_B the Boltzmann constant, $\hbar$ Planck's constant divided by 2π. There are other relaxation processes which arise from interactions which do not produce a net shift of the resonance but which obey

$$\frac{1}{T_1} \propto T. \tag{3}$$

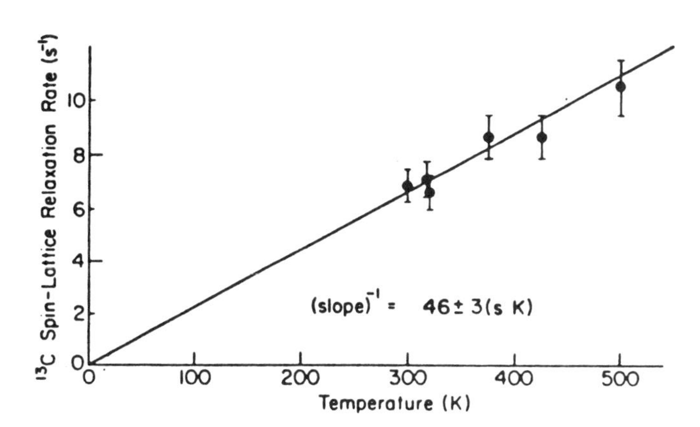

Figure 2. ^{13}C spin-lattice relaxation rate, $1/T_1$, vs. temperature, T for ^{13}CO adsorbed on Pt demonstrating that $T_1 T = \text{const}$ (46 ± 3 s K) in agreement with the Korringa relation, demonstrating that the ^{13}C nuclei are relaxed by the conduction electrons.

For the Knight shift K we take the difference between the frequency of ^{13}CO on the metal from its value in the appropriate metal carbonyl molecule. That gives for ^{13}CO on Pd a shift of 310 $\pm$ 30 ppm, leading to a prediction that

$$T_1 T = 43 \pm 7s \cdot \text{ K}. \tag{4}$$

The experimental value obtained by Shore[12] is

$$T_1 T = 29 \pm 3s \cdot \text{ K}. \tag{5}$$

Thus, for Pd we conclude that indeed the shift and the spin-lattice relaxation arise from interactions with conduction electron spins on the carbon atom.

For CO on Pt, the observed ^{13}C T_1 is even faster than predicted by Equation 2. We have discussed the detailed mechanisms responsible, plus information we learn about the mixing of CO-metal wave functions elsewhere[13].

III. THE STRUCTURE OF SIMPLE MOLECULES

When a molecule adsorbs on a surface, it may attach either by van der Waals forces ("physisorption") or by forming chemical bonds with the surface ("chemisorption"). In the latter case, since new bonds are formed, old ones must be broken, the structure of the molecule must certainly change, at least to a minor degree. NMR has proven to be a useful technique for determining the structure of simple adsorbed molecules. We have studied the structure of acetylene (HCCH) and ethylene (H_2CCH_2), and measured the bond length of CO.

The principle we have utilized is to measure the nuclear magnetic dipolar coupling between adjacent nuclei. The idea is simple. The dipolar coupling between adjacent nuclei can be represented classically by the dipolar magnetic field, H_d, a nucleus of spin S produces on a nucleus of spin I. If the static field, H, lies along the z-direction H_d is given by

$$H_d = \gamma_I \gamma_s \frac{\hbar^2}{r^3} \frac{(3\cos^2\theta - 1)}{3} m_s \tag{6}$$

where m_s is the eigenvalue of S_z, the z component of the angular momentum of spin S, $(m_s = S, S-1, ..., -S)$, γ_I and γ_s are the respective nuclear gyromagnetic ratios, r the distance between I and S, and where θ is the angle the internuclear vector $\vec{r}$ makes with the static magnetic field. (See Ref. 9 for a discussion of Equation 6). For $S = 1/2, m_s = \pm 1/2$, giving rise to two possible values of H_d for a given r and θ. Then the resonance condition on the angular frequency, ω,

$$\omega = \gamma(H_0 + H_d) \tag{7}$$

gives rise to two absorption lines split in angular frequency, $\Delta\omega$, by

$$\Delta\omega = 2\gamma |H_d|. \tag{8}$$

This phenomenon first discovered by Pake[14] and the pair of resonance lines is commonly called a Pake doublet. In a powder sample, all values of θ occur, leading to a partial smearing of the doublet structure. In addition, the structure is washed out by two other line broadening effects. The first is the fact that the chemical shift is anisotropic, so that it depends on the orientation of a molecule with respect to $\vec{H}_0$. The second effect arises from the magnetic susceptibility of the metal, which leads to the metal possessing an induced electronic magnetization. This magnetization produces an extra magnetic field whose magnitude is different at different points on the surface of the metal particle. Consequently, the resonance frequency is further smeared.

As a result of these two additional line-broadening mechanisms, the Pake doublet is not seen in a direct measurement of the absorption spectrum. However, in a spin echo these magnetic broadenings are eliminated by the refocusing effect of the second pulse. It can be shown that the dipolar coupling between like nuclei, however, is not refocused. The envelope of the echo amplitude versus 2τ is the Fourier transform of the Pake doublet powder pattern. It oscillates with a period which corresponds to the frequency splitting $\Delta\omega$.

We now illustrate our methods with the case of ethylene (H_2CCH_2). From single crystal studies, it is known that this molecule gives up one H on adsorption, becoming a CCH_3 (ethylidyne) species attached by one C atom. We expect (and find) the same with our catalyst samples.

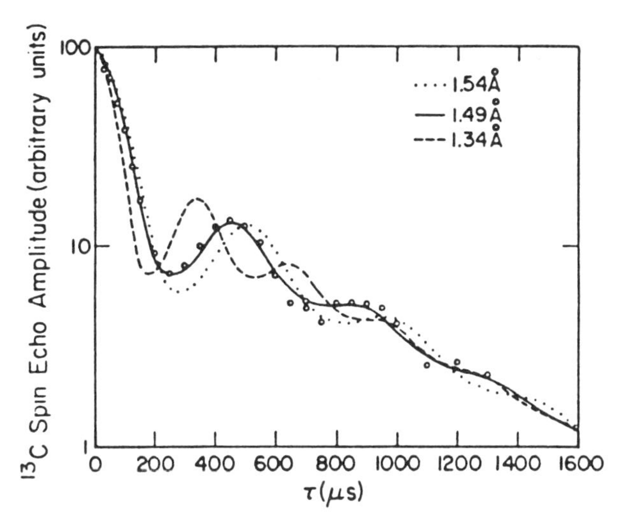

Figure 3. ^{13}C spin echo amplitude vs. pulse separation τ at 77 K for $^{13}C_2H_4$ on Pt clusters. The circles are data. The solid line is a fit with a C-C bond length of 1.49 Å; the dotted and the dashed lines are that of single bond length of 1.54 Å and double bond length of 1.34 Å, respectively.

Figure 3 shows the echo envelope for the ^{13}C resonance of H_2CCH_2 adsorbed on Pt obtained by P.-K. Wang et al.[15] The ethylene is enriched to 90% in ^{13}C. Theoretical curves for several bond lengths are shown, with the best fit indicating a length of 1.49 ± 0.02 Å, typical of a single bond. This fact suggests that we have a metal-$CHCH_3$ species, or a metal-CCH_3 species, or an H_2CCH_2 species in which both carbons attach to the metal. To choose among them, we perform a ^{13}C-^1H spin echo double resonance (SEDOR) in which we compare the ^{13}C spin echo formed with and without simultaneous ^1H spin flips. Crudely speaking, we note that flipping the ^1H is able to destroy the ^{13}C echo of any ^{13}C which is bonded to a ^1H. Thus, for either of the two species for which every C has a ^1H attached, we should be able to destroy the entire ^{13}C echo, whereas for the ethylidyne (CCH_3) species, we expect to be able to destroy only half the echo amplitude. Figure 4 shows the SEDOR data of P.-K. Wang et al together with several theoretical curves. The fraction of the echo destroyed by SEDOR is called the SEDOR fraction. We note that the data fit the structure CCH_3 with the CH_3 group rotating rapidly about the three-fold axis.

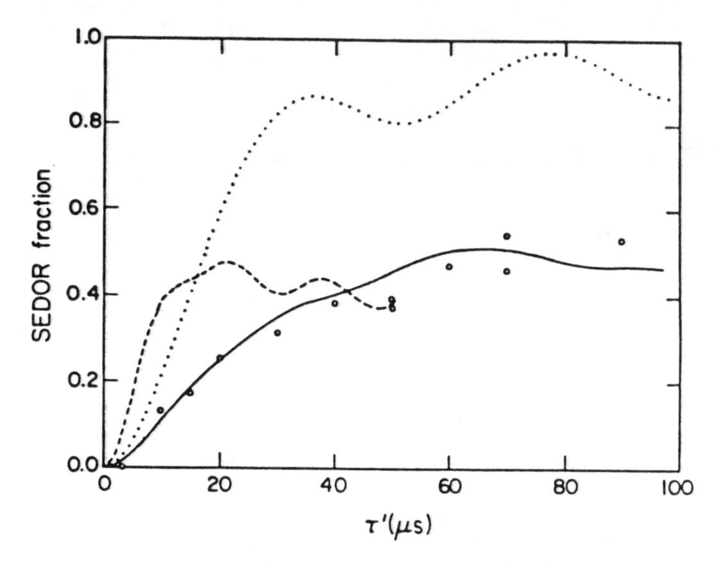

Figure 4. SEDOR fraction vs. time τ' at which the ^1H pulse is applied at 77 K for $^{13}C_2H_4$ on Pt clusters. The solid line is the theoretical prediction for the ethylidyne species (C-CH_3), with the methyl group rotating about the C-C axis, the dashed line is that of frozen C-CH_3, and the dotted line is that of CH-CH_3 with rotating methyl groups.

Using similar techniques, plus the method of multiple quantum coherence, P.-K. Wang et al. showed that about 3/4 of the acetylene molecules (HCCH in the gas) adsorb as CCH_2 molecules, with a C-C bond length midway between C-C single and double bonds. Thus, we surmise that both C atoms are bonded to the metal to some extent.

Susan Shore studied $^{13}C^{17}O$ adsorbed on Pd[12], using CO enriched to 90% with ^{13}C and 36% with ^{17}O. By means of ^{13}C-^{17}O SEDOR in which she observed the ^{17}O NMR, she measured the C-O bond length to be 1.20 ± 0.03 Å (See Figure 5).

IV. THE BREAK-UP ON HEATING

We can use data such as that of Figure 3 to observe breaking of the C-C bond which results from heating. P.-K. Wang et al. performed annealing

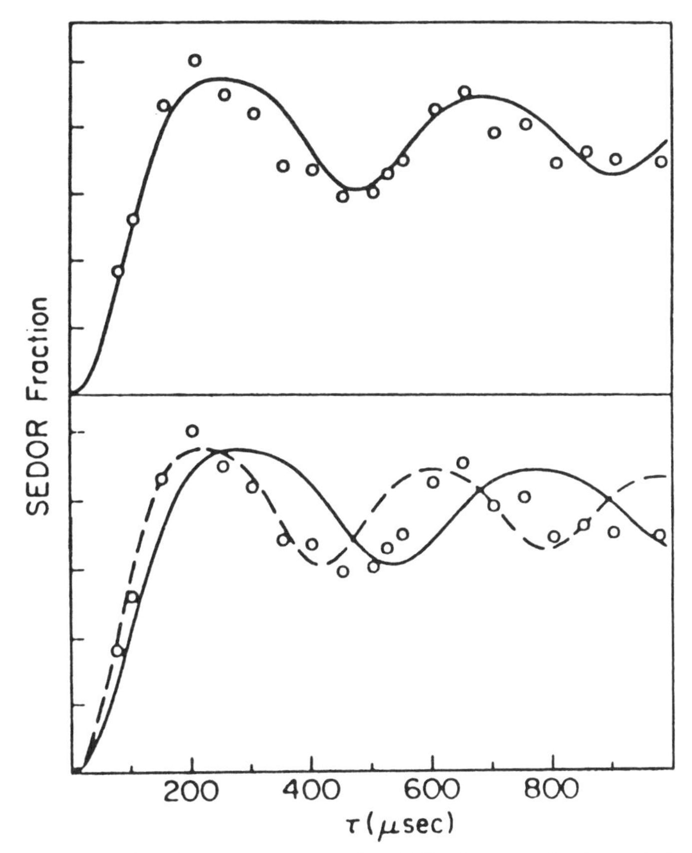

Figure 5. SEDOR fraction vs. τ for CO chemisorbed on Pd. (a) The data fitted by a bond length of 1.20 Å. (b) Predictions for bond lengths 1.15 Å (dashed line) and 1.25 Å (solid line). Each data point represents the sum of approximately 10,000 spin echoes.

experiments[17] in which they observed the ^{13}C resonance at 77 K, but in between measurements heated the sample to progressively higher temperatures. Since the oscillation in the echo envelope results from the dipolar coupling between ^{13}C spins of C atoms bonded to each other, breaking the bond causes that oscillation to disappear. Figure 6 shows the ^{13}C echo envelope for CCH_3 which has been annealed to progressively higher temperatures[17]. The curves all have the same echo amplitude for $\tau = 0$, but the curves for annealing to 479 K shows little or no oscillatory component. These curves can be expressed mathematically as a sum of two types of curve. One, corresponding to ^{13}C atoms bonded to other ^{13}C atoms, has an oscillation which dies out with τ, as seen in Figure 3. The other, corresponding to ^{13}C atoms not bonded to C atoms is a simple exponential decay (as in the 479 K curve). It arises from molecules whose C-C bond is broken (or molecules with a ^{13}C-^{12}C pair). We can fit the data of Figure 6 using such an analysis to determine the fraction of C-C bonds broken at each stage of the annealing. If one then expresses the bond breaking by a rate equation for the rate, ν, of C-C scission:

$$\nu = \nu_0 \exp(-E/k_B T) \tag{9}$$

P.-K. Wang et al. found that for $\nu_0 = 10^{13}$ Hz, E for ethylene on Pt and Ir was 36 kcal/mole, whereas for acetylene E was 37 kcal/mole on Ir but 53 kcal/mole on Pt. Thus, for acetylene we are seeing a dependence of C-C scission on the metal on which the acetylene has been adsorbed. (ν_0 could be measured by studying the time dependence of bond scission at fixed temperature, but we have not done such experiments. For a given ν, a change of a factor of 10 in ν_0 changes E by only 2 kcal/mole).

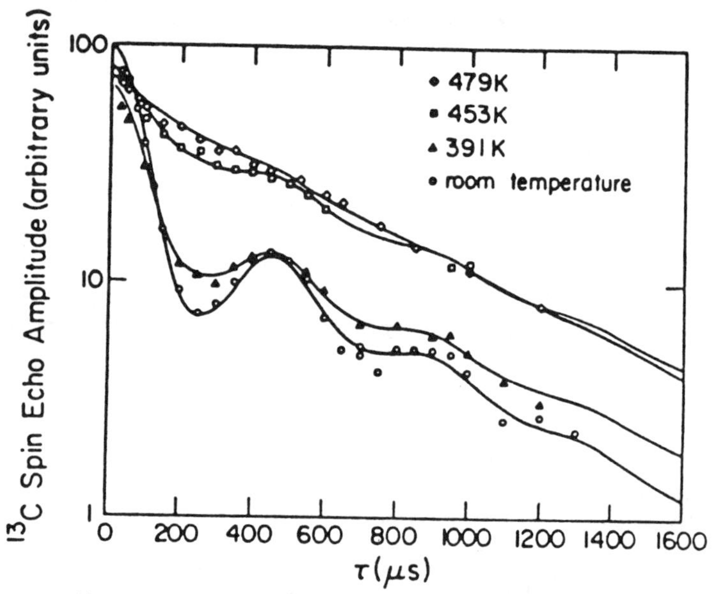

Figure 6. ^{13}C spin echo amplitude vs. τ at 77 K for $^{13}C_2H_4$ on Pt, after adsorption at room temperature (0), and after the sample has been heated to 391 K(Δ), 453 K($\square$), and 479 K ($\diamond$) for 3 hr. The disappearance of the oscillation after annealing at higher temperatures arises because of breaking of the C-C bonds.

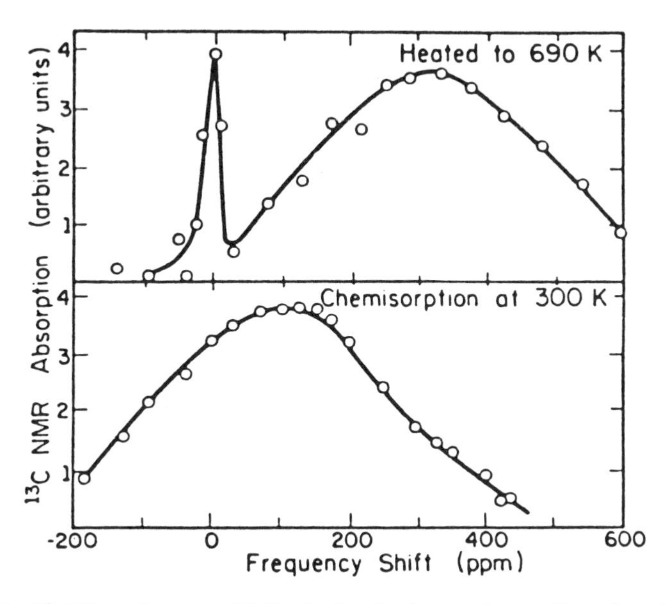

Figure 7. ^{13}C line shape at 77 K of adsorbed acetylene, after chemisorption at 300 K, and after the sample has been heated to 690 K. The solid lines are guides to the eyes.

After the C-C bonds have broken, the ^{13}C spectrum changes as shown in Figure 7. The sharp line with a shift close to 0 ppm is at the position of methane (CH_4). The broad line near 330 ppm arises from isolated C atoms[17]. We know they are isolated for the following reasons. (a) The fact that the ^{13}C echo envelope does not display a slow beat shows that the ^{13}C's are not bonded to other ^{13}C s. (b) By ^{13}C-^{1}H SEDOR, we have shown that less than 15% of the ^{13}C's are attached to H atoms. The fact that the C atoms are attached to the metal is shown by the fact that their spin-lattice relaxation rate $1/T_1$ is proportional to temperature, T, the signal of metallic relaxation mechanisms.

V. MOTION

NMR reveals motion of atoms or molecules in several ways. We will mention the effect on the width of the NMR line, a phenomenon called motional narrowing. As mentioned above, the ^{13}C NMR linewidth arises because of two effects, the ^{13}C chemical shift anisotropy and the local field at the metal surface arising from the magnetic susceptibility of the metal particles. These interactions cause the resonance frequency of a ^{13}C nucleus to vary with position on the particle surface. If the atoms are at rest there is a static broadening, $\delta\omega_0$ in angular frequency, which results. If, however, the atoms can move on the surface, the ^{13}C nuclear resonance frequency changes in time. For sufficiently rapid motion, one should now use some average frequency. If τ_0 is the time for the frequency to change an amount comparable to $\delta\omega_0$ (i.e. diffuse over a distance such as from the equator to the north pole of one of the metal particles), magnetic resonance theory[9] shows that the apparent line width $\delta\omega$ obeys the equation

$$\delta\omega = (\delta\omega_0)^2\tau_0. \tag{10}$$

Thus, since τ_0 varies with temperature so does $\delta\omega$. This equation describes the line width for motion which is fast enough that

$$\delta\omega_0\tau_0 \leq 1 \tag{11}$$

Figure 8 shows data of Shore et al for ^{13}CO on Pd[12]. The line shapes of the resonance are observed at a variety of temperatures. By 298 K, the full motional narrowing has set it.

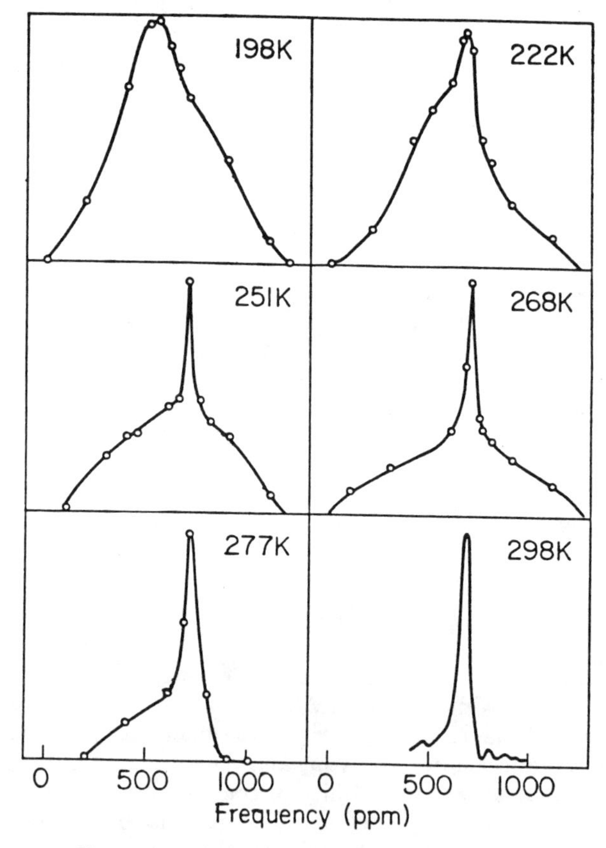

Figure 8. The ^{13}C line shape as a function of temperature for ^{13}CO adsorbed on Pd, showing the narrowing of the resonance line as the CO diffusion rate increases at higher temperatures.

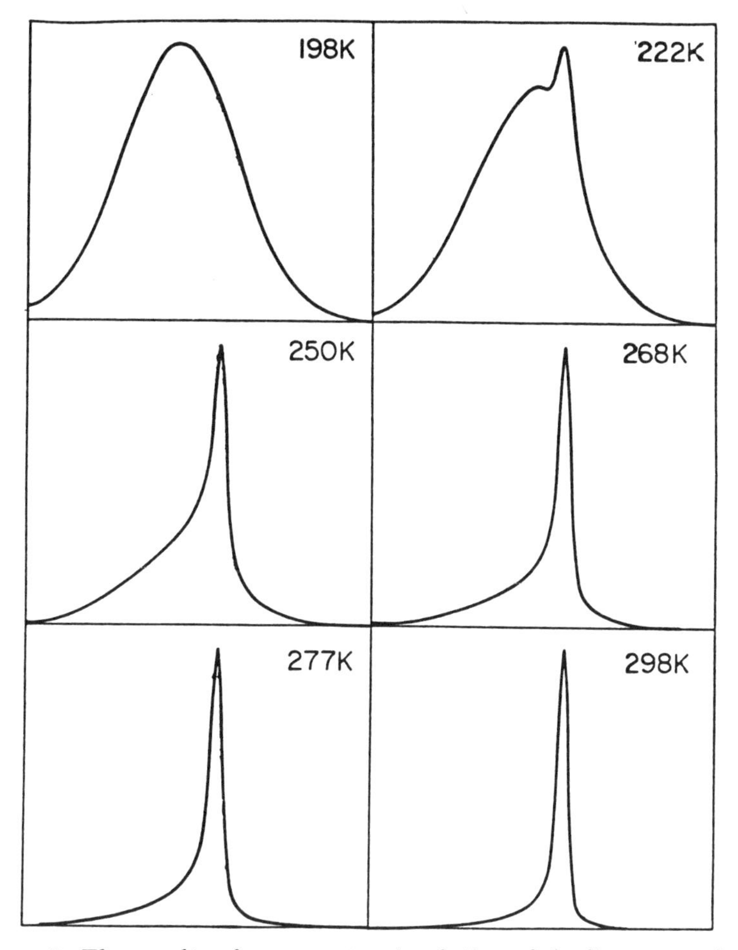

Figure 9. The results of a computer simulation of the line narrowing of Figure 8. The simulation assumes a diffusion energy of 6 kcal/mol and a particle-size distribution reasonable for our sample.

It is interesting to note that the narrowing is manifested by a narrow peak on a broad base (see for example 251 K). This effect arises because of the fact that we have a distribution in particle size, as I will now explain.

$\delta\omega_0$ is easily shown to be independent of particle size. So is the diffusion constant of atoms or molecules on the metal surface. But since τ_0 corresponds to diffusion from the equator to the north pole, it involves diffusion over a longer *distance* for larger particles than for smaller particles. As a result, the condition for motional narrowing, Equation 11, is satisfied at lower temperatures for smaller particles than for larger particles. Shore et al. have applied the motional narrowing criterion to a particle size distribution determined by electron microscope to obtain the theoretical curves of Figure 9. They find an activation energy for diffusion of CO or Pd of 6 ± 2 kcal/mole. This value is about 1/2 that found by Ansermet for diffusion of CO on Pt. P.-K. Wang, et al. found 7 ± 2 kcal/mole for C atoms on Pt, thus showing that the C atoms are very mobile.

In addition to observation of line narrowing, motion reveals itself in NMR studies of spin lattice relaxation times as well as in effects on the transverse relaxation time called T_2.

I am pleased to acknowledge that this research has been supported by the Department of Energy Division of Material Sciences under contract no. DE-AC02-76ER01198, and by grants from the Exxon Education Foundation.

REFERENCES

1. H. Frauenfelder, Helv. Phys. Acta 23, 347 (1950).
2. Howard Rhodes, Harold Stokes, Serge Rudaz, Claus Makowka, Po-Kang Wang, Jean-Philippe Ansermet, Zhiyue Wang, David Zax, Dale Durand, Chris Klug.
3. (a) C.P. Slichter in The Structure of Surfaces, eds. M.A. Van Hove and S.Y. Tong (Springer-Verlag, 1984). (b) P.K. Wang, J.-P. Ansermet, S.L. Rudaz, A. Wang, S. Shore, C.P. Slichter, and H.J. Sinfelt, Science 234, 35 (1986).
4. H.E. Rhodes, P.-K. Wang, H.T. Stokes, C.P. Slichter, and H.J. Sinfelt, Phys. Rev. B 26, 3559 (1982); H.E. Rhodes, P.-K. Wang, C.D. Makowka, S.L. Rudaz, H.T. Stokes, C.P. Slichter, and J.H. Sinfelt, Phys. Rev. B 26, 3569 (1982); H.T. Stokes, H.E. Rhodes, P.-K. Wang, C.P. Slichter, and J.H. Sinfelt, Phys. Rev. B 26, 3575 (1982).
5. C.D. Makowka, C.P. Slichter, and J.H. Sinfelt, Phys. Rev. Lett. 49, 379 (1982) and Phys. Rev. B 31, 5663 (1985).
6. J.H. Van Vleck, The Theory of Electric and Magnetic Susceptibilities, London, Oxford Univ. Press (1952).
7. J.R. Schrieffer and P. Soven, Physics Today 38 (4), 24 (1975).
8. G. Blyholder, J. Phys. Chem. 68, 2772 (1964).
9. C.P. Slichter, Principles of Magnetic Resonance, 2nd edition, corrected 2nd printing (Springer-Verlag, Berlin, 1980).
10. S.L. Rudaz, J.-P. Ansermet, P.-K. Wang, C.P. Slichter, and J.H. Sinfelt, Phys. Rev. Lett. 54, 71 (1985).
11. J. Korringa, Physica (Utrecht) 16, 601 (1950).
12. S.E. Shore, J.-P. Ansermet, C.P. Slichter, and J.H. Sinfelt, Phys. Rev. Lett. 58, 953 (1957).
13. J.-P. Ansermet, P.-K. Wang, C.P. Slichter, and J.H. Sinfelt, Phys. Rev. B 37, 1417 (1988).
14. G.E. Pake, J. Chem. Phys. 16, 327 (1948).
15. P.-K. Wang, C.P. Slichter, and J.H. Sinfelt, J. Phys. Chem. 89, 3606 (1985).
16. P.-K. Wang, C.P. Slichter, and J.H. Sinfelt, Phys. Rev. Lett. 53, 82 (1984).
17. P.-K. Wang, J.-P. Ansermet, C.P. Slichter, and J.H. Sinfelt, Phys. Rev. Lett. 55, 2731 (1985).

APERIODIC CRYSTALS: BIOLOGY, CHEMISTRY AND PHYSICS IN A FUGUE WITH STRETTO

Peter G. Wolynes
Noyes Laboratory
University of Illinois
Urbana, IL 61801

I. INTRODUCTION

Important unsolved problems often straddle several of the canonically defined intellectural disciplines. This situation occurs because successful ideas diffuse from one area of study to another rather quickly but when a key concept is missing a barrier to progress arises in many fields. Often at this stage a useful development is the introduction of what we might call a "soft concept." A "soft concept" is a poetic and thought-provoking idea, that is specific enough to suggest further work but vague enough not to prematurely limit the search for truth. Hans Frauenfelder, whose birthday we celebrate in this volume, loves to introduce such soft concepts and many of us value him and his work precisely for providing these ideas to us. This article is a review of a continuing exploration of a soft concept: the aperiodic crystal. The concept was not, in fact, introduced by Hans but it has been one with which he has often toyed in his studies of protein dynamics. Many of the ideas involved in the development of this concept have parallels with Hans's ideas of proteins and my contributions to this problem have been greatly stimulated by these parallels and by my interactions with him.

The problem for which the aperiodic crystal concept seems relevant are all tied up with the question of how a complex structure, one apparently requiring many bits of information to describe it, can be stable for long periods of time. This kind of question arises in many parts of biology, chemistry and physics, as well as other human endeavors. In biology we ask how the complexity of genetic information or of psychological memory can be stable. In physics, it was a truism that stable structures could be described by a few parameters[1] until disordered spin systems with complex and nonobvious order were uncovered[2]. In physical chemistry the clearest manifestation is the existence of glasses. Windows are sensibly stable for centuries yet at the molecular level the structure of glass seems hardly different from that of a swirling cup of water with its myriad configurations. The term aperiodic crystal is most apt for these structural glasses and my essay here will deal primarily with the study of the glass transition. Nevertheless I will try to make clear how there can be useful crossfertilization with ideas from many areas of study.

In this essay I hope to illustrate the allure of a soft concept. In doing this I ask the reader to keep in mind the forceful articulation of the aperiodic crystal notion in the book "What is Life?" In it Schrödinger said[3]: "In physics we have dealt hitherto only with periodic crystals. To a humble physicist's mind, these are very interesting and complicated objects; they constitute one of the more fascinating and complicated material structures by which inanimate nature puzzles his wits. Yet compared with the aperiodic crystal, they are rather plain and dull. The difference is of the same kind as that between an ordinary wallpaper in which the same pattern is repeated again and again in regular periodicity and a masterpiece of embroidery say a Raphael tapestry, which shows no dull repetition, but an elaborate coherent, meaningful design traced by the great master."

The analogy with artistic works is a particularly good one because it allows us to realize that structurally sound structures can be built on a simple global plan or may be based on an unfolding pattern in which only local rules of order may be apparent. Another analogy to which we will return is to a musical composition obeying only local rules of harmony but which, for great works, have a sense of inevitability.

II. GLASSES: OBSERVATIONAL AND STRUCTURAL PHENOMENOLOGY

A. Macroscopic Behavior

Although glassy materials are familiar to everyone it is not always realized just how ubiquitous the formation of glasses is. Glasses can be formed from materials with ionic, covalent, van der Waals or metallic bonding[4]. Many people believe any material can be produced in a glassy state. Nevertheless, crystallizaiton into ordinary periodic crystals does intrude for some of the simpler substances such as the condensed states of the rare gases. The crystallization problem seems to be separable from the phenomena of the glassy state and that is the stance taken here. Glasses may be formed in many ways, for example, by vapor deposition or by chemical reactions, but the most characteristic approach is the cooling of the melt. If crystallization is avoided glass formation invariably occurs (except for liquid helium!). The dynamics of motion slows down and eventually no motion occurs on the laboratory time scale. It is, of course, very familiar that molecular processes slow down with cooling, sometimes quite a bit. Chemical reactions involve surmounting an energy barrier so such processes gets much slower upon cooling usually following an Arrhenius law: $k = \omega_0 e^{-E/k_B T}$. A few glass-forming liquids show this behavior, such as pure silica. Here the puzzle is to explain why the energy barriers are there in the first place. More commonly the slowing down of transport is more dramatic than the Arrhenius law. Fluids at high temperature have very low "activation energies" for molecular motion. In fact computer simulation studies show there really are not discrete barrier surmounting events in ordinary simple liquids[5]. However at low temperature approaching the glass the activation energy increases sometimes reaching values of tens of electron volts! This behavior, shown for D^{-1} in Figure 1, is often fitted with an empirical form known as the Vogel-Fulcher law:

$$D = D_0 e^{-A/(T-T_0)} \tag{1}$$

Such a low if it held rigorously true would imply no motion below the temperature T_0. In fact there are always deviations from this law because as we approach T_0, the system can no longer approach equilibrium before the diffusion (or similarly viscosity) is measured then the diffusion is faster than expected. The Vogel-Fulcher law does reasonably well at describing duffusion or viscosity over sixteen orders of magnitude. Surprisingly, for a three parameter fit, it does not do extremely well. The behavior at low viscosity seems different, with a change around one poise or so.

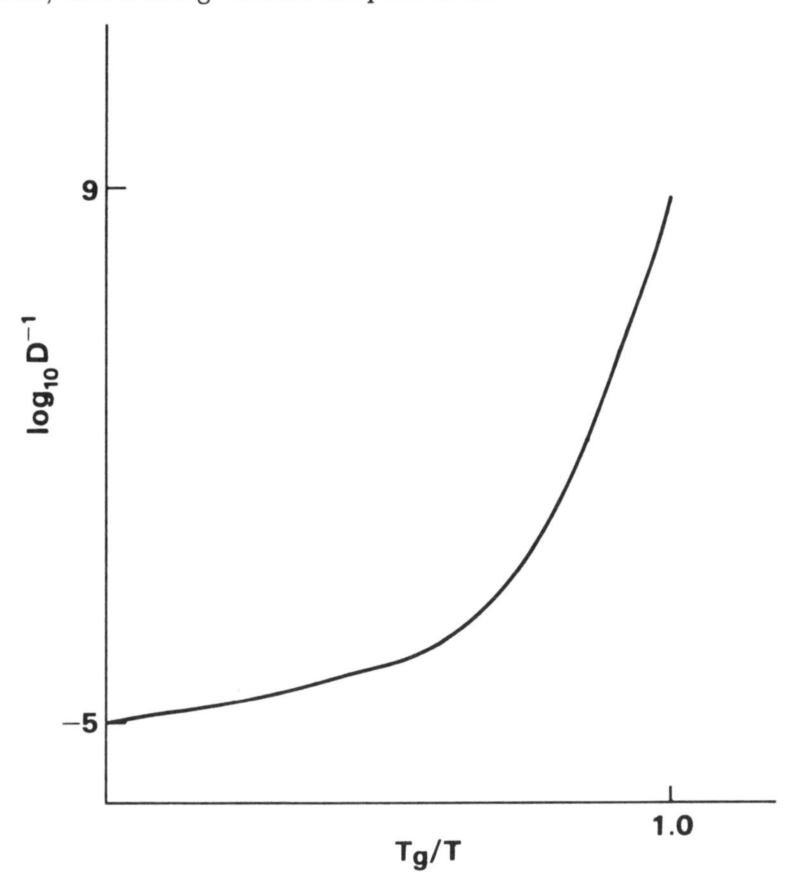

Figure 1. The logarithm of the inverse diffusion constant is plotted versus temperature, unstarting the Vogel-Fulcher behavior.

The appearance of a characteristic temperature in the Equation (1) suggests looking at the thermodynamic changes on forming a glass. As in Figure 2a for the volume, extensive quantities do not show discontinuities but rather, very nearly, discontinuities of slope. Similarly the heat capacity of a liquid is rather high but at the glass transition it falls to a value more typical of the crystalline state. The transition point is kinetically determined but for materials obeying Eq. (1) is relatively sharp. Materials that show large thermal anomalies show the dramatic departures from Arrhenius transport behavior. This has led to the useful classification of liquids into "fragile" and "strong" (e.g. SiO_2) categories by Angell[6]. The heat capacity (as shown in Figure 2b) is especially interesting because it is related to the entropy. Since the liquid heat capacity is higher than the solid, (although it has more entropy than a crystal at melting) it is losing entropy much more rapidly when it cools. In the usual case if the glass transition did not intervene the liquid entropy would fall below the crystal at a temperature T_K, which is not very far from the T_0 found in the kinetic data. The idea of a liquid with less entropy than a crystal is disturbing and this observation is known as the Kauzmann paradox[7]. In fact the crisis is rather more severe since a little below T_K, as noted ever earlier by Simon, there would apparently be a violation of the third law of thermodynamics[8]. There seems to be a connection (although it involves an extrapolation!) between thermodynamics and kinetics.

Another feature connected with glasses is that nonexponential decays are usually observed for supercooled liquids. This occurs much above the glass transition temperature, but it does signal the presence of structures with different time scales of persistence. In systems not showing large thermal anomalies the nonexponentiality is generally weak.

B. Macroscopic Structure

Because of the ubiquity of glass formation and because of the current inadequacy of our experimental tools for studying large systems lacking periodicity, many ideas for glassy structure have been put forward. The idea that an aperiodic system can have well-defined structural themes is both an old and a currently developing idea. The first person to make much headway with crystallographic thinking for liquids and glasses was Bernal[8]. His studies began with a beautiful experiment in which ball bearings were placed in an elastic bay and uniformly compressed. This is a crude simulation of what happens when a van der Waals fluid is cooled since the attractive forces cause a contraction of the fluid and the effective diameter of the particles increases upon cooling[9]. In fact Bernal ended up with a mechanically stable but noncrystalline packing i.e. a glass. By pouring some paint into the structure, allowing it to harden and then patiently disassembling the structure Bernal was able to investigate the local connectivity of the structure. He discovered the local packings and their statistics were very different from those in the periodic crystalline structures. For example there were a surprising number of near icosahedral local structures. The icosahedron is a very efficient local packing of spheres but the five-fold symmetry of icosahedra make the higher packing of such icosahedra very inefficient in a periodic structure.

The icosahedron can form large ordered structures. In various intermetallic compound it is the basis of periodic crystalline structures with huge unit cell as for example, $Mg_{32}(Al, Zn)_{49}$ a so-called Frank-Kasper phase[10]. We might imagine an aperiodic packing as the limit of such a structure with an infinite unit cell. In a somewhat different sense this situation has been realized

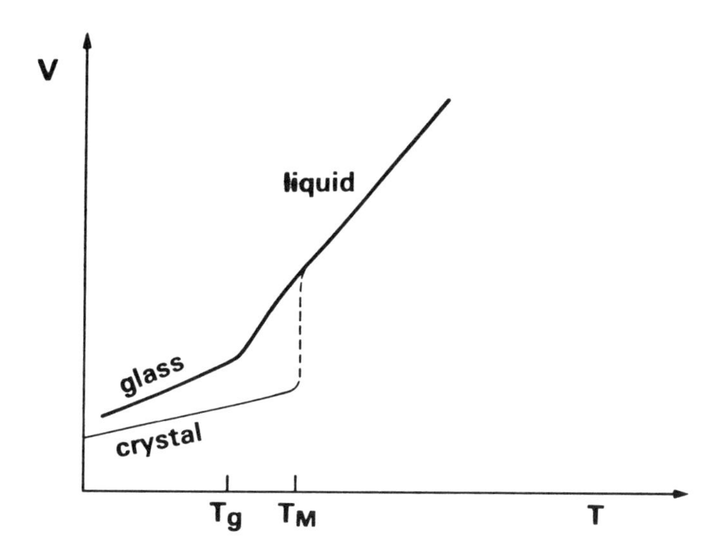

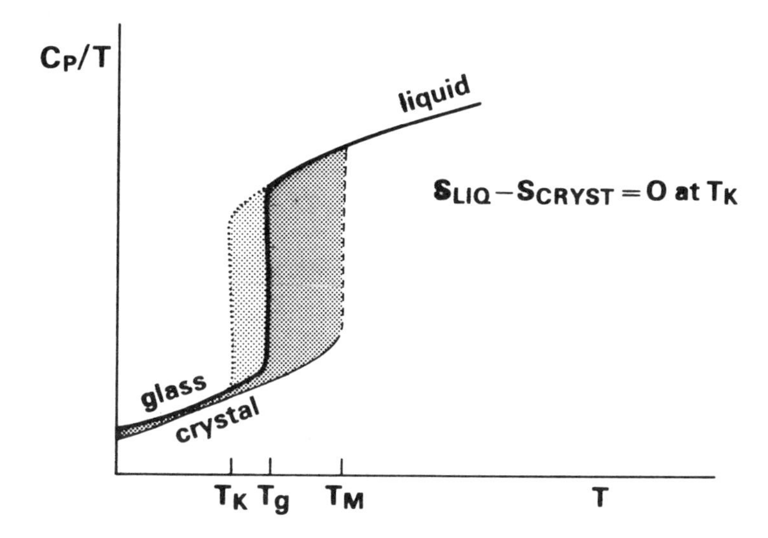

Figure 2. (a) The volume plotted versus temperature for the crystal, liquid and glassy states. (b) The heat capacity is plotted as C_p/T versus T. The shaded area is the entropy loss on cooling from the equilibrium melt. At T_K this entropy loss equals the entropy of fusion.

in the recent discoveries of quasicrystalline alloys. As Levine and Steinhardt[11] originally pointed out, these structures are related to the Penrose packing with five-fold symmetry[12], an illustration of which is shown in Figure 3. They are not precisely periodic structures but are different from glasses in that they are weakly periodic[13] or quasi-periodic, any given size region can be found repeated nearly but an even new pattern is formed on that scale[14]. The Penrose pattern does indeed Bragg diffract but at an infinite number of reciprocal vectors. It is experimentally distinguishable from glasses which have a continuous scattering pattern. The chemical similarity of the quasicrystalline alloys and metallic glasses suggests a connection however.

An appealing notion is that there may be a series of transitions, periodic, quasiperiodic, chaotic (glassy) like those occurring in the transition to turbulence. This idea has impelled many studies[15].

The packing problem for icosahedra is a geometrical one and many beautiful geometric ideas have been used to describe aperiodic packings. An important notion is that aperiodic packings may be related to periodic or regular packings in curved space. Just as a soccer ball can be covered with pentagons, icosahedra nicely pack in a positively curved space. Sadoc[16] has exploited this in describing many models of glasses in terms of distorted structures formed from uncurving and adding defects to a curved space packing, much as a flat map of the World can be produced with various projections. Nelson and others have developed many field theoretic models of glass structure based on the curved space notion[17,18]. Just as there are many space groups for periodic crystals but melting is a general phenomenon, the ubiquity of glasses suggests that a rather general approach to the glass transition needs to be found and these curved space ideas should be thought of as beautiful but particular examples of possible mechanisms.

III. NAIVE THEORIES OF THE GLASS TRANSITION: STABILITY ANALYSIS AND FRICTION CRISIS

A. Stability of an Aperiodic Crystal

One of the preliminary studies of any phase transition is to establish the limits of stability of the different phases. If the transition is of a classical continuous nature this stability criterion also locates the transition. If the transition is first order only a bound on the transition temperature is produced. In any case important clues to the nature of the transition can be found by such a stability analysis.

The instability of a periodic crystalline solid to thermally excited vibrations has been a starting point for understanding melting for a long time. In an obscure paper, Sutherland proposed that melting occurs when a critical value of thermal motion exists[19]. The idea that melting occurs when the mean square displacement of atoms from their sites reaches a critical value is now known as the Lindemann criterion[20]. At melting the root mean square displacement of an atom is roughly one tenth of the interparticle spacing. Is a Lindemann criterion relevant for the devitrification of glass?

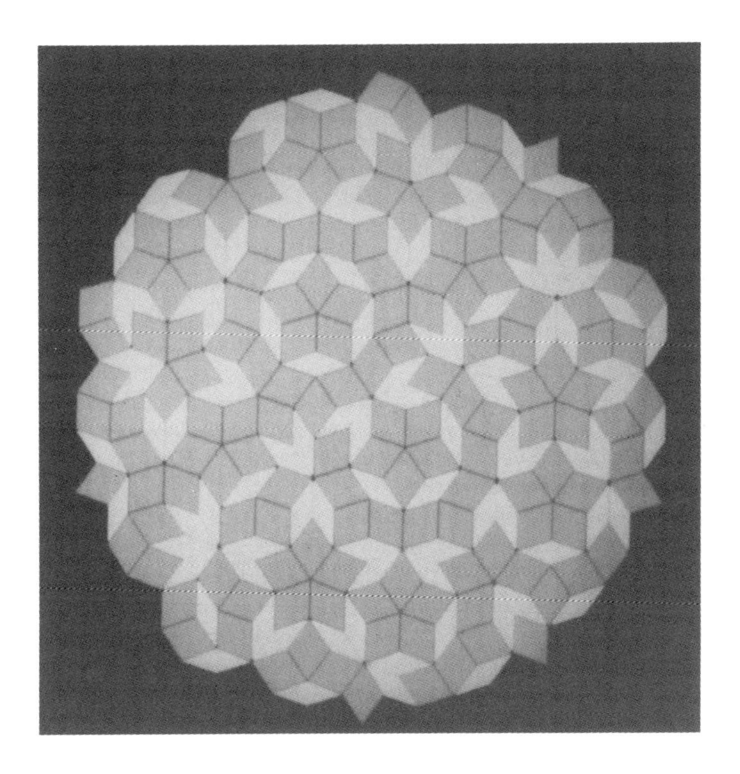

Figure 3. A portion of a Penrose tiling.

Information on the amplitude of thermal motion in glasses comes from several sources. Computer simulation studies of glass formation use the form of particle displacement versus time as a criterion for whether the system is a glass or a liquid. For a "glass" the $\langle r^2(t)\rangle$ increases rapidly to a plateau value while for a "liquid" $\langle r^2(t)\rangle$ first increases rapidly followed by a slower but continued increase given by the diffusion law $\langle r^2\rangle = 6Dt$, D being the diffusion constant. On the time scales of simulation ($\sim$ 100 psec. at the moment) this behavior of the simulated glass is just like that of a solid. The shift from one behavior to the other is rather abrupt on the simulation time scale. Laboratory data are rather sparse on the amplitude of thermal motions but the Mössbauer effect does come to the rescue. Using the Mössbauer effect directly on Mössbauer active nuclei dissolved in a glassy liquid[21] or through the use of Mössbauer scattering one obtains information on the recoilless fraction like that shown in Figure 4. In a crystal the recoilless fraction is directly a measure of the extent of thermal motion being essentially a Debye-Waller factor, $f = \gamma_0 e^{-q^2\langle r^2\rangle}$ where q is the wave vector of the gamma ray. At melting the Mössbauer signal generally disappears. Notice that the glass "Debye-Waller factor" follows a crystal-like behavior upon heating until T_g whereupon it usually begins to drop precipitously although, to be sure, continuously, until the Mössbauer signal disappears entirely at a somewhat higher temperature. A striking feature of the Mössbauer signal is that there is no appreciable broadening of the line until the higher temperatures where it begins to disappear. Thus the charge in the static structure not some sort of motional averaging effect.

Several static theories give a Lindemann criterion for the glass transition. Stoessel and Wolynes adapted a self-consistent phonon theory of Fixman[22] to study the stability of an amorphous lattice[23] based on a random close packing of hard spheres. Fixman's theory computes an effective potential for the thermal vibration about a site by averaging a Mayer f bond over the thermal vibrations of its neighbors. Fixman's approach is very much like an extension of the virial expansion for repulsive core systems to a state of broken translational symmetry. It is essentially exact near close packing and gives a very good rendition of the phase transition between the fcc hard sphere solid and liquid when combined with a liquid equation of state. Although many approximations were made by Stoessel and Wolynes the transition densities obtained by this approach are rather like those found for computer simulated glasses. Like the instability of a periodic crystal the mechanical instability sets in at a finite value of $\langle r^2\rangle$. Recently Hall, Mertz and Wolynes have carried out more elaborate calculations using the self-consistent phonon approach[24]. In these calculations the centers of vibration in the aperiodic lattice are allowed to readjust to minimize the free energy and different atoms are allowed to have different mean square displacements. Although the study is rather coarse it shows that aperiodic free energy minima exist and that the transition is very abrupt, seeming to affect the whole sample at once.

Another kind of theory giving a Lindemann criterion is based on the density functional theories of freezing[25]. These theories characterize the periodic solid by the presence of a density wave. Successful theories of the structure of liquids have been based on expressions for the free energy of forming an inhomogeneous spatially varying density. Thus the relative free energy of a

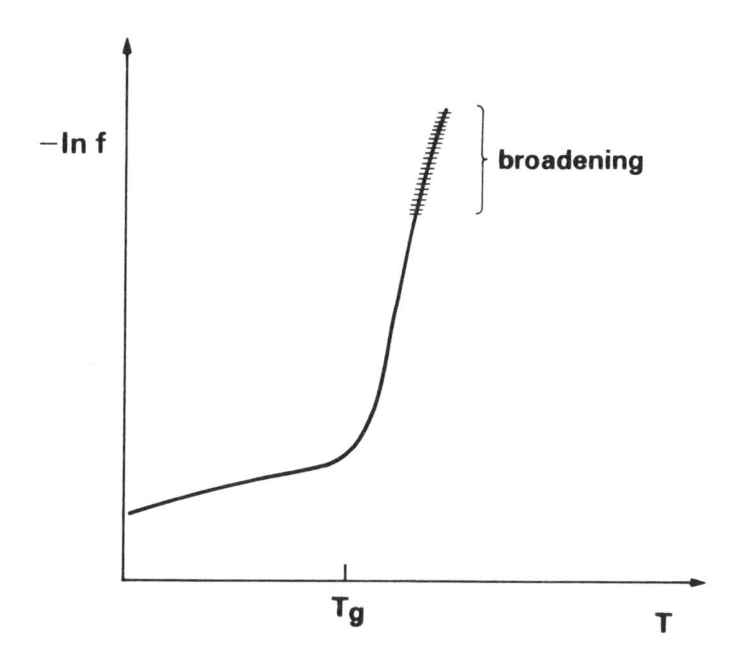

Figure 4. The logarithm of the recoilless fraction is plotted versus temperature for the crystal, liquid and glassy states. In the striped region broadening of the Mössbauer line is observed.

crystal and a liquid can be evaluated. Many approximate free energy functionals exist. The one used by Ramakrishnan and Youssoff in their seminal work[26] has been extensively investigated by others. This free energy consists of two parts, an ideal gas entropy term and an effective interaction term:

$$\begin{aligned}-\frac{1}{k_B T}F[\rho(r)] = &-\int d^3x\rho(x)\log\rho(x)\\ &+\int\int d^3xd^3x'(\rho(x)-\rho_0)c(x-x')(\rho(x)-\rho_0).\end{aligned} \tag{2}$$

This free energy functional is the first part of a systematic expansion about the uniform state. In practice, the direct correlation function $c(r)$ may be determined by demanding that the magnitude of small amplitude density fluctuations in the uniform state $\langle|\rho_k|^2\rangle$ is correctly given. The latter is obtainable from diffraction experiments. Alternatively a variety of approximation to $C(\boldsymbol{r})$ exists from ab initio liquid structure theory.

Although trial density waves in terms of an expansion in Fourier components were originally used, an alternative trial function is based on the picture of thermal vibrations about fixed sites:

$$\rho(r) = \sum_{\substack{\text{lattice}\\ \text{sites}}} \left(\frac{\alpha}{\pi}\right)^{1/2} e^{-\alpha(x-x_i)^2} \tag{3}$$

When used for periodic lattices this function in terms of a single parameter α which is inversely related to the thermal vibration amplitude gives an excellent agreement with the more complex Fourier expansions[27]. This form is very convenient for calculations on the possible freezing into an aperiodic lattice as well. Singh, Stoessel and Wolynes used this sort of trial function with the $\{x_i\}$ chosen to lie on a Bernal packing to show that such an aperiodic density wave can become locally stable at a high enough density[28]. The competition of effects in $F(\alpha)$ is clear. $\alpha = 0$ is favored by the ideal gas entropy but the interactions are minimized if the vibrational envelopes of atoms avoid each other i.e. at large α. Thus a secondary minimum occurs at a finite α, discontinuously. Like the self-consistent phonon theory, the density functional theory gives a Lindemann criterion and a discontinuous transition. This should not be too surprising because both of these theories have the same mathematical structure for the aperiodic crystal as for the periodic crystal.

B. Friction, Mode Coupling and the Glass Transition

An apparently different approach to the glass transition views it as a crisis arising from friction effects. This view is particularly worth examining because of the emerging consensus that dynamical friction plays an important role in the theory of chemical reactions in condensed phases[29].

In the low viscosity liquid regime, the mechanism for friction is essentially the same as in a gas, isolated binary collision events, which are like those in the gas but more frequent. As the temperature is lowered, or the density decreased, a new effect involving multiple collisions comes in. This can be seen in the form of the force correlation function pictured in Figure 5[30]. Linear response theory indicates that the friction on a moving particle in the fluid is essentially the integral of this force autocorrelation (some technicalities aside):

$$\zeta = \frac{1}{k_B T} \int_0^\infty \langle F_z(0) F_z(t) \rangle dt \tag{4}$$

The rapid initial decay of $\langle F(0)F(t)\rangle$ is due to the individual collisions. The longer decay has some contribution from hydrodynamic effects but a more important part comes from structural relaxation in the fluid. This structural relaxation contribution to dynamics has been well documented and quantitatively worked out for the hard sphere fluid at moderate densities by Kirkpatrick[31] and others[32]. Qualitatively we can see how it could lead to a friction crisis. If we think of the structure of the fluid being frozen it is clear that the initial amplitude of the structural relaxation $A(\rho)$ should increase slowly with density. On the other hand the structural relaxation time also increases with density. Carrying out the integral in, we have roughly:

$$\zeta = \zeta_0 + A(\rho)\tau_R \tag{5}$$

where ζ_0 is the collisional contribution and τ_R is relevant relaxation time. Since the structural relaxes by the diffusion of fluid particles τ_R should be inversely proportional to the diffusion constant. Using the Stokes-Einstein relation $D = k_B T/\zeta$ then implies a relation between τ_R and ζ:

$$\tau_R = B\zeta \tag{6}$$

If these equations are solved simultaneously we find a feedback relationship leading to a divergence:

$$\zeta = \frac{\zeta_0}{1 - BA(\rho)} \tag{7}$$

Thus if $BA(\rho)$ exceeds one the friction constant would diverge, leading to a vanishing diffusion constant e.g. a glass. Arguments like this can be traced back to Geszti[33] and Turski and Sjölander[34]. The argument is rather generic and similar to the theory of dielectric friction in complex, charged fluids[35]. Theories of the glass transition with a very similar flavor have arisen by taking the mode-coupling approach to dense fluid dynamics seriously[34,37,38,42]. These theories are essentially self-consistent perturbation theories of the dynamics of many body systems. They have had enormous successes in the description of dynamic critical phenomena[39], quantum localization[40] and in describing quantitatively the dynamics of moderately dense fluids[41]. Often these theories involve rather complicated mathematical manipulations which have, in my opinion, frightened some of the younger generation of physicists from studying them. In their application to the glass transition they must be pushed to the limit and the lack of a small parameter casts doubt on their validity, despite their powerful successes outside the glassy regime.

Insight into the mathematical structure of the glass transition predicted by mode-coupling theories has been gained by the ruthless simplification of the mode coupling theory proposed by Leutheusser[42]. Leutheusser eliminates all wave-vector dependence from the theories, replaces spatial integrals by single values and arrives at a deceptively simple integro-differential equation:

$$\ddot{\phi} + \zeta_0\dot{\phi} + \omega_0^2\phi + 4\lambda\omega_0^2 \int_0^t d\tau \phi^2(\tau)\dot{\phi}(t-\tau) = 0 \tag{8}$$

On physical grounds this equation is *too* general and has been criticized because it makes it seem that practically any physical system will exhibit a glass transition. Nevertheless its solutions have much in common with the more

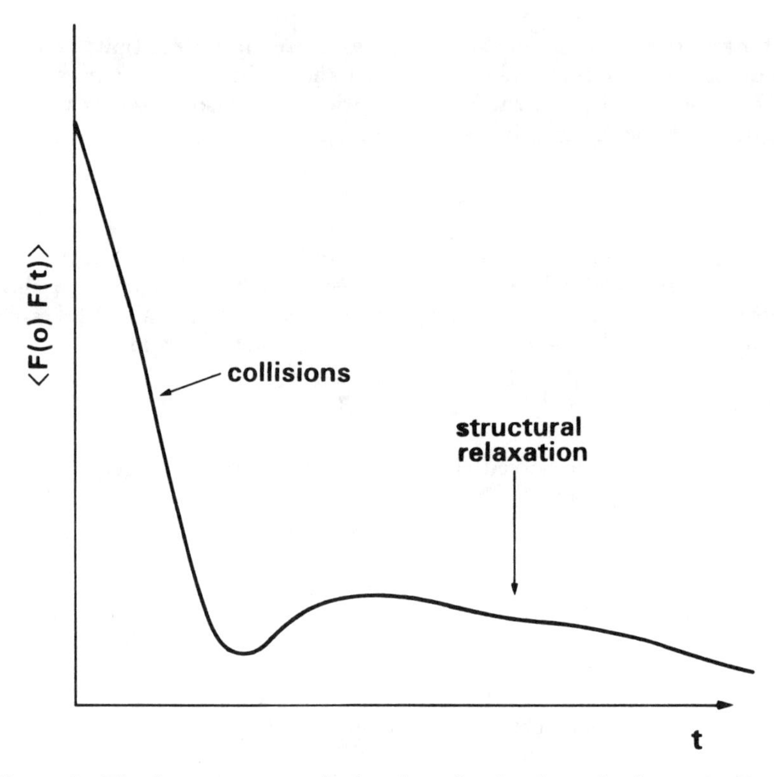

Figure 5. The force autocorrelation function is plotted schematically.

realistic but complex calculations. The Leutheusser equations show a continuous slowing down near their transition but also show a discontinuity in the long time limit of the autocorrelation function! (We will see shortly how this fits in with actual glassy behavior.)

Despite their different roots the mode coupling theories have a great deal in common with the static approaches outlined in the last section. Kirkpatrick and Wolynes have shown their relationship, at least for approximate versions of the theory[43]. (Their physical approximations are as ruthless as the mathematical ones made by Leutheusser.) The free energy functional allows one to find the force on a test particle at x caused by a density fluctuation:

$$F = \int d^3x' \nabla c(x - x') \delta\rho(x', t) \tag{9}$$

This then gives an expression for the force correlation in terms of the fluctuations of the density surrounding the particle:

$$\langle F(0)F(t) \rangle = \int d^3x_1 d^3x_2 \nabla c(x - x_1) \nabla c(x - x_2) \delta\rho(x_1, 0) \delta\rho(x_2, t) \tag{10}$$

If the density fluctuations do not decay in time then $\lim_{t\to\infty}\langle F(0)F(t)\rangle = k$ is nonzero. The zero frequency friction coefficient would diverge but the finite friction coefficient would have a pole

$$\zeta(\omega) = \frac{1}{k_B T}\int_0^\infty \langle F(0)F(t)\rangle e^{i\omega t}\,dt = \frac{k}{i\omega} + \zeta'. \tag{11}$$

This means at low frequency the "friction" force is not proportional to velocity but to displacement: k is an effective spring constant and the motion of the test particle is confined. Thus we can demand self-consistency. If the test particle is confined to a harmonic well so must the bath particles. The long time limit of the density correlations will be a Debye-Waller factor times their initial correlation. Thus one obtains a self-consistent condition using a Fourier representation.

$$k = \frac{1}{2}\int \frac{d^D g}{(2\pi)^D} C(q) n h(q) e^{-q^2/2k}. \tag{12}$$

Here D is the dimension of space and $S(q) = nh(q) - 1$ is the static structure function.

A very similar result is obtained by the use of the static density functional theory. At high dimension the theories exactly agree and, it would seem, are systematic in the sense that $1/D$ is a small parameter. Unfortunately the forms of hard sphere packings in very high dimension are unknown[44] but the transition likely occurs at a realizable density. We see there is an intimate connection between the friction crisis envisioned by the dynamic theories and the existence and stability of aperiodic crystal packings.

The naive theories discussed in the previous sections certainly do not exactly describe the glass transition in the laboratory. They compare rather well with the computer simulations but we know these are limited by time scale. What the theories do suggest is a change in dynamical behavior. Above the transition temperature (or volume) predicted by these theories the fluid is strongly fluctuating, no individual local structure lasts much longer than a vibrational period. Below the transition temperature individual local structures become metastable and can begin to live quite long compared to a vibrational period.

IV. DIVERSITY, STABILITY AND ENTROPY

A. Spin Glasses and Structural Glasses: Symmetry and Randomness

The naive theories of glasses fail to take into account that there may be *many* aperiodic crystal structures. Thus unlike the case of periodic crystals we must understand the statistics of these structures and the transitions between them. While the diversity of aperiodic crystal states seems manifest in that they apparently require many bits of information to describe them it has been difficult to understand this diversity quantitatively from simple theory.

A great deal has been learned about structural glasses by computer cimulation techniques. Powerful ideas about the statistics of energy minima have been learned this way in the studies of Stillinger and Weber[45] and much of this should carry over to the free energy minima discussed here. In fact, a program to do this is underway[24]. Purely analytic models can give us insight into the interplay of structural diversity and stability which is the basic issues here. What is required is a simpler system in which the molecular details are suppressed (although they are important!) but the collective features are preserved. No totally satisfactory model like this has been found. The glass problem now lacks a simple model to play the role that the Ising magnet analogue did for the liquid-gas transition (at least a partially satisfactory analogy does exist — the Potts glass[43,46,47,48]. Before describing this "exotic" system I'd like to introduce some general ideas about spin glasses.

The idea of a spin glass was invented to deal with transitions observed in disordered magnetic alloys[2]. If magnetic impurities, e.g., manganese ions, are dispersed in a metallic but nonmagnetic matrix, e.g., copper, the spin-spin interactions between the impurity spins are of oscillating sign because they are mediated by the electron gas of the host. Since the impurities do not lie on a regular lattice the interactions are sometimes ferromagnetic and sometimes antiferromagnetic. It is not clear that any simple order such as parallel or antiparallel alignment of the spins should be the ground state and thus one might have expected the system to be only weakly affected by the interactions and exhibit no phase transitions. Instead experiments showed that below a reasonably sharp temperature the spins retained their alignment for long periods of time and associated with this there were hysteresis effects and weak thermal anomalies. This coexistence of extreme sluggishness and a disordered state suggested a connection with the structural glasses and hence the name. The parallels between glass and the usual spin glass have always been tantalizing but on closer examination there have always been significant differences.

The most studied spin glass is the Ising spin glass. The real Hamiltonian of the experimental system is abstracted rather severely to a model with just a few features:

$$H = \sum_{i,j} J_{ij} S_i S_j. \tag{13}$$

Here the $\{S_i\}$ are Ising spins and J_{ij} are fixed but random variables mimicking the interactions between the impurity spins whose locations are fixed. The J_{ij} may be short or long ranged but have zero mean and a sizable mean square.

One feature of this Hamiltonian that seems common to glasses is that it exhibits *frustration*[49]. With only two choices, up or down, for each of the Ising spins it is impossible simultaneously to satisfy the conflicting demands of each of the interactions in the model. This conflict is much like the competing demands in glasses of satisfying the local packing constraints while still creating a space-filling structure. The curved space ideas of Sadoc and Nelson make this clear.

Two features of the Ising Spin Glass Hamiltonian are very different from those of a glass, its randomness and its symmetry. The spin glass has same randomness already put it, by hand, as it were. The Hamiltonian for a glass is just the interactions between the particles which may be simple or complicated but is the same throughout the sample. Thus the randomness of glassy configurations must be self-generated. The situation is reminiscent of the theory of turbulence[50]. One question is how does it originate but we may also ask given the experimental fact of its existence what are its consequences for dynamics and stability. The problem of the origin of chaos here is mathematically much more complicated than the well investigated problems of the dynamics of systems with a few degrees of freedom because it lacks their one dimensional character (i.e. there is no analog of time). Studies of one dimensional regular systems have shown how chaos might arise[15b] but in one dimension it has been shown that the chaotic states can never be stable[15c]. Renormalization group arguments for spins on hierachical lattices which are regular but frustrated suggest that chaotic structures can be thermodynamically significant[51]. There are also suggestions from the theory of tilings[52] and cellular automata[53]. Certain tiles called Wang dominoes can be constructed which act as computers. The rules of association of these tiles are such that, given a certain seed structure (a program), an arbitrary computation can be carried out. Since Turing's theorem says there are computations that never end, some seeds for such filings could generate as random a structure as possible! Clearly the origin of randomness in the context of glasses is a deep question which will require considerable study.

Even if the randomness is granted, the symmetry of Ising spin glass is rather special and different from any, one can easily see, in glasses. The Ising spins are symmetric between up and down. This symmetry in the case of regular systems leads to continuous phase transitions for Ising systems. The simple arguments for the glass transition described in the last section lead, at least in one sense, to a discontinuous transition. This reflects the lack of symmetry of the liquid-glass system. This is why these naive theories while modeled on similar early treatments of spin glasses[54] give very different results.

There are disordered spin models which lack the Ising symmetry and they have been shown (in mean field theory) to have very different transitions. The similarity of these systems suggests that while they may be less studied they are more generic[55]; they also should be better analogies to structural glasses as first emphasized by Kirkpatrick and Wolynes[43]. Several spin glasses lacking this symmetry have been introduced. The first is the p-state Potts glass. Here the spin variables point at the corners of a hypertetrahedron with p vertices. The Hamiltonian for the Potts glass PG is

$$H = \sum J_{ij}(\delta_{S_i S_j} - 1). \tag{14}$$

For $p = 2$ this is just the Ising system but for p greater than two the up-down symmetry is not present. Mean field theory for the regular Potts ferromagnet gives a first order transition. Another system lacking the Ising symmetry is the p-spin interaction model[56,57] whose Hamiltonian is:

$$H_p = \sum_{i,j,k} J_{ijk\cdots p} S_i S_j S_k \ldots S_p \tag{15}$$

in which triplets on higher numbers of Ising spin interact at a time. $\{J_{ijk\ldots p}\}$ again are random variables. The limit as $p \to \infty$ may seem rather complicated but in fact it is simple. Essentially for $p = \infty$, the energies of different spin configurations are independent random variables. This random energy model (REM) introduced by Derrida[57] is exactly soluble and the solution has a simple structure[58] (to which we will return shortly). The REM captures a great deal of the physics of the mean field theories of these nonsymmetric spin glasses. In mean field all these systems show similar transitions. We will call them generically Potts glasses.

B. Sophisticated Mean Field Theories for Potts Glasses

The naive mean field theories of glasses which we discussed earlier have in common with the early theories of spin glasses that they characterize the frozen state by a single parameter. In the case of glasses it is the Debye-Waller factor the long time limit of the density correlation function.

In the case of spin glasses the parameter is the Edwards-Anderson order parameter, q, which is the long time limit of the spin autocorrelation function. This single parameter does not do justice to the complex diversity of structures that a glass or spin glass can take on. Two approaches to this diversity have been taken for the traditional spin glass in the mean field limit. One of these is to explicitly construct solutions of minimum free energy for the spin glass and analyze their statistics. This approach was embarked upon by Thouless, Anderson and Palmer (TAP)[59] and we now know essentially all the results can be obtained this way. A royal road to deal with the diversity was also invented – the replica approach[2,60,61]. Despite its lack of physical immediacy the replica approach has led to a great appreciation for the phase structure of the traditional spin glass in the mean field limit. First, the mean field Potts glass will be discussed from the replica perspective and then from the TAP viewpoint, wherein the physical nature of the transition becomes clear. In all cases I will sketch only the results and refer the reader to the literature for the algebraic details which can be quite messy.

The replica approach starts by considering several copies of the system[62]. Below the transition one can see that the different copies may fall into different states, if there is more than one thermal ground state. Thus there are many order parameters; $q_{\alpha\beta} = \frac{1}{N}\sum \bar{S}^i_\alpha \bar{S}^i_\beta$. Now, technically, the replica approach relies on the "replica trick" in which one considers a copies of the system and finally takes the limit $n \to 0$! This construction actually gives an overlap function $q(x)$, representing the fact that different copies may be in different states that may or may not be related to each other. The limit as $x \to 1$ of $q(x)$ is the Edwards-Anderson parameter $q = q_{\alpha\alpha}$ and $q(x)$ also allows one to characterize the thermal ground states statistically through a probability the state typically, distribution $P(q = q^{\alpha\beta})$ which tells how close they are to each other[62]. Thermodynamic quantities depend on averages over $P(q)$.

Parisi, who introduced the $q(x)$ idea, solved the Ising spin glass in the mean field limit[62]. His solution shows that the Edwards-Anderson orderparameter, $q(1)$ rises continuously from zero. This is a reflection of the second order nature of the transition arising from the Ising symmetry. Furthermore, although $P(q)$ contains a delta function part $\delta(q - q(1))$, there is also a continuous part showing there are gound states which are related to each other to arbitrary degrees. Carrying this analysis further the thermal states of the Ising spin glass were shown to exhibit the property of ultrametricity[63] that is they could be thought to be the twigs at the surface of an evolutionary tree! This ultrametric property brought to mind analogies to many other problems in biology and in optimization.

The replica method when applied to the Potts glass gives a very different result. Gross, Kanter and Sompolinsky[64] showed that there was a critical temperature below which $q(1)$ discontinuously became nonzero, when the number of Potts directions exceeds four. This is reminiscent of the naive mean field theories of the structural glasses. The overlap probability function also is rather simple:

$$P(q) = (1 - \bar{x})\delta(q - q(1)) + (\bar{x})\delta(q) \tag{16}$$

where $1 - \bar{x} = o(T - T_c)$ below the transition. States are either strongly related to each other or are entirely unrelated. There doesn't seem to be much need for evolutionary trees! Just below the transition most of the state are entirely unrelated. Because of this although $q(1)$ behaves discontinuously at the transition the free energy which contain averages over $P(q)$ does not have a discontinuous slope. Thus the mean field Potts glass exhibits no latent heat or volume change at the transition. Instead there is a discontinuity of the heat capacity at the transition. In a thermodynamic sense the transition is second order. The similarity of these properties to the ideal structural glass transition is very seductive.

The analogy is strengthened by consideration of the p-spin interaction spin glasses. The replica method applied to this model gives the same kind of transition as for the Potts glass. In exploring this model with the replica technique, Kirkpatrick and Thirumalai discovered two characteristics temperatures[46]. At the lower temperature the variational equations for $q(x)$ were completely satisfied and this gave the thermodynamic transition just discussed. At a higher temperature the variational equation for $q(1)$ was satisfied but not the equation for $\bar{x}$! The presence of two characteristic temperatures is not totally unexpected for a first order transition in which there is a limit of stability and a transition point. When mode coupling theory was applied to the 3-spin model by Kirkpatrick and Thirumalai they found a transition at the higher temperature. The mode coupling equation for the 3-spin model has exactly the same form as the Leutheusser equation abstracted from the dense fluid mode coupling theory.

The meaning of the characteristic temperatures and indeed the mechanism of the thermodynamic transition is greatly clarified by carrying out the mean field theory by studying the statistics of the free energy minima obtained from the TAP equations. Kirkpatrick and Wolynes did this for the Potts glass[47] using a technique developed by de Dominicis and Young for the Ising spin glass[65]. Their analysis showed that at the higher temperature T_A individual free energy minima were locally stable. However there is a thermodynamically large number of them. The configurational entropy S_c exactly makes up for the free energy cost of being confined to a single minimum so no thermodynamic phase transition occurs. This configurational entropy decreases with temperature and vanishes at T_K which the same as the thermodynamic transition found by the replica approach. Thus the Kauzmann paradox for the Potts glass is indeed resolved by a phase transition.

The mystery of this phase transition mechanism is simplified when the random energy model is considered. The average energy spectrum of the random energy model is a Gaussian but there are a finite albeit large number of states. So the actual energy spectrum should be thought of as a series of spikes *roughtly* approximating the smooth Gaussian. At any temperature a band of these states contributes to the canonical partition function the position of the band being determined by a balance of the density of states favoring energies near zero and the Boltzmann factor favoring the low energy states. As the temperature is lowered the contributing band gets lower in energy and samples sparser regions of the energy spectrum. Finally the discreteness of the spectrum begins to appear, the number of state available is small and one must settle into the lowest state with a vanishing entropy. At this point the Edwards-Anderson order parameter discontinuously charges from zero to one since the system is energetically confined to this small band of states.

A mechanism much like the Random Energy Model was also found in the Gibbs-Dimarzio theory of the polymeric glass transition[66]. Some clues to this kind of behavior are seen in computational studies of the hard sphere glass[24]. The free energy quenches can be statistically analyzed. A stick diagram of the free energies per particle to which they quench is shown in Figure 6. At low densities although many different structures are found they have essentially the same free energy per particle but at high densities there is a big spread. Thus in thermal weighting the configurational entropy will be decreasing dramatically.

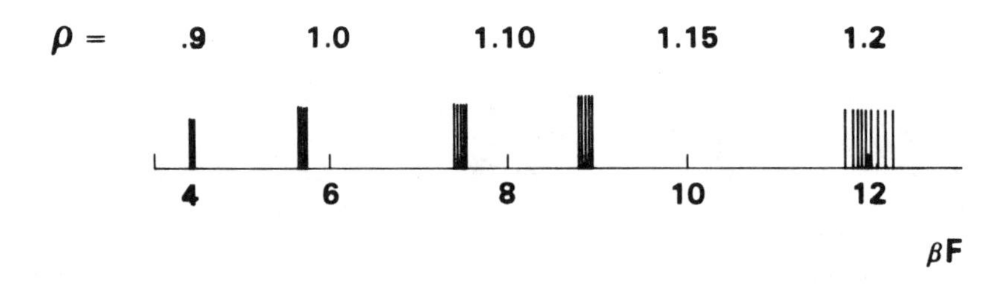

Figure 6. The free energies per particle for different minima are plotted. The densities are in units of $\rho\sigma^3$.

The entropy crisis mechanism is a beautiful way of reconciling complexity with stability. Although a structure appears to require many bits to describe it, in fact, only a few states out of the many complex ones are picked out by the thermal average. The vanishing configurational entropy means the structure actually does not need so many bits of description. Again we can liken this to the behavior of a computer, the output is often very complicated but may be the result of a randomly written and short program.

C. Speculation Beyond Mean Field Theory

Mean field theory is strictly true only for systems with infinite range interactions. The understanding of the Potts glass and its connections with real structural glasses with short range interactions requires some consideration of the effects beyond mean field theory. This is especially true because of the peculiar nature of the transition. It is especially difficult because the theory of short-ranged Ising spin glass is still somewhat controversial[67,68,69] and how it is related to the mean field limit is more obscure. Furthermore we have argued that the transitions we have discussed here have something of the character of first order transitions and there are still open questions about these when one deals with metastable situations[70]. Our speculations here will start with a "droplet" picture which takes much from the theory of first order phase transitions. It relies on the mean field theory to understand the phase structure of each droplet. Then we will discuss some scaling hypotheses[71] motivated stylistically by the droplet picture and by studies of Huse and Fisher of the Ising spin glass where the droplets are particularly fuzzy. Below T_A the individual TAP states (i.e. local free energy minima) are dynamically stable for the infinite range Potts glass. When a long but finite range is introduced we can look more carefully at stability by first examining fluctuations about a TAP solution and the correlation length for these fluctuations[47]. This sort of analysis for the Ising spin glass showed that the states were always marginally stable, at all temperatures below the critical temperature there are unstable fluctuation modes and infinite range correlations. The analysis of Kirkpatrick and Wolynes for the Potts glass indicates a different behavior. At T_A, to be sure, there is marginal stability but not at lower temperatures. T_A behaves just like a spinodal or limit of stability of a first order transition. The simple analysis would give a correlation length diverging like $(T - T_A)^{1/4}$. Spinodals are particularly tricky even for ordinary first order phase transitions. A divergent correlation length means a vanishing surface tension[72] but that means droplet of other phases (TAP states here) can form easily, thus the singular behavior is wiped out by transitions to those states when close to T_A. Thus there may be some remnant of a change in behavior at T_A but we expect no singularity. This is in harmony with the findings of Hall and Wolynes for the structural glass[73]. They noted that the distance between energy minima, as calculated by Stillinger and Weber, are of the order of the aperiodic crystal Lindemann distance i.e. the thermal vibrations at T_A. Thus the energy barriers at T_A are of order $k_B T$. Presumably the fluctuations that wipe out the singularity below T_A also have a leveling effect above T_A so the crisp dynamic singularities obtained by the mode coupling theories should be smeared. Recently some elaborations of the mode coupling theories have resulted in smooth transitions. These theories put in effects that act like the activated transitions discussed here but whether they are the same or not is far from clear.

What is the structure of the finite range Potts glass below T_A? Above, T_k, A given TAP state is unstable to droplet configurations[42,71]. Consider a region of size R. For the finite range model introducing one of the other TAP states in this region can cost only a surface energy. The finite correlation length suggests a finite surface tension, σ, but this could be modified by wetting effects. There is a macroscopic number of TAP states for $T_k < T < T_A$ so any of them could be introduced. This gives an entropic contribution $-k_BTS_cR^d$, where d is the dimension of space. The free energy cost is hence

$$F(R) = -k_BTS_cR^d + (d-1)\sigma R^{d-1}. \tag{17}$$

Thus for R^* of order σ/k_BTS_c it is preferable to disorder the structure. We would end up with a kind of Mosaic structure. Translated into the language of the aperiodic crystal we would have compact regions with relatively good bonding surrounded by walls with weaker bonding. This is a little like the old "amorphon" picture[74]. This would be consistent with the Mössbauer results shown earlier. Warming from T_g, the constrained regions shrink leaving more "connective" tissue which is more fluid and doesn't contribute to the recoilless fraction. Significant broadening occurs only when a droplet lifetime become as short as the gamma ray coherence time. The droplet picture also gives a picture of the dynamics between T_A and T_K. Within a pure TAP state or a TAP droplet of the Mosaic any other TAP state can be nucleated the same free energy expression then gives a barrier for this process which diverges as T_K is approached. The barrier height goes like S_c^{1-d}. Thus as the entropy crisis approaches relaxation times diverge. This argument, of course, assumes that the domain walls of the Mosaic are pinned in some fashion like that of a soap froth.

The droplet speculations indicate that there would be a divergent length at T_K, the droplet size, even though there are no soft modes. This length R^* would go like $S_c^{-1} \propto (T-T_K)^{-1}$. Not surprisingly for a mean field theory with low order corrections this divergence would not satisfy the hyper scaling relation $\nu d = 2 - \alpha$. The theory indicates that there is a discontinuity in the heat capacity thus $\alpha = 0$, the scaling relation implying a divergent length like $\xi_K = (T-T_K)^{-2/d}$, as pointed out for the Potts glass by Gross et al[64]. The exponent has a simple interpretation. The continuity of the entropy density across the transition implies that just below T_K there are still a large but not macroscopic number of states. Below T_K the logarithm of the number of thermally sampled state diverges like $(T-T_K)^{-1}$. If we are uncertain about the temperature of the system by $\pm\Delta T$ about T_K we see then that a sample of radius $\Delta T^{2/d}$ is needed to tell if we are above or below T_K. If below the number of states appropriate to a small region will level off at this volume, if above it will continue to grow exponentially.

The extreme degeneracy below the transition may plausibly act to lower the surface energy of drops since their surface can be wetted. A similar effect occurs in the ferromagnetic Potts model[75]. If this happens we can modify the surface energy to be $\sigma_1 R^{(d-1)y}$. If we demand R^* to be ξ_K this gives $y = \frac{1}{2}d/(d-1)$, the surface energy is comparable to the spread in energies of the thermally sampled TAP states themselves. It is amusing that this modification of the droplet picture would give barriers scaling like $S_c^{-1} \approx (T - T_K)^{-1}$, precisely the Vogel-Fulcher or Adam-Gibbs behavior[76].

Since the dynamic range of configurational entropy is not large in experiment it is rather hard to distinguish one scaling form from another so perhaps one should take these arguments as being indicative of a scenario for the dynamic slowing in structural glasses. Because of the exponential scaling of time scales the asymptotic ideal glass transition (if there is one!) will be almost as difficult to reach by experiment, as the length scale of the now popular string theories of elementary particles is to probe directly. A completely convincing development of the ideas presented here will require the calculation of system specific, non-universal properties in order to be experimentally tested. Certainly the classification of various liquids into "strong" and "fragile" categories is not in conflict with the droplet or scaling picture and one's chemical intuition about surface energies in these systems.

V. THE APERIODIC CRYSTAL IN OTHER AREAS OF BIOLOGY, CHEMISTRY, PHYSICS AND MUSIC

Clearly the aperiodic crystal is still a soft concept, even in the area of glasses and spin glasses. Nevertheless at a celebration for Hans Frauenfelder it is not inappropriate to speculate further how the present ideas may couple to some of those parts of science he holds dear.

A. Biomolecules

The notion of an aperiodic crystal has its roots in the study of biomolecules. Several of the ideas used for glasses were modeled on notions used there, in fact. The idea of a fixed protein structure is an important one, but also one with its limits, as Hans has so often pointed out. The idea of using the Debye-Waller fctor and the Mössbauer effect as a probe of order has been useful both for glasses and biomolecules[77]. Also Hans' idea of substates probably appeared in the biomolecular context before its complete incarnation in the physics of spin glasses.

Is there a glass transition in proteins? Hans has argued that there is evidence for it in the Mössbauer data. The notions presented here emphasize the importance of more thermodynamic measurements to see if this is related to the more conventional glass transition. Stein has put forward a spin glass model for this transition[78] which has been quite stimulating. The considerations here suggest that we should not try to slavishly map these problems on to an Ising spin glass. Potts glasses or the Random Energy Model (also used by Stein) seem to be better analogues.

This has much to do with the search for ultrametricity in proteins. The Potts glass has only trivial ultrametricity. In fact a preliminary study by Elber and Karplus[79] does indeed find that the minimum energy structures of protein are either strongly related or apparently unrelated to each other. Nevertheless Hans and his group have found kinetic evidence for tiers of states in myoglobin[80]. I feel these states are not those found in the molecular dynamics simulation and do not arise from a purely statistical many body effect. They are more likely the result of evolution. Either the tier structure has been engineered for a functional purpose or else it is a kind of fossil record of the evolutionary process itself. I am not sure this would displease Hans.

Where else might the aperiodic crystal concept be useful in the study of biomolecules? One area being explored is in protein folding[81]. Here it may be that the avoidance of glassy dynamics, that is long-lived misfolded intermediates, may be the name of the game. In any case the connection of glassy ideas with optimization problems may come forecefully through in the practical problem of folding proteins on the computer. The Potts analogy may be a little disheartening since recent studies have shown the connection of the 3-spin interaction glass with the so called golf-course problem[82,83]. In a golf course problem one has no guidance as to where to go you simply keep trying for a hole-in-one (or two or three, etc.) This is connected with the unrelatedness of the minima. Thus we will need something better than simulated annealing[84] but, of course, the practitioners knew this already.

Another area worth exploring is allostery. Here precisely the problem is how a system with potentially many states behaves as if it had only a few (entropy crisis!). To date models of allostery have been rather particular and, one might say, anatomical in character[85,86]. Nevertheless there may be cases where a collective viewpoint will be useful.

B. Chemical Reactions

Hans' great interest in chemical reactivity has done much to stimulate thinking about reaction, dynamics in condensed phases[82,29], especially about the role of viscosity. One of the perplexing features of his data has been that the dependence of rates on viscosity is weaker than expected on the basis of Kramers theory. This has been rationalized by invoking the spatial[88] or temporal[89] dependence of friction. Since many of the experiments were done in supercooled systems the present picture suggests that some care might be necessary because the concept of linear friction has been pushed to its limit for the motion of the solvent molecules. The interplay between hydrodynamics of the macromolecule and the expected nonlinear microphysics needs investigation.

The current picture also suggests investigations on reactions in which the solvent plays a more intimate role, for example, electron transfer or S_N2 reactions. Here successful theories are based on the linear response of the environment[90]. If the configurational entropy is small this linear response is no longer guaranteed simply by the law of large numbers. It would be interesting to investigate linear free energy relations for these reactions in the supercooled liquid. There may be either anomalies in the relation or in the coefficients of these relationships.

C. Elementary Particles and Cosmology

Starting with his early experiments on parity conservation[91], Hans has continued to take a lively interest in the physics of elementary particles. The broken symmetry concept, hinted at in those kind of experiments, has escaped from condensed mater physics and infected large parts of particle physics and cosmology. The aperiodic crystal picture has borrowed somewhat from these areas[18] but as yet the glass transition ideas haven't exerted much influence reciprocally. Still it is, almost inconceivable (to an outsider such as myself) that some glassy phenomenon didn't rear its head in the early universe or be at work in the nether reaches of space-time[92].

D. Music

Computer generated music or "musique aleatoire" might well benefit by using aperiodic crystal or glassy ideas. Most people would agree it could not be hurt.

Hans' taste in music lies in the older human generated kind. An important form is the fugue. The connection of fugues with aperiodic crystals is clear. They must obey local rules (not be too dissonant) but they must also exhibit a long range structure whose nature can be quite complex and hard to predict. Many fugues have a stretto section. In the stretto the theme enters in the several voices in rapid, overlapping succession, often only a fragment of the theme is actually expressed. I find the stretto one of the most exciting parts of a fugue, partly for itself, partly for the resolution which usually follows. Current research on aperiodic crystal problems is a little like a fugue with stretto. Similar ideas have been announced in rapid succession in many fields, often answering each other's questions.

Hans' favorite composer is Beethoven. Beethoven is not known primarily for his fugues. In his diary Beethoven remarks: "On the whole, the carrying out of several voices in strict relationship mutually hinders their progress."

Certainly a cautionary remark for interdisciplinary research! Although Beethoven did not write many fugues he did write some great ones. A great fugue with stretto can be found in the finale of his piano sonata, Op.110. Perhaps Beethoven's real attitude towards fugues is better captured in another quote: "To *make* a fugue requires no particular skill; in my study days I made dozens of them. But the fancy also wishes to assert its privileges, and today a new and really poetic element must be introduced into the traditional form."

This attitude could also characterize Hans' style of doing science. Happy Birthday, Hans!

ACKNOWLEDGEMENTS

In addition to thanking Hans Frauenfelder for his continuing interest in these strange things, I'd also like to thank the many participants at the UCSB Institute for Theoretical Physics program on Relaxation and Reaction Dynamics in Complex Systems with whom I discussed much of this material. The work on the Potts glass analogy was done in a very fun collaboration with Ted

Kirkpatrick. I would also like to thank the members of my research group who have worked on glassy problems — Joe Bryngelson, Randall Hall, John Mertz, Yashwant Singh and James Stoessel. My work was supported by the National Science Foundation at the University of Illinois under grant CHE 84-18619. The computational studies cited were carried out at the National Center for Supercomputing Applications in Urbana and supported by the Materials Research Laboratory under grant NSF DMR 83-16981. The helpful support of the Guggenheim Foundation is acknowledged. I am also grateful for the hospitality and support of the Institute for Molecular Science, Okazaki, Japan, where this manuscript was written.

REFERENCES

1. L.D. Landau and E.M. Lifshitz, Statistical Physics (Pergamon, London, 1980).
2. P.W. Anderson in *Ill-Condensed Matter*, eds. R. Balian, R. Maynard and G. Toulouse (North-Holland, Amsterdam, 1979).
3. E. Schrödinger, *What is Life?* (Cambridge University Press, 1944).
4. G.S. Grest and M. Cohen, Adv. Chem. Phys. 48, 455 (1982).
5. B.J. Alder, D.M. Gass and T.E. Wainwright, J. Chem. Phys. 53, 3813 (1970); For a review from the experimental viewpoint see J. Jonas, Acc. Chem. Res. 17, 74 (1984).
6. C.A. Angell, in Proceedings of the Workshop on Relaxation Effects in Disordered Systems, ed. K. Ngai and T.K. Lee (McGregor and Werner, New York); C.A. Angell, A. Dworkin, P. Figuiere, A. Fuchs and H. Szwarc, J. de Chimie Phys. 82, 773 (1985).
7. W. Kauzmann, Chem. Rev. 43, 219 (1948).
8. F. Simon, Z. Anorg. Algemein. Chemie. 203, 217 (1931).
   J.D. Bernal, Proc. Roy. Soc. London Ser. A280, 299 (1964). The role of the icosahedron was also emphasized by F.C. Frank, Proc. R. Soc. (London) Ser. A251, 43 (1952).
9. D. Chandler, J.D. Weeks and H.C. Andersen, Science 220, 787 (1983).
10. F.C. Frank and J.S. Kasper, Acta. Crystallop. 11, 184 (1958); 12, 483 (1959).
11. D. Levine and P. Steinhardt, Phys. Rev. Lett. 53, 2477 (1984). The materials were discovered by D. Shechtman, I. Blech, D. Gratias, and J. Cahn, Phys. Rev. Lett. 53, 1951 (1984).
12. R. Penrose, J. Inst. Math. Its. Appl. 10, 266 (1974).
13. S. Aubry, J. de Physique 44, 147 (1983).
14. M. Gardner, Scientific American 236, 110 (1977).
15. a) D. Ruelle, Physica 113*A*, 319 (1982); b) P. Bak, Rep. Prog. Phys. 45, 587 (1982); c) M.E. Fisher and D.A. Huse, in *Melting Localization and Chaos*, eds. R. Kalia and P. Vashishta (Elsevier, 1982).
16. M. Kleman and J.F. Sadoc, J. Phys. Lett. 40, L569 (1979). J.F. Sadoc, J. Noncryst. Solids 44, 1 (1981).
17. D.R. Nelson, Phys. Rev. *B*28, 5515 (1983).
18. D.R. Nelson, Phys. Rev. Lett. 50, 982 (1983).
19. W. Sutherland, Philos. Mag. 30, 318 (1890).

20. F.A. Lindemann, Z. Phys. 11, 609 (1910).
21. F.J. Litterst, Nucl. Inst. and Meth. 199, 87 (1982); D.C. Champeney and D.F. Sedgwick, J. Phys. 5, 1903 (1972).
22. M. Fixman, J. Chem. Phys. 51, 3270 (1969).
23. J.P. Stoessel and P.G. Wolynes, J. Chem. Phys. 80, 4502 (1984).
24. R. Hall, J. Mertz and P.G. Wolynes, to be submitted.
25. For a review see A.D.J. Hagmet, Science 236, 1076 (1987).
26. T. Ramakrishnan and M. Yussouff, Phys. Rev. *B*19, 2775 (1979).
27. P. Tarazona, Mol. Phys. 52, 81 (1984).
28. Y. Singh, J.P. Stoessel and P.G. Wolynes, Phys. Rev. Lett. 54, 1059 (1985).
29. H. Frauenfelder and P.G. Wolynes, Science 229, 337 (1985) contains a pedagogic review in this area.
30. B.J. Alder, in *Molecular Dynamics Simulations of Statistical Mechanical Systems*, ed. G. Ciccotti and W.G. Hoover (North-Holland, Amsterdam, 1986). Rather detailed work on precisely the correlation function discussed, has been carried out by B.J. Berne and his research group.
31. T.R. Kirkpatrick, Phys. Rev. Lett. 53, 1735 (1984); J. Noncryst. Solids 75, 437 (1985); T.R. Kirkpatrick and J. Nieuwoudt, Phys. Rev. *A*33, 2651, 2658 (1986).
32. H. van Beijeren, Phys. Lett. *A*105, 191 (1984).
33. T. Gestzi and J. Kertesz, *Liquid Metals*, ed. R. Evans and D.A. Greenword (Bristol Inst. Phys.) (1976). T. Gestzi, J. Phys. *C*14, 5805 (1983).
34. A. Sjölander and L.A. Turski, J. Phys. *C*11, 1973 (1978).
35. See P.G. Wolynes, Ann. Rev. Phys. Chem. 31, 345 (1980) for a review of early work. The idea was first introduced in its modern form by R.W. Zwanzig, J. Chem. Phys. 38, 1603 (1963).
36. U. Bengtzelius, W. Götze and A. Sjölander, J. Phys. *C*17, 5915 (1984).
37. T.R. Kirkpatrick, Phys. Rev. *A*31, 939 (1985).
38. S.P. Das, G.F. Mazenko, S. Ramaswamy and J.J. Toner, Phys. Rev. Lett. 54, 118 (1985).
39. K. Kawasaki, in *Phase Transitions and Critical Phenomena*, eds. C. Comb and M. Green (Academic, New York, 1976) volume 5a.
40. D. Vollhardt and P. Wölfle, Phys. Rev. Lett. 45, 842 (1980); For a review see T.R. Kirkpatrick and J.R. Dorfman in Fundamental Problems in Statistical Mechanics VI. ed., E.G.D. Cohen (North Holland, Amsterdam, 1985).
41. For a review see W. Götze, *Proceedings of the NATO Advanced Summer School on Amorphous and Liquid Material.*
42. E. Leutheusser, Phys. Rev. *A*24, 2765 (1984); Z. Phys. *B*55, 235 (1984).
43. T.R. Kirkpatrick and P.G. Wolynes, Phys. Rev. *A*35, 3072 (1987).
44. C.A. Rogers, *Packing and Covering* (Cambridge Press, London, 1964).
45. F.H. Stillinger and T. Weber, Phys. Rev. *A*25, 978 (1982). For a review see their article in Science 225, 983 (1984).
46. T.R. Kirkpatrick and D. Thirumalai, Phys. Rev. Lett. 58, 2091 (1987).
47. T.R. Kirkpatrick and P.G. Wolynes, Phys. Rev. *B*36, 8552 (1987).
48. R. Kree, L.A. Turski and A.Z. Pelius, Phys. Rev. Lett 58, 1656 (1987).
49. G. Toulouse, Comm. in Phys. 2, 115 (1977).

50. P.C. Martin, J. de Physique 37, C1 (1976).
51. S.R. McKay, A.N. Berker and S. Kirkpatrick, Phys. Rev. Lett. 48, 767 (1982).
52. H. Wang, Scientific American 213, 98 (1979).
53. S. Wolfram, Phys. Rev Lett. 54, 735 (1985).
54. S.F. Edwards and P.W. Anderson, J. Phys. *F*5, 965 (1975); *F*4, 1927 (1976).
55. D. Sherrington, in *Heidelberg Colloquium on Glassy Dynamics, 1986*, eds. J.L. van Hemman and I. Morgenstern (Springer, Berlin, 1987). D. Sherrington, Prog. Theor. Phys. Suppl. 87, 180 (1986).
56. E. Gardner, Nucl. Phys. B [F514], 747 (1985).
57. B. Derrida, Phys. Rev. *B*24, 2413 (1981).
58. D.J. Gross and M. Mezard, Nucl. Phys. *B*240, 431 (1984).
59. D.J. Thouless, P.W. Anderson and R.G. Palmer, Philos. Mag. 35, 593 (1977).
60. The trick is to write $\langle \log\ Z \rangle = \underset{n \to 0}{\mathrm{Lt}} \left\langle \frac{Z^n - 1}{n} \right\rangle$. The average of Z^n has the structure of interacting copies of the system. This was used as early as 1959 by R.H. Brout, Phys. Rev. 115, 824 (1959).
61. The checkered history of the replica approach to spin glasses is reviewed in D. Chowdhury, *Spin Glasses and other Frustrable Systems* (World Scientific, Singapore, 1986).
62. G. Parisi, Phys. Rev. Lett 43, 1754 (1979); Phys. Rev. Lett. 50, 1946 (1983).
63. R. Rammal, G. Toulouse and M. Virasoro, Rev. Mod. Phys. (1984).
64. D.J. Gross, I. Kanter and H. Sompolinsky, Phys. Rev. Lett. 55, 304 (1985).
65. C. de Dominicis and A.P. Young, J. Phys. *A*16, 2063 (1983).
66. J.H. Gibb and E. DiMarzio, J. Chem. Phys. 28, 373 (1958).
67. W.L. McMillan, Phys. Rev. *B*31, 340 (1985); *B*29, 4026 (1984).
68. D.S. Fisher and D.A. Huse, Phys. Rev. Lett. 56, 1601 (1986); Phys. Rev. *B*35, 6841 (1987).
69. A.J. Bray and M.A. Moore, J. Phys. *C*17, L463 (1984); Phys. Rev. *B*29, 340 (1985). A very thoughtful review by these authors can be found in the *Heidelberg Colloquium on Spin Glasses*, ed. by J.L. van Hemmen and I. Morgenstern (Springer, Berlin, 1987).
70. J.D. Gunton, M. San Miguel and P.S. Sahni, in *Phase Transitions and Critical Phenomena* Vol. 8, eds. C. Domb and J.L. Lebowitz (Academic Press, New York, 1983), p. 269.
71. T.K. Kirkpatrick and P.G. Wolynes, work in progress.
72. B. Widom, J. Chem. Phys. 43, 3892 (1965).
73. R. Hall and P.G. Wolynes, J. Chem. Phys. 86, 2943 (1987).
74. See M.R. Hoare, Ann. N.Y. Acad. Sci. 271, 186 (1976).
75. B. Deridda and M. Schick, J. Phys. *A*19, 1439 (1986).
76. G. Adams and J.H. Gibbs, J. Chem. Phys. 43, 139 (1965).
77. H. Frauenfelder, G.A. Petsko and D. Tsernoglou, Nature 280, 558 (1979).
78. D.L. Stein, Proc. Natl. Acad. Sci. USA 82, 3670 (1985).
79. R. Elber and M. Karplus, Science 235, 318 (1987).

80. A. Ansari, J. Berendzer, S.F. Bowne, H. Frauenfelder, I. Iber, T.B. Sauke, E. Shyamsunder and R.D. Young, Proc. Natl. Acad. Sci. USA 82, 5000 (1985).
81. J.D. Brynegelson and P.G. Wolynes, Proc. Natl. Acad. Sci. USA, 84, 7524 (1987).
82. E. Baum, Phys. Rev. Lett. 57, 2764 (1986); 59, 374 (1987).
83. G. Baskaran and D.L. Stein, Phys. Rev. Lett. 59, 373 (1987).
84. S. Kirkpatrick, C.D. Gelatt and M.P. Vecchi, Science 220, 671 (1983).
85. M.F. Perutz, Nature 228, 726 (1970).
86. A. Szabo and M. Karplus, J. Mol. Biol. 72, 163 (1972); B.R. Gelin and M. Karplus, Proc. Natl. Acad. Sci. USA 74, 801 (1974).
87. D. Beece, L. Elisenstein, H. Frauenfelderll, D. Good, M.C. Marden, L. Reinisch, A.H. Reynolds, L.B. Sorensen and K.T. Yue, Biochem. 19, 5147 (1980).
88. B. Gavish, Phys. Rev. Lett. 44, 1160 (1980).
89. R.F. Grote and J.T. Hynes, J. Chem. Phys. 73, 2715 (1980).
90. The seminal work is R.A. Marcus, J. Chem. Phys. 24, 966 (1956); Disc. Far. Soc. 29, 21 (1960).
91. H. Frauenfelder, R. Bobone, E. von Goeler, N. Levine, H.R. Lewis, R.M. Peacock, A. Rossi and G. de Pasquali, Phys. Rev. 106, 386 (1957); V. Yuan, H. Frauenfelder, R.W. Harper, J.D. Bowman, R. Carlini, D.W. Macarthur, R.E. Mischke, D.E. Nagle, R.L. Talaga and A.B. McDonald, Phys. Rev. Lett. *L*57, 1177 (1986).
92. N.H. Christ, R. Friedberg and T.D. Lee, Nucl. Phys. *B*202, 89 (1982); *B*210 [F56], 310, 337 (1982).

EQUILIBRIUM AND NON-EQUILIBRIUM DYNAMICS IN PROTEINS

Robert H. Austin
Department of Physics
Princeton University
Princeton, NJ 08544

INTRODUCTION

Part of this paper won't be the same as the talk I gave at Hans' Birthday Blowout. There are two reasons for that. Reason One...oh hell, Reason Zero is that I was wrong, Martin Karplus. Reason One is that I have recently been acutely aware that many biophysicists simply do not understand Hans' contributions to biophysics, and I have to make a stronger case for that. Reason Two is related to what makes Hans such a great advisor to young, budding graduate students. I'll get to the punchline later in this manuscript, before that those of you who don't know about Hans need some education.

Hans is one of those cosmopolitan, sophisticated Europeans that we read all about in the New Yorker (I even saw him wear an ascot once). Strangely enough, he lives smack dab in the ethnic center of the midwest (the ethnic food of the midwest is cheese whiz, of course). The University of Illinois is a great State University that gets its good share of totally naïve little nerds like myself coming to graduate school, clueless as to how to survive in the hardball real world that lies outside the states that begin with vowels. Places like Harvard or Princeton, if they ever let us in, would chew us up and leave behind a demoralized, smoking shell. Hans taught me how to survive, but he did it gently, giving me that most precious of all commodities: his time.

Hans was there at 4 AM as the data came in, Hans helped take the raw data off the oscilloscope screen of the original log time base recorder, which I and my dear friend Laura Eisenstein built in a hectic few weeks. Hans yelled at us when we were sloppy, praised us when things looked right, taught us how to play hard as well as how to work hard. So Hans, here's to you and your kindness in taking the kids off the farm and teaching them how to live. We are so grateful that you stayed with us.

Now, in keeping with Hans' teaching, it is time to argue with him, but hopefully in a manner that he will appreciate. Hans of course isn't without his critics in biophysisics, and one reason for that criticism is that Hans thinks like a physicist. That is, he likes to build simple physical models that attempt to explain in a general way a large group of data. That's the way we are taught. At the risk of insulting many out there, the non-physicist way to do science can tend towards building detailed microscopic models for each case, and I think that there is resentment by those folks towards Hans' sweeping generalities. Well, here I will continue in the noble Frauenfelder mode and stress that we need to push on further into the non-equilibrium aspects of protein dynamics in order to fully realize what I view as Hans' goal of a simple and unified picture of protein action. That is, while myoglobin may indeed be the hydrogen atom of biophysics, we must go on using the principles Hans has established in the more complex molecules since often in the increased complexity new physical principles emerge unexpectedly from the complexity. Laura Eisenstein, who also as I mentioned helped raise me, was proceeding in that direction when we so tragically lost her. Laura, this is for you too.

1. A SIMPLE PICTURE OF EQUILIBRIUM FLUCTUATIONS

Equilibrium fluctuations are transitions between those states in a system which are readily accessible due to thermal fluctuations. Thus, the free energy of the states must lie within approximately k_bT of the ground state, where k_b is boltzmann's constant and T is the temperature in Kelvin. Examples of such fluctuations might be the motions which give rise to the Debye-Waller factors in X-ray diffraction of protein crystals, those motions seen in the ubiquitous motion pictures of computer simulations of protein motions.

Proteins are complex 3-d structures with a large number of short range interactions. As Dan Stein and Hans have pointed out, locally the environment looks random and possibly *frustrated*[1,2] due to the conflicting interaction terms among the various amino acids with electrostatic contributions terms, hydrogen bonding and Van der Waals adding to the stew. Stein in an ansatz compared the energy conformational distribution to that found in the Edwards-Anderson spin glass model. The predictions of such an analysis are:

(1) There exist many local conformational minima of near degenerate energy.

(2) There is a distribution of energy barriers among the minima.

These considerations lead us naturally to expect that the conformational distribution of thermally accessible states in a protein is a continuum of states. I don't think that we should press the issue too strongly about what energy distribution fits the data best. Young and Bowen[3] have certainly achieved better fits than Stein did with a gaussian distribution, but I think most (but not all, see Refs. 4 and 5) of us agree that the basic concept of a distribution of conformational states that Hans has so energetically forwarded is correct.

All these musings came after the fact: it was Hans' classic experiments[6,7] in the early 1970's using low temperature recombination kinetics that revealed exactly the kind of continuous energy distributions that we now "expect" from condensed proteins (Figure 1). Attempts have been made to fit these kinetics with a finite series of exponential processes, but we hope that it is clear from the above considerations that a distribution of rates makes far more sense physically. Certain eminent people seem particularly resistant to this idea....

Since the photolysis experiments of Frauenfelder there has been a large number of other experiments which have convincingly demonstrated the validity of Hans' earlier insight. One of the most important was the measurement of the Debye-Waller effect (temperature dependent intensity in X-ray scattering) by Petsko, Frauenfelder and Ternoglou[8], while recently I with Dan Stein and Joseph Wang have used rare earth emission lifetimes to also demonstrate the basic correctness of the conformational distribution idea (Figure 2)[9].

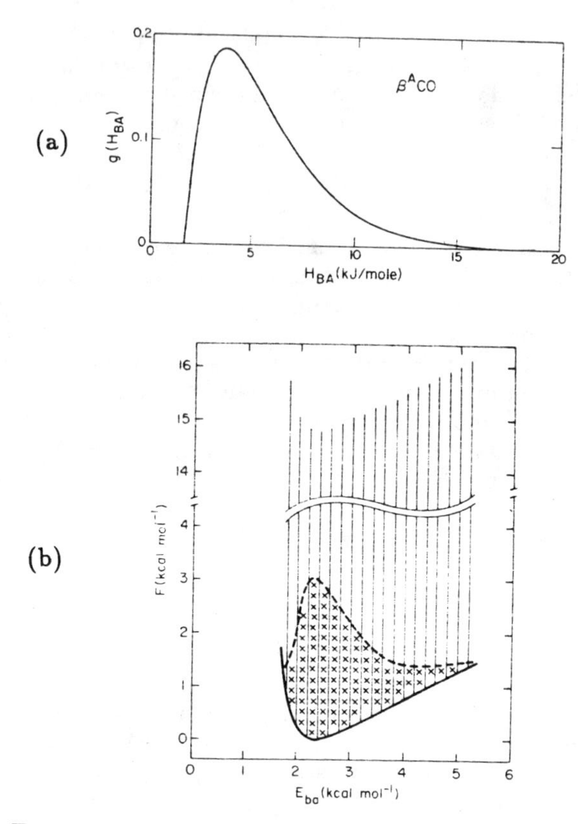

Figure 1. Energy level distribution. (a) A schematic of the distribution of activation energy levels expected in a protein. (b) The energy surface that can give rise to the energy level distribution in part a. The original version of this figure can be found in Ref. 6, originated by Hans Frauenfelder long before it became fashionable.

2. ARE EQUILIBRIUM FLUCTUATIONS ENOUGH?

However, it is the non-linear events, the events characterized by free energy changes much greater than k_bT, that make life work. In the non-linear events a large chunk of energy is input to the system and the system responses remarkably in a *directed* manner along non-equilibrium paths to accomplish a task.

Examples of such behavior are proton pumping across membranes, the R-T switch in hemoglobin and at a higher level muscle contraction and nerve cell firing. The question is, what of the insights gained from the equilibrium studies can be applied to understand the more important non-equilibrium events? Should we continue to study the equilibrium events or should we conserve our efforts for non-equilibrium events?

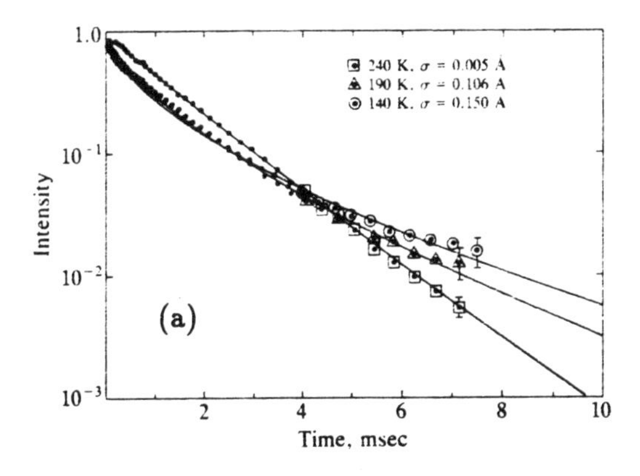

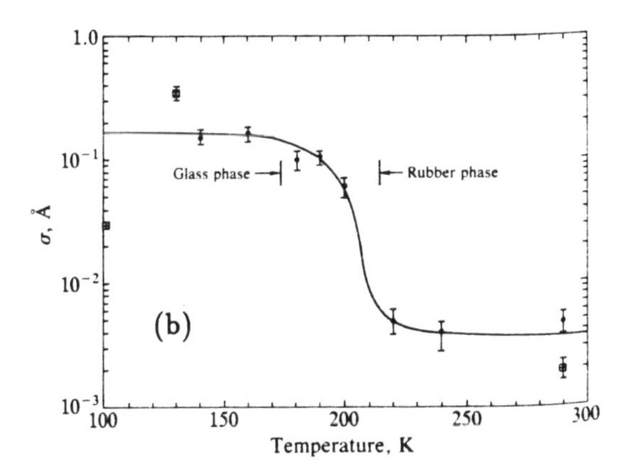

Figure 2. Terbium Luminescent Lifetime Distribution. (a) The decay of excited terbium atoms bound to the protein Calmodulin. The three curves are for three different temperatures, above, at and below the glass transition temperature of the solvent, glycerol/water. (b) A plot of the effective width of the distribution of protein coordinates vs. temperature, as determined from the terbium lifetime distribution. Note the sudden change in the effective distribution at the glass transition of the solvent.

Equilibrium fluctuations are not the stuff of life! I would expect that any condensed polymer would show somewhat similar effects as we have observed in the low temperature recombination experiments, which is a frozen-in distribution of activation energies. As I mentioned above we must ask ourselves what will happen when the engine burns some fuel? As it is, we are sort of measuring the specific heat of the engine without letting it run.

Now the basic problem in biophysical studies of proteins is arranging yourself to be there at the precise instant the engine burns its fuel, and figuring out the proper probe to study the event. These are non-trivial problems, and to a large extent unsolved at present. Fortunately, or unfortunately, the heme proteins have an ideal initiation and response system built in: the photolysis event and the subsequent absorption changes. It is unfortunate in that this is one of the few systems that is so conveniently accessible to experiment, and so there is a grave risk of concentrating exclusively on heme proteins and ignoring all others. In fact, I have been told that there have been more papers written on hemoglobin than there have been on Quantum Mechanics! Knowing all too well the overly prolific publishing traits of chemists and the extremely cautious publishing traits of physicists (except those working on high T_c materials!), I think this fact may be true. I recently got a Chinese fortune cookies that summed up the problem quite nicely. It said: "Quantity is the enemy of Quality". There must be a latter day Confucius employed at the Central Chinese Cookie Agency.

3. NON-LINEAR EVENTS IN HEME PROTEINS

So, since we have to look for the lost key under the light until somebody invents the flashlight, let's take advantage of what nature has given us and explore the heme proteins a bit. What do we mean by a "non-linear event" in a protein? We infer by this that the protein upon perturbation does not relax back to its equilibrium state but instead evolves with time to a new equilibrium state. In that sense, it acts like a switch with hysterisis. When carboxy myoglobin is perturbed by the absorption of a photon, it undergoes evolution in the sense that the resultant deoxy state is different than the carboxy state, but the differences seem to be confined to the heme with little or no purturbation of the protein structure and no major salt bridges or hydrogen bonds seem to be broken[10].

A much more spectacular structural rearrangement, which can be called non-linear, occurs in hemoglobin; the deoxy state is in the so-called tertiary "T" configuration while the fully loaded hemoglobin Hb·4CO is in the "R" configuration[11]. Thanks to the X-ray diffraction work of Perutz, we know that there are both local and remote structural changes that occur upon photolysis in hemoglobin. Locally we know that the F helix which is coordinated via His F8 to the iron undergoes substantial structural rearrangements upon photolysis, and that further changes remote from the heme also occur.

This bond modification can be viewed as the consequence of a general rotation of the $\alpha_1\beta_2$ dimer relative to the $\alpha_2\beta_1$ dimer. A consequence of this rotation are the following changes upon making the R→T switch:

(1) The hydrogen bond between Tyr $\alpha_1 42$ and Asp $\beta_2 99$ is formed.

(2) There are positive contact interactions between Trp $\beta_2 37$ and Tyr $\alpha_1 140$, Asp $\alpha_1 94$ and Arg $\alpha_1 92$.

(3) Tyr $\beta_2 35$ hydrogen bonds to Asp $\alpha_1 126$.

From the known differences in the binding free energies of Hb in the R and the T states we know that approximately +0.2 eV of free energy is stored somewhere in the hemoglobin T structure. On a chemical scale, this really isn't very much energy, hence the long struggle in hemoglobin to find what and where exactly are the critical components. Three questions can be asked:

(1) What is the rate at which this "purposeful" energy is transmitted to the structure?

(2) What is the mechanism by which the energy transmission is accomplished?

(3) Do the experiments of Hans Frauenfelder shed any light on the physics of this process?

Let's address the last question first by discussing some things we have learned about Hb dynamics using Hans' ideas. First, I have shown that the the distribution of activation energies changes with the R-T switch. Experiments were done using the low temperature recombination kinetics of carp hemoglobin, a hemoglobin that can be locked into the R or the T state under suitable conditions of effectors and pH[12]. We certainly do see changes in the activation energy spectrum of R and T state hemoglobin, as we show in Figure 3.

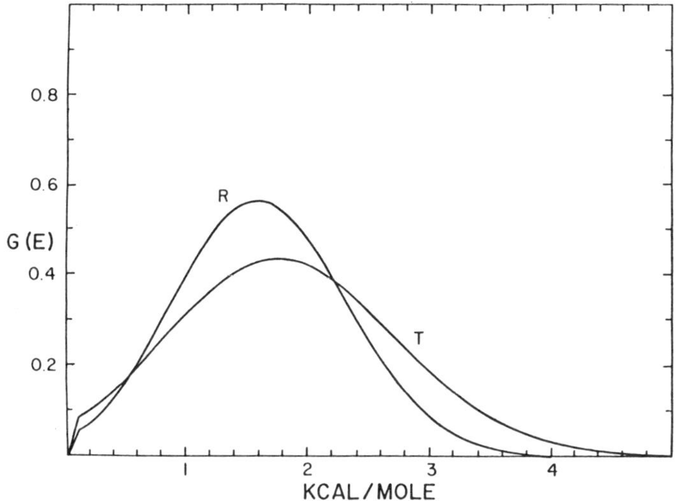

Figure 3. The activation energy spectrum, fitted to Stein's gaussian model, for R and T state carp hemoglobin[12].

The next obvious question: can we interpret the changes in activation energy spectrum with the known energy of cooperativity? Uortunately, not all of us agree on how to connect the energy of a protein conformation with the activation energy for rebinding. John Hopfield and Noam Agmon[13] have devised a beautiful model for the connection between protein conformational structure and activation energy, and Champion in this volume will introduce a model that incorporates Agmon-Hopfield plus other elements to attack this problem, but up to now the results are controversial. To my mind, the linear strain model proposed by these colleagues is a fleshing out of the original proposal by Hans that the conformations that give rise to the distribution lie along a nuclear coordinate energy. Hopfield and Agmon proposed that the actual distribution of conformational states could be calulated by assuming that the protein made a conformational change between (for example) the carboxy and the deoxy states:

$$V(x) = \frac{1}{2}kx^2 - \frac{1}{2}k(x - x_o)^2 \tag{1}$$

where k is the spring constant for the potential surface and x_o is the shift in the mean coordinate position of the potential surface. I won't go into the full details of this model, the reader is invited to consult Agmon and Hopfield's orginal papers or see Champion's article in this volumn. Suffice it to say that the net energy stored in the conformational change is given by:

$$\frac{1}{2}kx_o^2 \tag{2}$$

From our fits of the R-T differences we arrive at numbers indicating that the stored conformational energy difference in the carp hemoglobin is on the order of 5 kcal/mole (0.2 eV), on the order of the known energy of cooperativity. This result should be viewed as a triumph for Hans and his ideas: here it is possible to actually go from the distribution of energies to a real energy of biological interest, and it shows that it is probably in the structure of the protein, not at the iron site, that the conformational energies are stored. This shouldn't be surprising, since we just catalogued all the various hydrogen bonds and salt bridges that are changed in the R-T switch.

4. TENATIVE ATTEMPTS AT WATCHING CONFORMATIONAL FLOW

But, as we stressed in the introduction, there is more to the problem than just saying that the energy is stored in the conformation. We would like to know how the energy flows within the protein.

In order to look directly at the bond making events in the switch it would be very nice to find if there existed an optical window where only the hydrogen bonds that are altered can be monitored, since we don't want to look at the heme. If we were to be so lucky, then it would be possible to observe the rate of the distant bond making/breaking events independently of the heme absorption changes. The aromatic amino acids in hemoglobin and myoglobin have prominent absorption bands at 280 nm that overwhelm the heme absorption. *Thus, one would guess that the UV is a good region to look for heme-remote structural perturbations.* However, as we shall see later, there still is heme absorbance perturbations in the region and they must be handled carefully.

Difference spectra in the UV region are quite difficult due to the fact that many molecules can absorb strongly in this region, and hence the difference protocol must be carefully controlled. There have been several papers which have measured oxy/deoxy UV difference spectra in heme proteins[14,15] and also some work on the effect of inositol hexaphosphate (IHP) on methemoglobin in the UV spectral range[16]. Unfortunately, in the case of the oxy/deoxy difference spectra, *none* of the groups are in agreement! Most of the discrepancies can be explained by shifts in the baseline of the spectra. As we will show, our spectra can be validated by observing the isosbestic points via transient absorption spectroscopy. We feel quite confident from the agreement between our transient data and our static data that the difference curves shown here are the correct curves (Figure 4).

The myoglobin and hemoglobin difference spectra in the UV are not identical. Let us recall that the myoglobin spectra are simply deoxy vs. carboxy curves, while the hemoglobin curves are T-state deoxy vs. R-state carboxy, and thus changes between the hemoglobin curves and the myoglobin curves can be assumed to consist *to some extent* of the non-linear R-T changes that occur only for hemoglobin. There are indeed spectral changes in the UV in myoglobin, but since we believe that only local, heme-related changes occur in myoglobin upon photolysis to the deoxy state the differences seen in the UV can be ascribed presumably strictly to the heme group.

It is well known that the Soret region of the absorption spectra shows variations with the R$\rightarrow$T state, so it is not surprising that the UV region also does. The following differences are apparent: first, the hemoglobin difference spectrum shows a characteristic structure at 290 nm which is not present in the myoglobin spectra. Further, the hemoglobin curves indicate considerably greater changes in the 290-320 region of the spectrum than does myoglobin. We can *hope* that these changes are due to purturbations of the aromatic amino acids and not heme-related changes. Do we have more than hope to work on?

Perutz and his coworkers[15], used the mutant hemoglobins, Hemoglobin Kempsey (replacement of Asp-99 β by Asn) and Hemoglobin NES-des-Arg-Tyr, to determine that the small perturbations at 290 nm are due to the Tyr-42 and Tyr 35 residues, while the so-called *c* and *d* bands from 295 nm to 320 nm are due to the Trp β_2 37 aromatic. Note that in keeping with the intensity of the Trp absorption spectrum and redshift that we would expect any Trp modifications to appear prominently at the longest wavelengths. Indeed, hemoglobin Hirose, which is lacking the Trp $\beta_2$37 shows full loss of a prominent band between 290 nm and 310 nm[17].

5. UV RESONANCE RAMAN RESULTS—EARLY RESULTS

From the above evidence, it is clear that the UV should be a great region to explore non-local dynamics of the protein structure. Tom Spiro of the Chemistry Department at Princeton came to me and suggested a collaboration. This was particularly interesting to me since Professor Spiro has been a pioneer in the development of UV transient Raman spectroscopy. He had evidence that the resonantly enhanced tryptophan region at 853 cm^{-1} showed clear evidence for a "T" conformation 10 nsec after photolysis, in spite of the initial "R" configuration of the protein[18] as we can see in Figure 5.

Using 200 nm light as the resonantly enhanced excitation, the Raman line at 853 cm^{-1} is associated with the tryptophan group. Note that these data

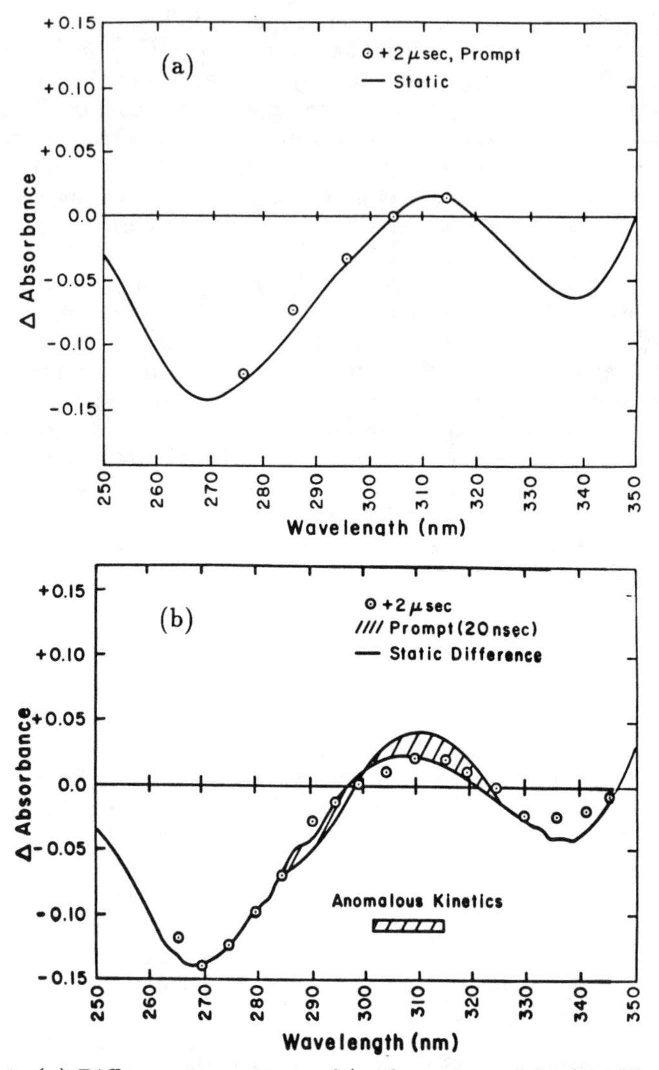

Figure 4. (a) Difference spectrum of (carboxy-myoglobin)— (deoxy-myoglobin) in the aromatic region. The solid line is the static difference spectrum, while the ⊙ are taken from transient kinetics either at 20 nsec after photolysis or 2 μsec after photolysis. (b) Difference spectrum of (carboxy-hemoglobin)— (deoxy-hemoglobin) in the aromatic region. The solid line is the static difference spectrum, the cross-hatched region is the altered absorbance seen approximately 20 nsec after photolysis, and the ⊙ is the difference spectrum seen at 2 μsec after photolysis.

indicate a splitting of the 850 cm^{-1} into two lines in both the deoxy Hb state

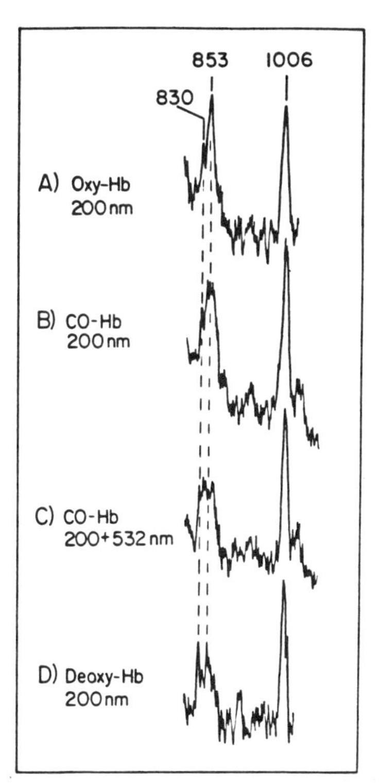

Figure 5. Early UV Raman scattering data[18]. The pre-Park transient UV resonance Raman (UVRR) spectra of hemoglobin. The scattering wavelength was 200 nm.

and the photolyzed Hb species, while the oxy species is shown as single peak. The implication from this is that the tryptophan signal strongly resembles the "T" state within 7 nsec after photolysis. *This really was quite shocking!!.* It implied that the transmission of the structural changes occured very fast, not at the leisurely 20 μsec that is believed to be the structural switch times as determined from numerous kinetic studies.

6. MYOGLOBIN TRANSIENT KINETICS

Sperm whale myoglobin served as our control since it does not have any of the large structural changes that occur in hemoglobin, hence we expect the UV region to be "quiet" and not show any unusual transients. We observed no unusual absorbance changes in the UV compared to the visible spectrum. Figure 6 shows this most clearly: the transient difference spectra observed at 20 nsec and 2 μsec overlaid quite well the static difference spectra, indicating that any structural relaxation is either very small and/or very fast on our time scale. Thus, the evidence here is for only local, heme-related changes that are prompt and not diffusive in nature.

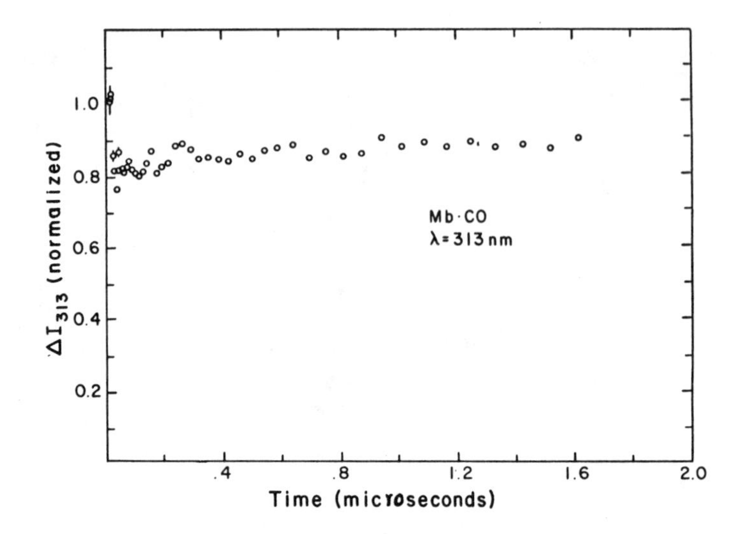

Figure 6. Mb Transient UV Kinetics. The transient absorption of Mb at λ=313 nm as a function of time. Note that no fast kinetics in the nanosecond to microsecond region are seen. We can assume that the small change in absorbance seen here is due to heme contributions and not changes in aromatics such as tryptophan or tyrosine.

7. HEMOGLOBIN TRANSIENT KINETICS

The simple picture seen in the case of myoglobin is considerably more complicated in hemoglobin. As I mentioned in the introduction, I suppose that this could be painted in a negative light concerning the applications of Hans' work since extrapolations of the conformational dynamics to room temperatures predict a simple time averaging of the internal states which gives rise to an exponential recombination time. This is emphatically NOT seen in hemoglobin.

The R$\rightarrow$T switch also occurs during rebinding, and this has associated with it spectral perturbations that are independent of the overall ligand rebinding absorbance changes. The spectral resolution of these changes has been elegantly addressed in the paper by Hofrichter, et al[19]. Hofrichter, et al., were able via singular value decomposition techniques to isolate 3 transient spectra. The largest spectral component, called SVD 1, is due solely to carboxy/decarboxy differences, and thus is a marker of only the ligation state of the protein (the simple population of deoxy hemes). The second spectral component, SVD 2, is identified as due to *quaternary* changes and represents about 6 % of the total signal, integrated over visible wavelengths. Fortunately, SVD 2 has an isosbestic at 438 nm, so that if measurements are made at that wavelength there is no contribution due to quaternary effects. Thus correction for deoxy concentration measurements were made as close to the isosbestic point as was technically feasible (due to the availability of 436 narrow band dielectric filters), that is, at the quaternary isosbestic point, and used to "normalize" the UV transient absorption changes. Since the population changes are in effect removed from the UV curves, it should be then possible to compare the time course of the SVD 2 curve, the quaternary spectrum, with our UV kinetics.

Figure 7 shows the transient 313 nm kinetics. Similar kinetics, but with different amplitudes, are seen as a function of wavelength as we showed in Figure 4.

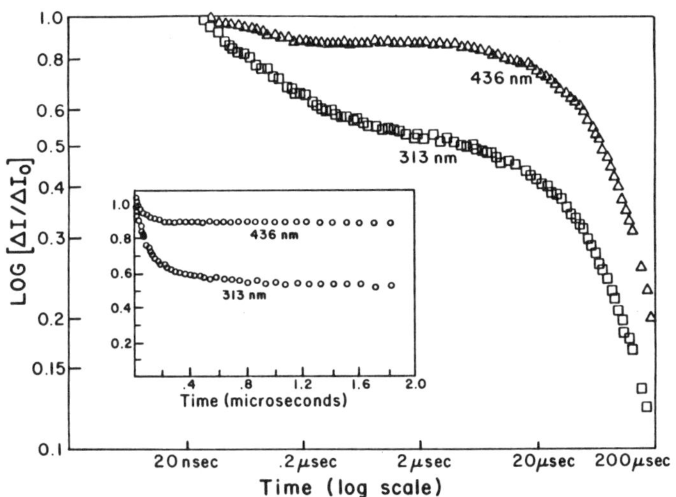

Figure 7. Hemoglobin Transient Kinetics. The small box insert shows that transient intensity changes for human hemoglobin at 313 nm and 436 nm plotted on a linear scale at short times (0 to 2 μsec). The large frame shows the transient *absorbance*, normalized to unity, for 313 nm and 436 nm plotted on a log-log scale. Note that the 313 nm fast component is both qualitatively different in time response and quantitatively larger in amplitude than the 436 nm data.

There are several interesting features to the 313 nm data. There is a large, *possibly* multi-exponential relaxation of an initial prompt absorbance change to a baseline at approximately 2 μsec, followed by a small relaxation to another baseline at 200 μsec. Normalization of the 313 nm data by the 436 nm data results in the plot of Figure 8.

IF the UV curves reflect "merely" the quarternary heme perturbations, then the time constants of the UV and the visible processes should be identical. IF the heme absorbance changes stop at 350 nm, then one can observe in the UV only the aromatic rings and their perturbations. But, that is a big IF, and certainly not true in general, since even myoglobin shows an absorbance change at 305 nm. The UV difference spectra seem to indicate that one can indeed see predominately the aromatic spectra, and the UVRR also seem to indicate that the aromatics have a *prompt* spectral change. What more could one ask for? The structural switching seems to have prompt components in

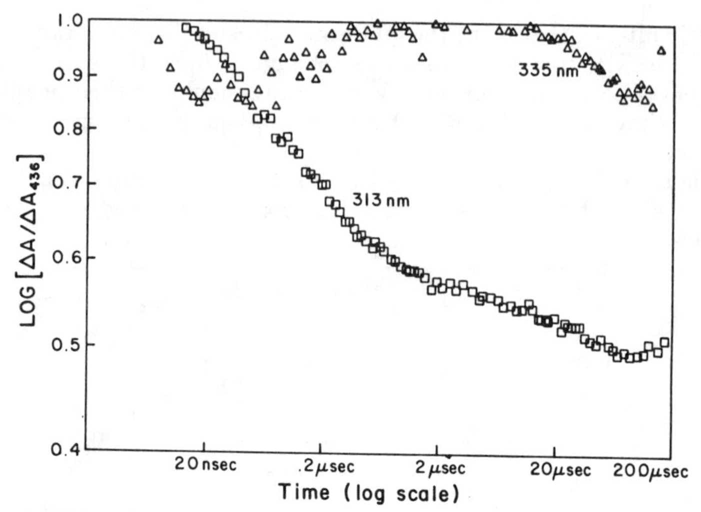

Figure 8. The Corrected Transient Kinetics of Hemoglobin at 313 nm. The corrected transient absorbance kinetics of hemoglobin plotted on a log δ intensity versus log time scale at 313 nm and 335 nm. Normalization of the UV data by the 436 nm data allows for correction of the geminate recombination that occurs.

the UV, leading to rosy visions of sexy things like solitons and rigidly coupled motions. Martin Karplus scowled darkly at this point.

8. THE DEATH OF AN IDEA

Up to this point we have been suffering from the weakness that pervades such studies of kinetics: we had indirect evidence for prompt perturbations at a distance, but it was still possible for heme changes to masquerade as changes in the UV absorbance. Enticed by the revolutionary aspects of this idea we grandly plowed on towards publication.

However, two things happened to save the day. One, I dragged my heels writing up the work due to (1) the publishing traits of physicists mentioned earlier, and (2) the lack of any grad student help at Princeton, the Land of String Theory and Clean-Handed Experimentalists. Two, Tom Spiro got a new post doc, Young Park, to try and repeat the earlier UVRR work.

Park and his colleague Chang Sue found that they could not reproduce the data, and in fact found out that the 853 cm^{-1} band is *insensitive to any R-T differences* (Figure 9), much less a strained transient configuration!

So, what happened? How could a paper purporting to show a clear effect turn out to be wrong? Appearently the "effect" seen was the result of wishful thinking and preselection out of a set of noisy data plots with spikes in the right places.

It's time to talk about Hans again, and the sociology of basic research. As I said, Hans was always there with us in the trenches taking the data,

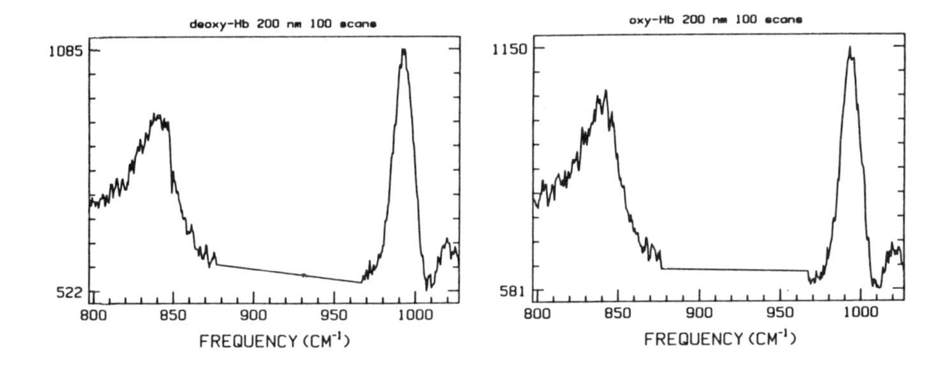

Figure 9. Recent UV Resonance Raman Results of Park and Sue. High statistics UV-Raman difference spectra at 853 cm^{-1} of the same spectral region displayed in the previous UVRR figure. Note the lack of significant difference between the R and the T state.

plotting it up, worrying about it. Today, with these large research groups that churn out papers like so many cars, often the Professor only sees the finished, polished, massaged plots from the draftsman. I don't think that such an approach works in the long run. I read in *Nobel Dreams*[20], the devastating exposé of hard-ball tactics in high energy physics, that one can only trust the data from new graduate students, those that have been there too long learn what the advisor really wants and deliver accordingly. We can question sometimes the interpretation of Hans' data, but the data themselves are always firm, as I have learned.

Although my UV absorbance is correct and still stands, I can't base any interpretations on the now changed Raman work. I believe that there still are prompt structural changes occurring in the aromatics. I guess I'll just have to wait for the Eaton-Hofrichter juggernaut to do careful experiments down here in the UV. There could well be some element of information lying in those curves, but for now they will have to lie fallow. We seem to be back to the orginal idea of Gibson[21] that the R-T switch takes place *structurally* in about 20 μsec.

9. REPRISE

I have shown, I hope in an honest manner which is sure to infuriate all, how an attempt to discover the way that energy is transmitted through the hemoglobin structure faded way because of questions concerning supporting data.

There are several facts that I believe still stand:

(1) The transient absorption changes seen in hemoglobin are indicative of conformational changes remote from the heme site that are "prompt".

(2) You can fit the transient absorption changes seen in hemoglobin to a series of exponentials, but the truth is more likely to be a continuous flow of processes and a distribution of rates.

(3) Studies confined to the heme site in hemoglobin are looking in the wrong spot if the answer to the mechanism of allosteric action is being sought.

(4) Computer dynamical modeling will not reveal the source of these dynamics.

More general principles that ignore the details still remain to be revealed. One of them, which has been clearly expressed by Hans, is the concept of hierarchical diffusion and ultrametricity. An ultrametric space can be formed by the set of configurations that a complex structure forms by repeated folding decisions. The clearest example of this in biological systems is the evolutionary tree. The distance between any two species A and B is defined as the number of generations one must go back until a common ancestor can be found. Ultrametricity then states that, for any 3 species A, B and C, the distance AC is less than the smaller of AB or BC. For a protein, we could imagine a polymer with links that folds by rotating the links through fixed angles. The distance between any two final configurations could be defined as the number of rotations you would have to make to arrive at a common ancestral configuration. Since the growth of the configurational space is like the evolution tree, the space of polymer configurations should be an ultrametric one. This is of course speculative and not all would agree[22] but no clear analysis has been done yet.

Hans has suggested that the dynamics of protein relaxation may be a configurational diffusion in an ultrametric space [23]. Is it possible that the switch from the R state to T state proceeds by such a configurational diffusion? I don't know, but I do leave you with this time-honored log-log plot of the UV relaxation signal I have struggled over (Figure 10). Although the fit shown is a 3 exponential fit, done to please the chemists, to my eye the decay preceding the final switch looks like a power law. Could we be seeing here the conformational diffusion of the hemoglobin structure as a continuous flow? Do proteins perhaps always flow in a hierarchial scheme between discrete final conformations? Maybe someday this will turn out to be one of the more important applications of Hans invigorating ideas to biophysics.

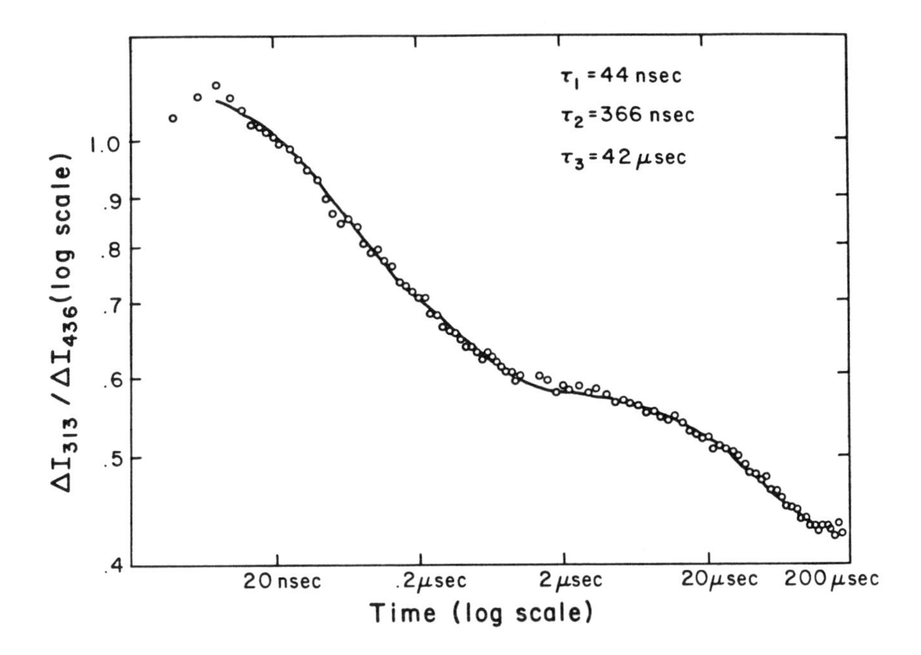

Figure 10. Fitting Kinetics to the UV Absorbance Change. A fit to the decay of the UV transient signal in hemoglobin, using a 3 exponetial fit.

REFERENCES

1. D.L. Stein, Proc. Natl. Sci. USA 82, 3670 (1985).
2. A. Ansari, J. Berendzen, S.F. Bowne, H. Frauenfelder, I.E.T. Iben, XII, T. Sauke, E. Shyamsunder and R.D. Young Proc. Natl. Acad. Sci. USA 82, 5000 (1985).
3. R.D. Young and S.F. Bowne, J. Chem. Phys. 81, 3730 (1984).
4. R.F. Goldstein and W. Bialek, Comments Mol. Cell. Biophys. 5, 407 (1986).

5. L. Powers and W. Blumberg, Biophys. J.54, 181 (1988).
6. R.H. Austin, K.W. Beeson, L. Eisenstein, H. Frauenfelder and I.C. Gunsalus, Biochemistry 14, 5355 (1975).
7. R.H. Austin, K.W. Beeson, L. Eisenstein, H. Frauenfelder, I.C. Gunsalus and V.P. Marshall, Phys. Rev. Lett. 32, 403 (1974).
8. H. Frauenfelder, G.A. Petsko and D. Tsernoglou, Nature 280, 558 (1979).
9. R.H. Austin, D.L. Stein and J.Wang, Proc. Natl. Acad. Sci. USA 84, 1541 (1987).
10. R.E. Dickerson and I. Geis, *Hemoglobin* (Benjamin/Cummings, Menlo Park, CA, 1983).
11. J. Monod, J. Wyman and J. Changeux, J. Mol. Biol. 12, 88 (1965).
12. W.G. Cobau, J.D. LeGrange and R.H. Austin, Biophys. J. 47, 781 (1985).
13. N. Agmon and J.J. Hopfield, J. Chem. Phys. 79, 2042 (1983).
14. R.W. Briehl and J.F. Hobbs, J. Biol. Chem. 245, 544 (1970).
15. M.F. Perutz, A.R. Fersht, S.R. Simon and G.C.K. Roberts, Biochemistry 13, 2174 (1974).
16. E.R. Henry, D.L. Rousseau, J.J. Hopfield, R.W. Noble and S.R. Simpson, Biochemistry 24, 5907 (1985).
17. Y. Yanase, T. Fijimura, K. Kawasaki and K. Yamoaka, Jap. J. Hum. Genet. 13, 40 (1968).
18. S. Dasgupta, R.A. Copeland and T.G. Spiro, J. Biol. Chem. 261, 1960 (1986).
19. J. Hofrichter, J.H. Sommer, E.R. Henry and W.A. Eaton, Proc. Natl. Acad. Sci. USA 77, 5608 (1983).
20. G. Taubes, *Nobel Dreams* (Random House, New York, NY, 1986).
21. A.L. Tan, R.W. Noble and Q.H. Gibson, J. Biol. Chem. 247, 2493 (1973).
22. R. Elber and M. Karplus, Science 235, 318 (1987).
23. H. Frauenfelder, Ann. N.Y. Acad. Sci. 504, 161 (1985).

NEUTRINO OSCILLATION EXPERIMENTS WITH REACTORS AND WITH THE SUN

R.L. Mössbauer

Physik-Department, Technische Universität München
D-8046 Garching, West Germany

ABSTRACT

Neutrinos continue to exhibit a number of unexplained features and unknown properties. Amongst these, the question of the neutrino restmass remains to be one of the most challenging problems of nowadays physics. Thus far, only upper limits on neutrino masses have resulted from direct measurements as well as from searches for neutrino oscillations. Neutrinos provide instant information on the solar fusion reactions. The reasons for the shortage by a factor of three in the high energy portion of the solar neutrino flux observed in the Davis experiment are unknown at present. The "European Gallex Collaboration" will perform a solar neutrino experiment aiming at the low energy portion of the solar neutrino spectrum, which is directly related to the known solar luminosity. The new experiment due to the large distance involved will be highly sensitive to neutrino masses.

THE NEUTRINO AND ITS UNKNOWN PROPERTIES

The neutrino was hypothetically introduced more than 50 years ago by Wolfgang Pauli, in an effort to explain the continuous nature of the electron spectrum observed in radioactive beta-decay. It was only in 1956, when Cowan and Reines[1] succeeded for the first time in observing the new particle directly via the reaction

$$\bar{\nu}_e + p \rightarrow n + e^+ \tag{1}$$

For purposes of weak interactions, quarks and leptons are divided into distinct weak isospin doublets or generations:

$$\begin{array}{cc} \textit{Quarks} & \textit{Leptons} \\ \begin{pmatrix} u \\ d \end{pmatrix}\begin{pmatrix} c \\ s \end{pmatrix}\begin{pmatrix} t \\ b \end{pmatrix} & \begin{pmatrix} \nu_e \\ e \end{pmatrix}\begin{pmatrix} \nu_\mu \\ \mu \end{pmatrix}\begin{pmatrix} \nu_\tau \\ \tau \end{pmatrix} \end{array}$$

In each doublet, the charge decreases by one unit from top to bottom. The enormous relative mass difference between the members of the different lepton generations is a consequence of the spontaneous symmetry breaking, which makes gauge invariance a hidden symmetry and provides the gauge bosons W and Z with mass.

Observations of neutrinos are very difficult, because these particles are subject only to the weak interaction, with gravitational effects being completely negligible. By consequence, many of the properties of the neutrinos, even today, are still unknown: We do not know why there are several generations of neutrinos, nor do we know how many such generations exist. We do not know whether the flavor quantum number, which may be associated with each neutrino generation, is exact or only approximate. An exact quantum number would imply the existence of an as yet unknown symmetry. We have only limited information on the stability of the neutrinos. We do not know whether neutrinos are Dirac or Majorana particles, i.e. we do not know whether we are obliged to distinguish between particles and antiparticles. This latter question might finally be settled by experiments on double beta decay, however the present evidence is still inconclusive. We do not know whether the universe is filled with a neutrino background radiation (with intensity and distribution approximately equal to the photon background radiation), which supposedly exists as a remnant from the big-bang creation of the universe. There appears at present no possibility to measure this very intense neutrino background radiation because of its very low energy. And finally, we do not know whether neutrinos are particles donated with mass. There is no conclusive experimental evidence for the existence of massive neutrinos. Theoretical predictions concerning neutrino masses are very flexible. The standard minimal electro-weak theory assumes the neutrino mass to be zero, but non-minimal electroweak theories can readily cope with finite masses. Grand unified theories (GUT) can be constructed, which make $m_\nu = 0$ possible, but $m_\nu \neq 0$ is more likely. Present experimental limits on neutrino masses are given in Table 1:

Table 1: Summary of experimental limits for neutrino rest masses

Neutrino	Method	Limit	Ref
$\bar{\nu}_e$	^{3}H - decay	$m_{\nu_e} < 18$ eV	2
ν_μ	$\pi \rightarrow \mu\nu_\mu$ at rest	$m_{\nu_\mu} < 270$ keV	3
ν_τ	$\tau \rightarrow 3\pi\nu_\tau$	$m_{\nu_\tau} < 70$ MeV	4

There has been for a number of years a claim by a group at the ITEP at Moscow for the observation of an electron neutrino mass[5], but the experimental value is in contradiction with the results of Ref. 2 and the analysis has been questioned in the literature. A search for the mass of the electron neutrino in a range of a few eV or below is very difficult to perform, since possible excitations in atoms or solids are likewise of the order of eV and a mass contribution is very difficult to distinguish from such excitations.

There exists also a cosmological limit on neutrino masses which states, that the sum of the masses of neutrinos of all flavors should be smaller than about 100 eV. This cosmological limit applies to stable neutrinos and is obtained by attributing the entire possible mass in the universe to neutrinos.

WEAK INTERACTIONS AND WEAK MIXING

It has been known for some time that weak interactions proceeding via charged currents give rise to a coupling between different quark generations, in contrast to strong interactions, where strangeness conservation prevents such a coupling. The weak interaction mixing in the approximation where only the first two quark generations couple is given by:

$$\begin{pmatrix} d' \\ s' \end{pmatrix} = \begin{pmatrix} cos\theta_c & sin\theta_c \\ -sin\theta_c & cos\theta_c \end{pmatrix} \begin{pmatrix} d \\ s \end{pmatrix}$$

The mixing matrix involving the Cabbibo angle θ_c couples the weak interaction eigenstates on the left side of the equation with the mass eigenstates on the right side. The particular form of the Cabbibo mixing matrix prevents the appearance of strangeness changing neutral currents (GIM-mechanism[6]). The extension of the weak interaction mixing to a coupling of 3 quark families involves the 3×3 Cabbibo-Kobayashi-Maskawa matrix[7].

The reason for the existence of weak interaction mixing in the hadronic sector is unknown. By analogy, one may assume that weak interaction mixing exists also in the leptonic sector, necessitating a distinction between weak interaction eigenstates ν_e, ν_μ, ν_τ and mass eigenstates ν_1, ν_2, ν_3. In a two neutrino approximation, involving for instance ν_e and ν_μ, we would have the relation

$$\begin{pmatrix} \nu_e(t) \\ \nu_\mu(t) \end{pmatrix} = \begin{pmatrix} cos\theta & sin\theta \\ -sin\theta & cos\theta \end{pmatrix} \begin{pmatrix} \nu_1 \\ \nu_2 \end{pmatrix}$$

With the neutrino mass eigenstates having a well defined time dependence,

$$\nu_i(t) = \nu_i(0)exp[-iE_it] \approx \nu_i(0)exp[-i(p_\nu + m_i^2/2p_\nu)t]$$

and assuming the initial condition $\nu_e(0) = 1$, it is straightforward to evaluate the probability for having at a distance L from the source the initially generated neutrino ν_e or the neutrino ν_μ, respectively:

$$P[\nu_e(0) \rightarrow \nu_\mu(L)] = \frac{1}{2}sin^2 2\theta \left[1 - cos\frac{\Delta m^2 L}{2E_\nu}\right] \tag{2a}$$

$$P[\nu_e(0) \rightarrow \nu_e(L)] = 1 - P[\nu_e(0) \rightarrow \nu_\mu(L)] \tag{2b}$$

where $\Delta m^2 = m_1^2 - m_2^2$, with m_i being the mass associated with a neutrino mass eigenstate. Equations 2 demonstrate the neutrino intensity variations which are to be expected if weak interaction mixing does exist. These intensity variations, apparently, are absent if the mixing angle θ is equal to 0. For finite mixing they would show an oscillatory behavior depending on the distance L between source and detector, on the mass parameter Δm^2 and on the energy E_ν of the neutrinos emitted by the source. Illustrative examples for neutrino oscillation lengths are shown in Table 2.

Table 2: Neutrino oscillation lengths in vacuum for different values of the mass parameter Δm^2 and of the neutrino energy E_ν.

Δm^2 $\downarrow$	Sun (300 keV)	Reactor (4 MeV)	Meson factory (20 MeV)	High Energy accelerator (1 GeV)
1 eV^2	0.75 m	10 m	50 m	2.5 km
10^{-3} eV^2	750 m	10 km	50 km	2500 km
10^{-10} eV^2	$7.5 \cdot 10^6$ km	10^8 km	5×10^8 km	

NUCLEAR REACTORS AS NEUTRINO SOURCES

Nuclear reactors are copious sources of electron antineutrinos $\bar{\nu}_e$. The fission processes in these reactors are associated with the generation of nuclei in highly excited states. These nuclei are very neutron rich and therefore undergo radioactive decays, where neutrons are converted into protons, with the emission of electrons and electron-antineutrinos. Reactors, therefore, are powerful sources of neutrinos for oscillation experiments. With neutrino energies in the range of a few MeV, reactors according to Eqs. 2 are particularly well suited for studying small mass parameters Δm^2. Such experiments employ Eq. 2 in efforts to search for the reduction of neutrino flux (disappearance) due to oscillations into other flavors. Accelerator experiments performed at high energies, by contrast, cannot be extended into the range of very low mass parameters. High energies, however, offer the possibility to perform appearance experiments according to Eq. 2. Such experiments are particularly sensitive to mixing angles θ.

We have performed in the framework of an international cooperation a search for neutrino experiments at nuclear reactors. A summary of these experiments is shown in Table 3.

Table 3: Neutrino oscillation experiments at Gösgen and Grenoble

#	Site	Reactor Thermal power [MW]	Distance L[m]	Collaboration	Ref.
I	ILL	57	8.8	Caltech-ISN-TUM	8
II	Gösgen	2806	37.9	Caltech-SIN-TUM	9
III	Gösgen	2806	45.9	Caltech-SIN-TUM	9
IV	Gösgen	2806	64,7	Caltech-SIN-TUM	9

The experiments utilize the detection reaction in Eq. 1. A scheme of the detection principle and of the general assembly of the detector is shown in Fig. 1. Fig. 2 shows the measured positron spectra at the nuclear power reactor at Gösgen. Neutrino energies are related to the positron energies through $E_e = E_{e_+} + 1.804\ MeV$. Fig.3 shows the results of our data analysis, based on a two neutrino model. Our measurements are compatible with the absence of neutrino oscillations. They may nevertheless be used to establish ranges of oscillations parameters, where neutrino oscillations can no longer exist, as contrasted with ranges were such oscillations may still be possible. The measurements exclude the areas to the right of the curves. Oscillations would still be compatible with our experiments in the parameter ranges to the left of the curves.

THE SUN AS SOURCE OF NEUTRINOS

Nuclear reactors are copious sources of electron-antineutrinos. The sun, by contrast, is a copious source of electron-neutrinos. These neutrinos originate in the solar fusion processes, where protons are fused into ^4He-nuclei. The sun is practically transparent for solar neutrinos and these neutrinos, by consequence, arrive at the earth about 8 minutes after their creation. Neutrinos are our only source of information on the reactions proceeding in the solar core and provide therefore a unique possibility to study such reactions inside stellar matter. The photons associated with the solar luminosity, by contrast, undergo extensive interactions on their way from the interior of the sun to the surface, with traveling times being of the order of a million years.

Fig.4 gives a survey of the ^4He-fusion processes in the solar core. The production of each ^4He-nucleus requires the conversion of two protons in two neutrons and therefore is associated with the creation of two ν_e. More than 98 % of the solar luminosity is associated with the pp-fusion to deuterons. The rather rare sidebranch affiliated with ^7Be and ^8B provides only minute contributions to the solar luminosity. The entire solar neutrino spectrum, based on the so-called standard theoretical model of the sun, is presented

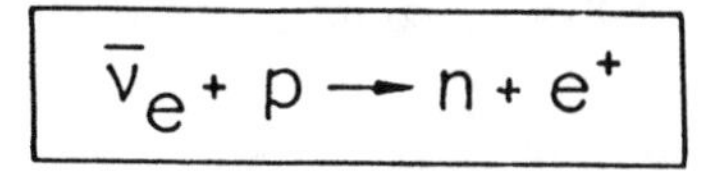

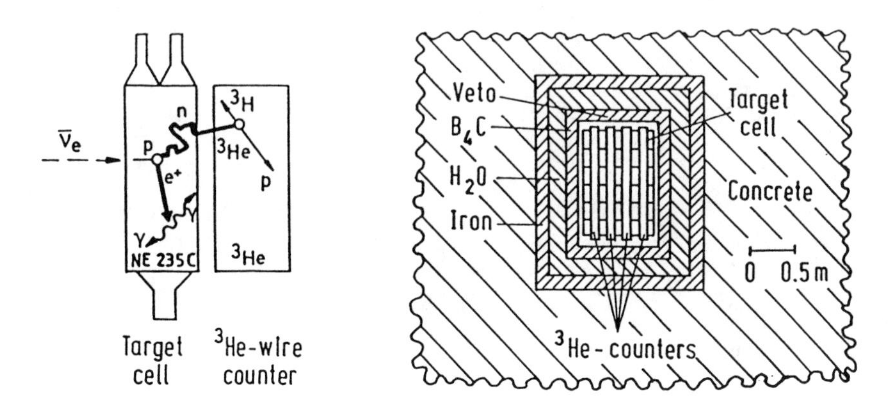

Figure 1. The left side of the figure shows the principle of the neutrino detection, using the reaction $\bar{\nu}_e + p \rightarrow n + e^+$: An electron antineutrino $\bar{\nu}_e$ coming from the nuclear reactor hits a proton in the target cell filled with a liquid scintillator and generates a neutron (n) and a positron (e^+). The positron, which slows down in the target cell, generates light which is observed by the photomultipliers mounted at both ends of the target cell. The neutron slows down in the target cell and drifts into an adjacent ^{3}He-wire chamber where it becomes absorbed and generates an electric pulse. The combined detection of a positron and a neutron is indicative of a neutrino absorption event. The right side of the figure shows schematically the composition of the actual detector, consisting of 30 target cells and 4 intercalated ^{3}He chambers. The central detector unit of about one m^3 size is surrounded by active and passive vetos schielding against cosmic radiation[9].

in Fig.5. This spectrum shows also the low-intensity contributions from the CNO-cycle.

The solar luminosity according to Fig.5 is essentially determined by the proton-proton fusion process. With roughly 26 MeV being liberated in the generation of each ^4He-nucleus, there exists a quantitative connection between solar luminosity and number of neutrinos arriving at the earth. Experimental studies of the solar neutrino flux were performed by R. Davis[11] for more than 15 years, in an effort to check the predictions of the standard solar model, which

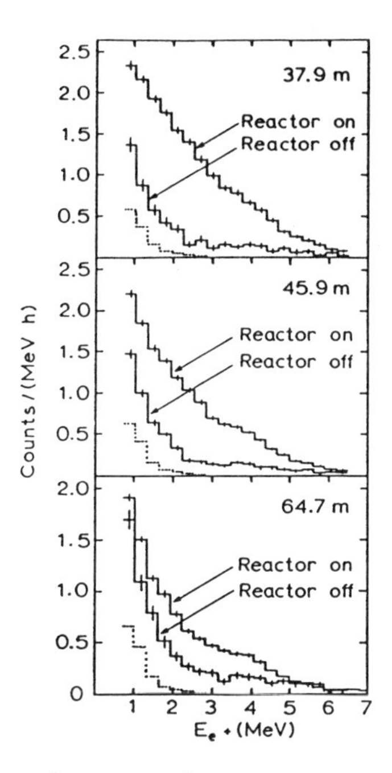

Figure 2. Experimental spectra for reactor-on and reactor-off periods after application of all selection criteria[9]. The errors shown are statistical. The contributions of the accidental background are indicated by the dashed curves. Typical measurement periods for each reactor-on spectrum have been one year.

is characteristic for all principle series stars. The terrestrial measurements of the solar neutrino flux were performed in the Homestake goldmine in South Dakota, using the reaction

$$\nu_e \; + \; ^{37}Cl \;\; \rightarrow \;\; ^{37}Ar \; + \; e^-$$

The detector employed a tank filled with 615 t of perchlorethylen. Within this huge detector, the solar neutrinos produced the conversion of less than one ^{37}Cl nucleus to an ^{37}Ar nucleus per day. The extraction of this excessively small number of ^{37}Ar atoms and their counting by means of their electron capture radioactivity was an extraordinary experimental achievement. Expressed in the solar neutrino unit SNU corresponding to one neutrino capture per sec in 10^{36} nuclei, the measurements gave a value of $(2.1 \pm 0.3) SNU$, as compared to the standard model prediction of $(5.8 \pm 2.2) SNU$, revealing a deficit of solar neutrinos by roughly a factor of three.

The measured deficit in solar neutrinos is very disturbing. It may originate from two different sources:

1. There may be something wrong with our understanding of the high energy branch of the solar neutrino spectrum. The ^{37}Cl capture process used for

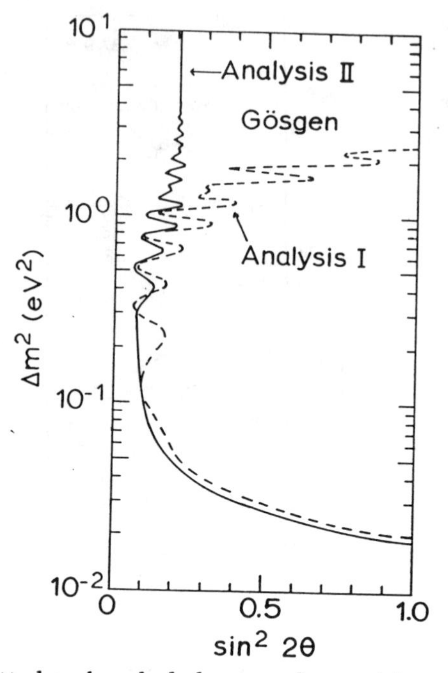

Figure 3. Permitted and excluded ranges for neutrino-oscillation parameters $\Delta m^2 = m_1^2 - m_2^2$ and $\sin^2 2\theta$, obtained for a two-flavor neturino model[9]. Excluded are in both types of analyses with 90% convidence the parameter ranges to the right of the curves; oscillation parameters to the left of the curves would still be compatible with the experimental results. Analysis I relies exclusively on measured neutrino intensities at three different distances between reactor core and neutrino detector (37.9 m; 45.9 m; 64.7 m). Analysis II employs additional information, in particular a knowledge of the reactor neutrino spectral distribution, the absolute value of the integrated flux and the absolute value of the detector efficiency.

the detection of the solar neutrinos has a rather high energy threshold of 0.81 MeV and therefore according to Fig.5 is essentially measuring the ^8B-part of the solar neutrino spectrum, this part being only a very low intensity side branch of the solar fusion process. The intensity of this branch is rather sensitive to the temperature of the interior of the sun and to the size of the fusion core. The predictions of the standard solar model may be in error concerning this side branch, though this is very unlikely to the extent necessary.

2. Neutrinos might oscillate during their passage from the interior of the sun to the earth. One might naively assume that the ν_e become evenly distributed over the three neutrino flavors ν_e, ν_μ and ν_τ while traveling from the solar surface to the earth. The ^{37}Cl-detector, being sensitive only

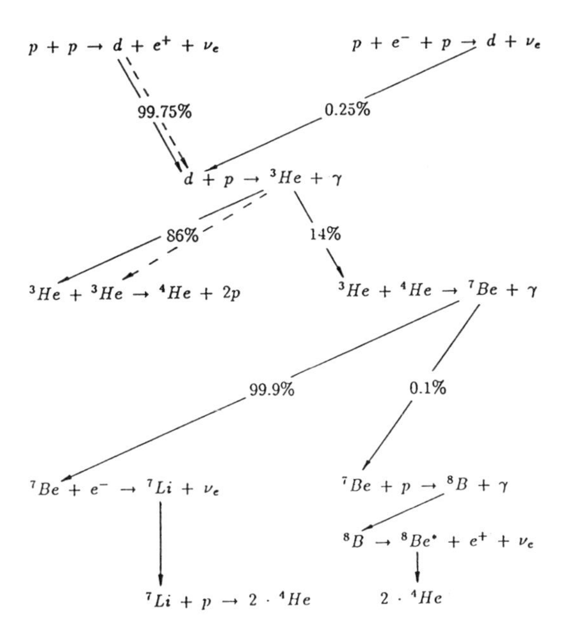

Figure 4. pp-reaction cycle[10]. The CNO-cycle, playing only a minor role, is not shown.

to ν_e, would then register only one third of the neutrino intensity created in the sun. This simple picture, however, would necessitate maximum neutrino mixing and it is hardly possible to grasp such an assumption. There exists, however, another possibility, the effective conversion of ν_e into other types of neutrinos within the solar matter.

NEUTRINOS OSCILLATIONS IN MATTER

The possibility of neutrino oscillations in matter was pointed out by Mikheyev and Smirnov[12,13], based on a previous theoretical analysis by Wolfenstein[14]. The MSW-effect is based on the fact that electron neutrinos in the sun may interact with the solar electrons via charged current reactions in addition to the neutral current reactions which are possible for neutrinos of all flavors, as indicated in Fig.6. The additional charged current interaction in

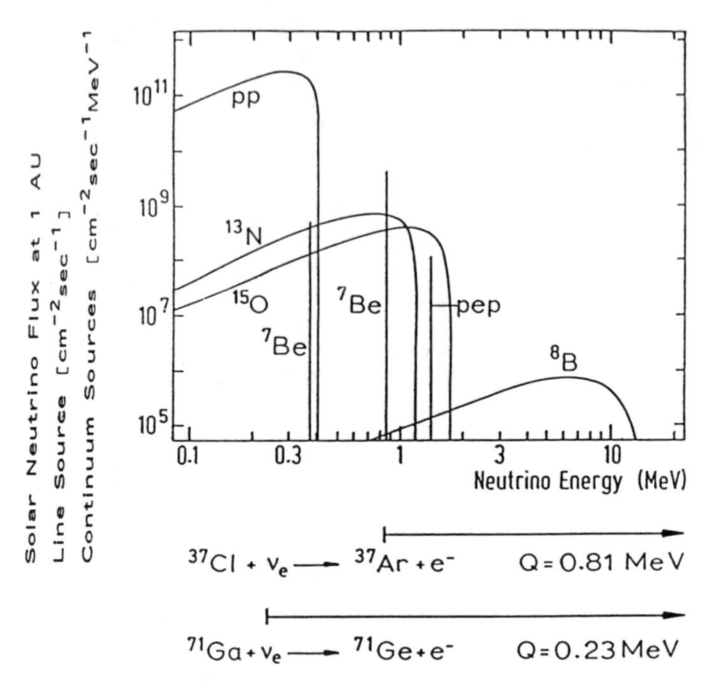

Figure 5. Solar neutrino spectrum according to the standard model of the sun, showing monochromatic and continuous components. The astronomical unit (AU) corresponds to the mean distance between sun and earth. Neutrino ranges pertinent to the chlorine experiment (threshold 814 keV) and to the gallium experiment (threshold 233 keV) are indicated.

matter confined to ν_e causes in a two-neutrino approximation the following relations between neutrino eigenstates ν_e, ν_μ, mass eigenstates ν_1, ν_2 in vacuum and mass eigenstates ν_{1m}, ν_{2m} in matter:

$$\begin{aligned}\begin{pmatrix}\nu_e\\ \nu_\mu\end{pmatrix} &= \begin{pmatrix}cos\theta_\nu & sin\theta_\nu\\ -sin\theta_\nu & cos\theta_\nu\end{pmatrix}\begin{pmatrix}\nu_1\\ \nu_2\end{pmatrix}\\ &= \begin{pmatrix}cos\theta_m & sin\theta_m\\ -sin\theta_m & cos\theta_m\end{pmatrix}\begin{pmatrix}\nu_{1m}\\ \nu_{2m}\end{pmatrix}\end{aligned} \tag{3}$$

Alternatively, we may write

$$\begin{aligned}\begin{pmatrix}\nu_{1m}\\ \nu_{2m}\end{pmatrix} &= \begin{pmatrix}cos\theta_m & -sin\theta_m\\ sin\theta_m & cos\theta_m\end{pmatrix}\begin{pmatrix}\nu_e\\ \nu_\mu\end{pmatrix}\\ &= \begin{pmatrix}cos(\theta_m-\theta_\nu) & -sin(\theta_m-\theta_\nu)\\ sin(\theta_m-\theta_\nu) & cos(\theta_m-\theta_\nu)\end{pmatrix}\begin{pmatrix}\nu_1\\ \nu_2\end{pmatrix}\end{aligned} \tag{4}$$

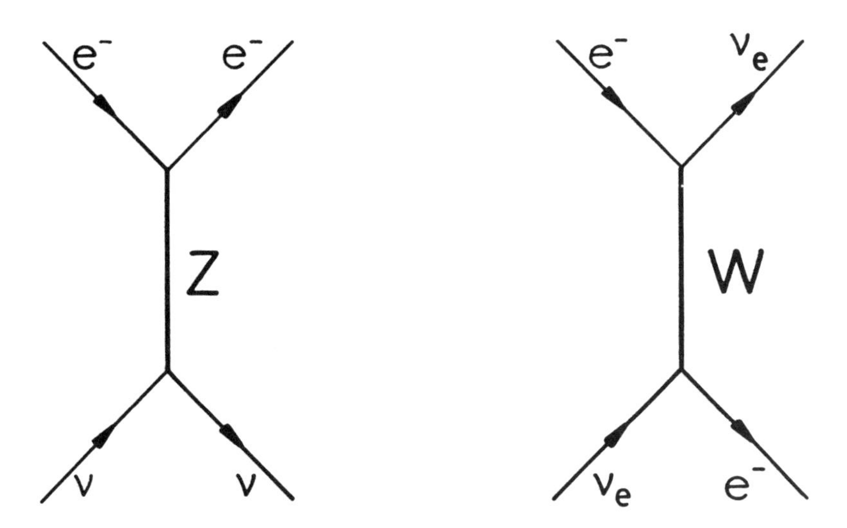

Figure 6. Interaction of electrons and neutrinos by exchange of neutral Z and charged W particles. Lepton number conservation at each vertex in the charged current diagram apparently is possible only for ν_e.

where θ_v and θ_m are the mixing angles applying to vacuum and matter, respectively. The matter oscillation angle θ expressed in terms of the ratio ℓ_v/ℓ_o is given by:

$$tan2\theta_m = \frac{sin2\theta_v}{cos2\theta_v} - \frac{\ell_\nu}{\ell_0}, \tag{5}$$

where the quantity

$$\frac{\ell_v}{\ell_o} = \frac{2\sqrt{2}G_F E_\nu N_e}{m_2^2 - m_1^2}$$

contains in the numerator the relevant charged current interaction, which is proportional to the electron density N_e in the sun. We may now anticipate the following succession of events which is illustrated in Fig.7:

1. A ν_e is created in the solar core were it interacts with electrons in a region of high density, which we may approximate by $N_e \rightarrow \infty$. Assuming a normal mass hierarchy, $m_1 < m_2$, we have

$$cos2\theta_v \ll \ell_v/\ell_o \; and \; \theta_m = \pi/2$$

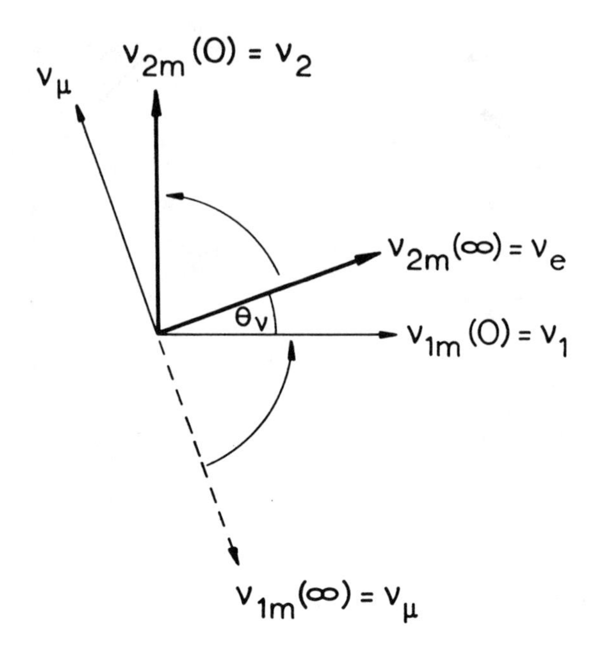

Figure 7. Weak interaction state factors ν_e and ν_μ and their turns, as the electron neutrinos ν_e move from the solar interior with high electron density (infinity) to the solar surface with 0 electron density (0). The neutrino mass eigenstates ν_1 and ν_2 in vacuum are also indicated.

According to (4) we then obtain for the state vectors

$$\begin{pmatrix} \nu_{1m}(\infty) \\ \nu_{2m}(\infty) \end{pmatrix} = \begin{pmatrix} 0 & -1 \\ 1 & 0 \end{pmatrix} \begin{pmatrix} \nu_e \\ \nu_\mu \end{pmatrix}$$

as illustrated in Fig.7.

2. Neutrinos may pass on their way from the solar interior to the surface through a region of electron density, where $cos2\theta_v = \ell_v/\ell_o$, corresponding to $\theta_m = \frac{\pi}{4}$ (compare Eq. 5). This resonance condition corresponds to a situation, where the terms in the main diagonal of the interaction matrix, i.e. the terms which leave the electron flavor unchanged, just balance. By consequence, a situation of maximum mixing arises, where ν_e mixes most effectively with ν_μ.
3. In continuing their travel, the ν_e will finally arrive at the solar surface, characterized by $N_e \to 0$ and a situation arises, where

$$cos2\theta_v \gg \ell_v/\ell_o \ and \ \theta_m = \theta_v$$

According to Eq.(4) we then have:

$$\begin{pmatrix} \nu_{1m}(0) \\ \nu_{2m}(0) \end{pmatrix} = \begin{pmatrix} 1 & 0 \\ 0 & 1 \end{pmatrix} \begin{pmatrix} \nu_1 \\ \nu_2 \end{pmatrix}$$

Fig.7 demonstrates, how this sequence of events would permit, assuming θ_v being very small, a most effective conversion of the ν_e initially created in the solar interior into the mass eigenstate ν_2 which effectively resembles ν_μ and travels unperturbed through the vacuum to the earth. In passing through the terrestrial matter, their might be within certain parameter ranges a conversion back into ν_e. In this case one would observe a higher flux of ν_e during the night and a higher average flux during winter periods. It is presently not possible to judge on the basis of the ^{37}Cl experiment whether the shortage of solar electron neutrinos is due to a misunderstanding of the solar model or due to an oscillatory behavior of the neutrinos within the solar matter. It therefore becomes crucial to study the main solar neutrino branch associated with the proton-proton fusion process, where the number of neu trinos can be readily related to the observed solar luminosity. A measurement of this low energy part of the solar neutrino spectrum requires a detector with a substantially lower threshold energy than used so far. The only feasible possibility appears at the moment the use of gallium as detector material. Such an experiment is actually in the preparatory state.

THE GALLIUM SOLAR NEUTRINO EXPERIMENT

The reaction

$$\nu_e + {}^{71}Ga \rightarrow {}^{71}Ge + e^-$$

has a threshold of only 0.23 MeV and according to Fig. 5 is well suited to measure a major portion of the pp-fusion process. This reaction will be used by the EUROPEAN GALLEX COLLABORATION[15] in their effort to measure the solar neutrino flux. The experiment will employ 30 t of gallium in the form of $GaCl_3$. Like the chlorine experiment, the gallium experiment will once more be a radiochemical experiment, with an expected reaction rate, where about one atom of ^{71}Ga would be daily converted to ^{71}Ge. The experiment will be performed in an underground neutrino laboratory in the center of the Gran-Sasso-highway tunnel some hundred twenty kilometers east of Rome, which provides a shielding of about 1200 meters of rock against cosmic radiation. The activated ^{71}Ge will be extracted within one day roughly every two weeks and will be measured via its radioactive decay back to ^{71}Ga, employing low level proportional counter techniques[16]. The extraction of the neutrino produced ^{71}Ge nuclei from the gallium tank has been successfully studied in a previous pilot experiment at BNL and extraction efficiencies in excess of 95 % should be possible. The procedure employs the addition of small amounts of inactive germanium as carrier to the tank solution. Upon activation, the employed $GaCl_3$ converts to volatile $GeCl_4$, which will be swept out from the solution by means of some auxiliary gas. After sweeping, the $GeCl_4$ is removed by gas scrubbers, extracted with CCl_4 and back-extracted with water.

Confined now to a small volume, a reduction to gaseous germane (GeH_4) is performed. After further purification, the germane admixed with Xenon is used as counting gas of a proportional counter. Special measures are taken to avoid possible interfering background reactions such as to facilitate the measurement of counting rates of less than one count per day. In order to remove ambiguities due to the complexity of the procedure and due to possible uncertainties in the reaction cross section, it is planned to calibrate the entire experimental procedure by means of an artificial neutrino source of one MCi ^{51}Cr, which will be placed in the center of the gallium tank. Details of this source experiment, in particular the possibility to engage a source depleted in ^{53}Cr causing undesired neutron absorption during the source production in a nuclear reactor, are still under study.

The European Gallex Collaboration will start taking data in late 1989. A measuring period of two years, involving some 50 extractions, should provide a statistical accuracy of about 10%, assuming a neutrino flux according to the standard solar model. Significant data should be available already after several months of measurements. Table 4 shows the expected rates according to the standard model, which besides the most relevant contributions from the pp-process will also contain contributions from the high energy portion of the solar neutrino spectrum. The latter part of the spectrum contributes very little to the solar neutrino flux, but adds substantially to the counting rates, because the detection efficiency increases roughly proportional to E_ν^2.

Table 4: Solar neutrino flux[10] and capture rate in ^{71}Ga (in units of SNU, corresponding to one neutrino capture per second per 10^{36} nuclei) for the various neutrino processes within the sun[16]. The notations gs and es refer to the ground state and the excited states in ^{71}Ge.

Source	Flux ($\nu_e/cm^2 sec$)	Energy Spectrum (MeV)	Capture Rate (SNU) gs	es
pp	6.09×10^{10}	$0 - 0.420$	70.3	– –
pep	1.50×10^{8}	1.440	2.5	0.7
^{7}Be	4.2×10^{9}	0.862(90%) 0.383(10%)	28.5	2.1
^{8}Be	4.2×10^{6}	$1 - 14$	1.2	11.5
^{13}N	4.4×10^{8}	$0 - 1.20$	2.5	0.2
^{15}O	3.4×10^{8}	$0 - 1.73$	3.4	0.6
Sum			108.4	15.1

Deviations of the measured neutrino rates from the expectation will have to be attributed essentially to the presence of neutrino oscillations in the solar

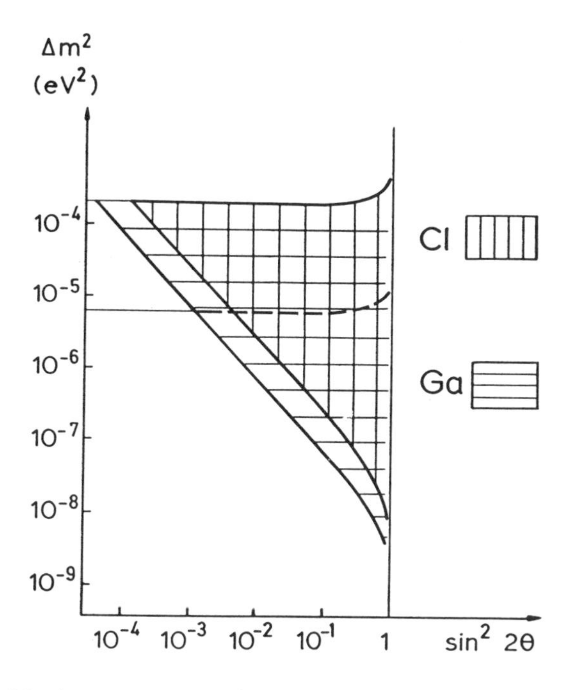

Figure 8. Maximum ranges of sensitivity of the chlorine- and the gallium- experiment for oscillation parameters $\Delta m^2 = m_1^2 - m_2^2$ and $\sin^2 2\theta$ due to matter oscillations. The horizontal limiting lines follow from the condition for the appearance of resonance (maximum mixing), assuming a maximum energy of the emitted neutrinos of $E_{\nu_e} =$ 14 MeV (solid line) and $E_{\nu_e} = 0.42$ MeV (dashed line). The inclined limiting lines follow from the condition of an adiabatic behavior of the electronic density changes across the resonance, for limiting energies of $E_{\nu_e} = 233$ kev (^{71}Ga) and 814 keV (^{37}Cl).

matter and therefore would provide direct information on neutrino masses and neutrino mixing. A coarse indication of the sensitivity of the chlorine- and gallium- experiments to neutrino oscillation parameters is shown in Fig.8.

A parallel effort to measure the low energy portion of the solar neutrino spectrum by means of a gallium detector is being planned by the Institute for Nuclear Physics of the Soviet Academy of Sciences in Moscow[17]. This experiment will use 60 t of gallium in metallic form, which might complicate the extraction procedure for the neutrino produced ^{71}Ge. The Soviet experiment will be performed in a underground neutrino laboratory in the Baksan-valley in the Caucasus.

There is substantial hope that the gallium experiments, replacing the distance between source and detector employed in terrestial experiments by the distance between source and earth, will substantially improve our knowledge on neutrino properties and thereby provide significant information for both elementary particle physics and astrophysics.

REFERENCES

1. C.L. Cowan, F. Reines, Harrison, Kruse, and Mc. Guire, Science 124, 103 (1956).
2. M. Fritschi, E. Holzschuh, W. Kündig, J.W. Petersen, R.E. Pixley, and H. Stüssi, Phys. Lett. 173B, 485 (1986).
3. B. Jackemann et al., Phys. Rev. Lett. 56, 1444 (1986) and R. Abela et al., Phys. Lett. 146B, 431 (1984).
4. ARGUS Collaboration, H. Albrecht et al., Phys. Lett. 163B, 404 (1985).
5. V.A. Lubimov et al. Phys. Lett. 94B, 266 (1980); S. Boris et al., Proc. Int. Europhysics Conf. on High Energy Physics Brighton, UK (1983), p. 386; S. Boris et al., Proc. 22nd Int. Conf. High Energy Physics (Leipzig, DDR, 1984) Vol. 1, p. 257; S. Boris et al., Phys. Lett. 159B, 217 (1985), V.A. Lubimov at VIth Moriond Workshop on Massive Neutrinos in Particle and Astrophysics (1986), p. 441.
6. S.L. Glashow, J. Iliopoulos, and L. Maiani, Phys. Rev. D2, 1285 (1970).
7. M. Kobayashi and K. Maskawa, Prog. Theor. Phys. 49, 282 (1972).
8. H. Kwon et al., Phys. Rev. D24, 1097 (1981).
9. G. Zacek, F.v. Feilitzsch, R.L. Mössbauer, L. Oberauer, V. Zacek, F. Boehm, P.H. Fisher, J.L. Gimlett, A.A. Hahn, H.E. Hendrikson, H. Kwon, J.L. Vuilleumier, and K. Gabathuler, Phys. Rev. D34, 2621 (1986).
10. J.N. Bahcall, W.F. Huebner, S.H. Lubow, P.D. Parker, and R.K. Ulrich, Rev. Mod. Phys. 54, 767 (1982).
11. R. Davis Jr, Proc. Neutrino Mass Miniconf. Telemark, Wisconsin (1980), ed. V. Barger and D. Cline, p. 38.
12. S.P. Mikheyev and A.Yu. Smirnov: Yad. Fiz. 42, 1441 (1985); Sov. J. Nucl. Phys. 42, 913 (1985); Nuovo Cimento 9C, 17 (1986).
13. A.Yu. Smirnov: Proc. VI. Moriond Workshop on Massive Neutrinos in Astrophysics and in Particle Physics, Tignes (1986) p. 355.
14. L. Wolfenstein, Phys. Rev. D17, 2369 (1978).
15. Members of the EUROPEAN GALLEX COLLABORATION: MPI-Heidelberg, KFK-Karlsruhe, TU-München, INFN-Milano, INFN-Rom, CEN-Saclay, CEN-Grenoble, Univ.-Nice, WIS-Rehovoth, BNL-Long Island/N.Y. (Speaker: T. Kirsten, MPI Heidelberg).
16. T. Kirsten, Proc. VI. Moriond Workshop on Massive Neutrinos in Astrophysics and in Particle Physics, Tignes (1986), p. 119.
17. I.R. Barbanov et al., "Solar Neutrinos and Neutrino Astronomy", Homestake (1984), AIP Conf. Proc. 126, 175 (1985).

THE PHYSICS OF EVOLUTION

Manfred Eigen
Max-Planck-Institut für Biophysikalische Chemie
Am Fassberg, D-3400 Göttingen, FRG

ABSTRACT

The Darwinian concept of evolution through natural selection has been revised and put on a solid physical basis, in a form which applies to self-replicable macromolecules. Two new concepts are introduced: 'sequence space' and 'quasi-species.' Evolutionary change in the DNA- or RNA-sequence of a gene can be mapped as a trajectory in a sequence space of dimension ν, where ν corresponds to the number of changeable positions in the genomic sequence. Emphasis, however, is shifted from the single surviving wildtype, a single point in the sequence space, to the complex structure of the mutant distribution that constitutes the quasi-species. Selection is equivalent to an establishment of the quasi-species in a localized region of sequence space, subject to threshold conditions for the error rate and sequence length. Arrival of a new mutant may violate the local threshold condition and thereby lead to a displacement of the quasi-species into a different region of sequence space. This transformation is similar to a phase transition; the dynamical equations that describe the quasi-species have been shown to be analogous to those of the two-dimensional Ising model of ferromagnetism. The occurrence of a selectively advantageous mutant is biased by the particulars of the quasi-species distribution, whose mutants are populated according to their fitness relative to that of the wild-type. Inasmuch as fitness regions are connected (like mountain ridges) the evolutionary trajectory is guided to regions of optimal fitness. Evolution experiments in test tubes confirm this modification of the simple 'chance and law' nature of the Darwinian concept. The results of the theory can also be applied to the construction of a machine that provides optimal conditions for a rapid evolution of functionally active macromolecules.

An introduction to the physics of molecular evolution by the author has appeared recently[1]. Detailed studies of the kinetics and mechanisms of replication of RNA, the most likely candidate for early evolution[2,3], and of the implications on natural selection have been given in Refs. 4 and 5. The quasi-species model has been constructed in Refs. 6 and 7 using the concept of sequence space. Subsequently various methods have been invented to elucidate this concept and to relate it to the theory of critical phenomena[8-19]. The instability of the quasi-species at the error threshold is discussed in Ref. 20. Evolution experiments with RNA strands in test tubes are described in Refs. 21 and 22.

REFERENCES

1. M. Eigen, Chemica Scripta 26*B*, 13 (1986).
2. M. Eigen and R. Winkler-Oswatitsch, Naturwissenschaften 68, 217 (1981).
3. M. Eigen and R. Winkler-Oswatitsch, Naturwissenschaften 68, 282 (1981).

4. C.K. Biebricher, M. Eigen and W.C. Gardiner, Biochem. 22, 2544 (1983).
5. C.K. Biebricher, M. Eigen and W.C. Gardiner, Jr., Biochem. 23, 3186 (1984); 24 (1985).
6. M. Eigen, Naturwissenschaften 58, 465 (1971).
7. M. Eigen and P. Schuster, Naturwissenschaften 64, 541 (1977); 65, 7 (1978); 65, 341 (1978).
8. M. Eigen, Adv. Chem. Phys. 33, 211 (1978).
9. C.J. Thompson and J.L. McBride, Math. Biosci. 21, 127 (1974).
10. B.L. Jones, R.H. Enns and S.S. Rangnekar, Bull. Math. Biol. 38, 15 (1976).
11. B.L. Jones, J. Math. Biol. 6, 169 (1978).
12. P. Schuster and K. Sigmund, Ber. Bunsenges. Phys. Chem. 89, 668 (1985).
13. J. Swetina and P. Schuster, Biophys. Chem. 16, 329 (1982).
14. J. Hofbauer and K. Sigmund, Evolutionstheorie und dynamische Systeme (Paul Parey, Berlin and Hamburg 1984).
15. R. Feistel and W. Ebeling, Bio Systems 15, 291 (1982); W. Ebeling, A. Engel, B. Esser and R. Feistel, J. Statist. Phys. 37, 369 (1984).
16. J.S. McCaskill, J. Chem. Phys. 80(10), 5194 (1984).
17. J.S. McCaskill, Biol. Cybernet. 50, 63 (1984).
18. D. Rumschitzki, J. Chem. Phys., in press.
19. I. Leuthäusser, J. Chem. Phys. 84, 1884 (1986).
20. M. Eigen, Ber. Bunsenges. Phys. Chem. 89, 658 (1985).
21. M. Sumper and R. Luce, Proc. Natl. Acad. Sci. USA 72, 162 (1975).
22. C.K. Biebricher, M. Eigen and R. Luce, J. Mol. Biol. 148, 369 (1981); 148, 391 (1981).

PRESSURE TUNING SPECTROSCOPY IN MODERN SCIENCE

H.G. Drickamer
School of Chemical Science, Department of Physics
and Materials Research Laboratory
University of Illinois
Urbana, IL 61801 USA

ABSTRACT

The optical, electrical, magnetic and chemical properties of condensed phases are determined by the characteristics of the energy levels available to the outer electrons of the entities which make up the phase. Because of different spacial characteristics the energy levels associated with different orbitals are perturbed differently by compression. The consequences of this pressure tuning are very broad. Here we consider two categories: the characterization of states, excitations or processes, and the testing of theories concerning electronic phenomena. Three examples in each category are outlined.

INTRODUCTION

The optical, electrical magnetic and chemical properties (collectively the electronic properties) of condensed phases depend on the characteristics of the energy levels available for the outer electrons on the atoms, ions or molecules which make up the phase. These properties depend most strongly on the normally occupied state-the-ground state, but also on the characteristics and relative energies of states accessible by thermal or electromagnetic excitation. From our viewpoint the basic effect of pressure is to reduce the volume and to increase the overlap among these outer orbitals. Since different types of orbitals have different spacial characteristics - different radial extent, angular momentum (shape) and different diffuseness or compressibility they are perturbed in different degrees by this overlap. This relative perturbation we call "pressure tuning."

One should mention that there is also a pressure tuning spectroscopy involving vibrational levels of molecules or crystals, which is active and exciting. In this paper I shall confine myself to electronic phenomena.

The consequences of pressure tuning are very broad, and any complete survey would extend to quite a number of categories. In this review I shall limit myself to two; the characterization of electronic states, excitations and properties and the testing of the application of models or theories of electronic phenomena. A variety of other topics are covered in previous reviews[1–7]. I shall outline three examples in each category.

I. CHARACTERIZATION OF ELECTRONIC STATES, EXCITATIONS AND PROPERTIES

The examples of characterization will, in part, be concerned with the balance between *intra*molecular and *inter*molecular interactions in perturbing an excitation. If one brings two molecular fragments together to form a molecule one establishes a bonding orbital which is stabilized in energy and an antibonding orbital which is destabilized. If one compresses this bond, both orbitals will, in general, increase in energy, but the antibonding orbital will increase more. Thus, the result of this *intra*molecular perturbation will be to increase the energy associated with a bonding to antibonding excitation (a blue shift in spectroscopic language).

The molecule will be interacting with its neighbors in a crystal or solution. For the type of molecules or large molecular ions we shall be concerned with in much of this paper, these interactions are predominantly of the van der Waals type. This is an attractive interaction of the form:

$$E \sim -\frac{\alpha_1 \alpha_2}{R^6}. \tag{1}$$

Here R is the distance between interacting entities, so this is a relatively short range interaction. The α_1 and α_2 are the polarizabilities of the interacting species. In general, although not always, a molecule is more polarizable when one of its electrons is in an excited state since the electron is less tightly bound. Thus the van der Waals forces will be stronger and there will be a tendency for a local decrease in volume of the system near an excited molecule even though the molecule itself may be slightly larger. Thus, for this *inter*molecular interaction, the effect of pressure will be to stabilize the excited state relative to the ground state and a shift to lower energy (red shift) of the excitation should occur. In the chance of our examples we shall see, among other things, the effect of various parameters on the balance between these two types of perturbations.

(A) First Example

The first example we take up involves fluorescence emission from two excited singlet states of different character. In addition to demonstrating the effect of type of excited state, and character of the medium on the balance discussed above, we shall demonstrate the effect of the change of the energy of the fluorescence on the fluorescence yield.

Figure 1 shows a schematic configuration coordinate diagram. The absoprtion and emission steps are vertical, in accordance with the Franck-Condon principle. The excited electron can descent to the lowest vibrational level giving off vibrational energy, and then emit fluorescence. Alternatively, it may return thermally to the ground state, by internal conversion, as shown, or, more probably, by intersystem crossing to a triplet state. Qualitatively the consequence will be the same. The rate of the thermal process will depend, in classical terms, on the height of the energy barrier. From a quantum mechanical viewpoint it will depend on the ability of the electron to penetrate the barrier by vibrational overlap. In either case, if the two wells move closer together in energy, the rate of the thermal process will increase, and the efficiency of fluorescence will decrease. If the potential wells move apart in energy, the opposite effect will obtain.

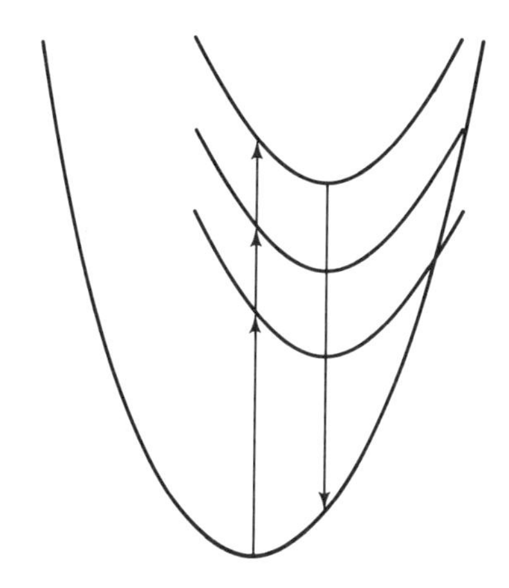

Figure 1. Schematic configuration coordinates diagram for fluorescence and non-radiative decay.

The molecule we discuss is a derivative of azulene[8] which has the unusual property of emitting from two different excited singlet states at the same time. S_2 lies at $\sim$ 25,000 cm^{-1} and S_1 at $\sim$ 15,000 cm^{-1} above S_0. S_2 is the normal π^* state or aromatic molecules, analogous to the ^1L_a state of anthracene in Platt's nomenclature. It is very polarizable and one would expect van der Waals forces to contribute strongly to a red shift of the $S_2 \rightarrow S_0$ emission. The S_1 state has nodes in the wave function which break up the conjugation of the π electrons. One would expect it to be certainly less polarizable than S_2 and possibly less than S_0. In any case, the difference in the van der Waals interactions between S_1 and S_0 will not be so dominant, so a blue shift for the $S_1 \rightarrow S_0$ emission is very possible.

In Figure 2 we exhibit the shift of the $S_2 \rightarrow S_0$ emission with pressure in polymethyl methacrylate (PMMA) and the change in integrated efficiency of fluorescence (area under the peak). The peak shifts red by $\sim$ 1500 cm^{-1} in 140 kbar and the efficiency drops by a factor of $\sim$ 30-40. In Figure 3 we show similar data for the $S_1 \rightarrow S_0$ emission, also in PMMA. Here we observe the expected blue shift, but at a rate which decreases noticeably with increasing

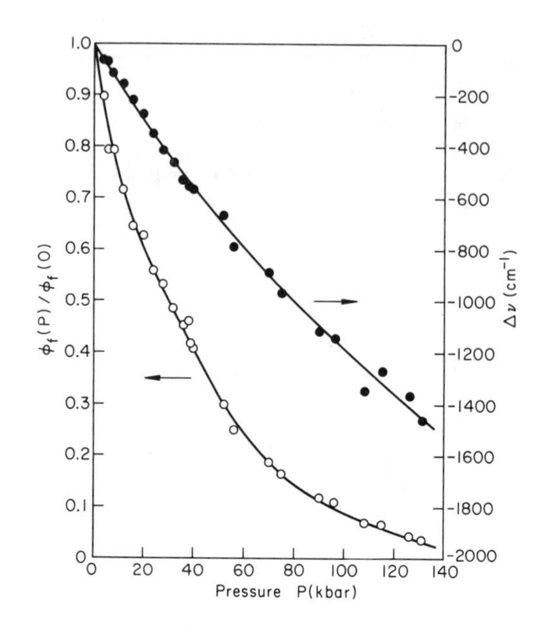

Figure 2. $S_2 \rightarrow S_0$ emission energy and fluorescence intensity change vs. pressure: azulene derivative in PMMA.

pressure. The intensity *increases* by a factor of $\sim$ 40 over 130 kbar which corresponds to our prediction.

In order to investigate the effect of the medium, we use polystyrene (PS) as a solvent which is much more polarizable than PMMA and so has stronger van der Waals attraction. For the $S_2 \rightarrow S_0$ emission (Figure 4) we observe a red shift of $\sim$ 3000 cm^{-1}, twice as large as in PMMA, with a correspondingly larger decrease in luminescent efficiency. The $S_1 \rightarrow S_0$ emission in PS (Figure 5) is especially interesting. At low pressures one observes a blue shift as in PMMA, but at higher pressures the decrease in blue shift observed in that medium becomes an actual red shift. The intensity precisely follows the trends predictable from the peak shift. This nicely illustrates the effect of "turning up" the van der Waals interaction.

(B) Second Example

The second example involves metal cluster compounds which form with metal-metal bonds connecting two or twenty or more metal atoms and which are stabilized by appropriate ligands. While high pressure studies of larger clusters have revealed interesting information, we confine ourselves here to binuclear clusters because one can treat them in simple order as pseudo-diatomic molecules and give a graphic description of the orbitals involved. Typical molecules are dirhenium or dimanganese decacarbonyl or molecular ions like

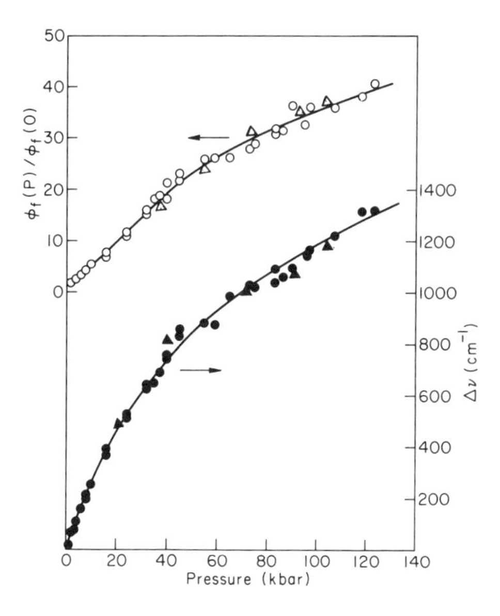

Figure 3. $S_1 \rightarrow S_0$ emission energy and fluorescence intensity change vs. pressure: azulene derivative in PMMA.

dirhenium or dimolybdenum octahalides[9,10]. One can label the intermetallic bonds as σ, π, or δ according to their angular momentum around the bond. Of course, there are different degrees of ligand mixing with orbitals or different symmetry and with antibonding vis a vis bonding orbitals, which one must consider in a quantitative treatment, but in the spirit of the present presentation one can treat these as second order considerations. In this spirit one can state that the highest occupied orbital is of δ type with π next and σ lowest. The antibonding orbitals are in inverse order of energy so that $\delta - \delta^*$ excitations typically lie near 15,000 cm^{-1}, the $\pi - \pi^*$ near 23,000–24,000 cm^{-1} and the $\sigma - \sigma^*$ peak near 30,000–33,000 cm^{-1}. Calcualtions[11] indicate that the electron density of the σ orbitals are very strongly localized between the metal atoms, so that *intra*molecular effects should dominate, while the δ orbitals are much more spread out, and the π orbitals associated with the metal bond are intermediate in extent. Unfortunately it is not possible to study all three excitations in the same molecule since not all molecules involve multiple bonds, and frequently charge transfer excitations disguise the higher energy excitations. Nevertheless a qualitative comparison among excitations is reasonable.

In Figure 6 is shown the $\sigma - \sigma^*$ excitation in crystalline dirhenium and dimanganese carbonyl. We see the expected shift to higher energy. Similar shifts are observed in various solvents.

In contrast, the $\delta - \delta^*$ excitations in the crystalline octahalides (Figure 7) exhibit large red shift indicating, as anticipated, that *inter*molecular forces are

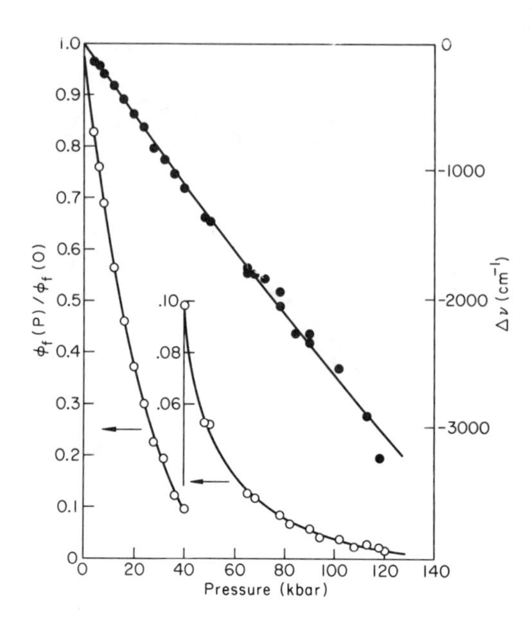

Figure 4. $S_2 \rightarrow S_0$ emission energy and fluorescence intensity change vs. pressure: azulene derivative in PS.

dominant. In the iodide there is an isomerization which causes large intensity loss which accounts for the scatter. As shown in Figure 8 the shift of the $\pi - \pi^*$ excitation is intermediate. (In this case in particular, for a quantitative treatment it is necessary to consider the difference in ligand mixing in the π and π^* orbitals.)

The above studies have all involved crystalline solids. It is of interest to examine solvent effects on the $\delta - \delta^*$ excitation. We considered two classes of solvents: dichloromethane, which has a modest dielectric constant (~ 7) and a relatively high polarizability providing an environment not unlike the crystal, and solvents like acetonitrile and nitromethane with dielectric constants near 40 and low polarizability. In Figure 9 we show the peak shifts in CH_2Cl_2. Not unexpectedly, they resemble the behavior observed in the crystal. In the constrast, Figure 10 exhibits the shifts observed in CH_3CN. There is a slight, but, we believe determinable red shift for the chloride, a blue shift for the bromide which starts very small and increases rapidly, and a large blue shift for the iodide. Divalent ions in a high dielectric constant solvent should be strongly solvated. It is reasonable to assume that the solvate shell is tightly bound and responds minimally to the van der Waals interaction. This is thus an increase in the effective value of R in equation (1) and a diminution in the importance of the van der Waals attraction. This effect is largest for the iodide with the largest ligands and least for the chloride with the smallest ligands.

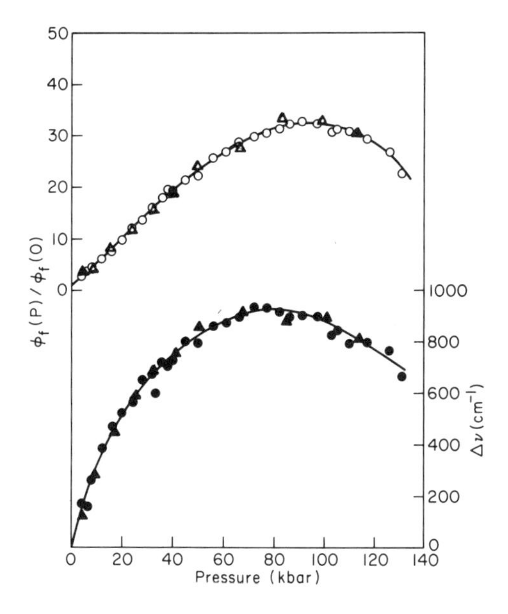

Figure 5. $S_1 \rightarrow S_0$ emission energy and fluorescence intensity change vs. pressure: azulene derivative in PS.

(C) Third Example

A final example of characterization involves the metal glyoximes. These are planar or essentially planar molecules containing a d^8 metal, Ni, Pd or Pt in the center. They crystallize in layers with alternate molecules at 90°. The interlayer distance for a given glyoxime is identical to ± 0.01 Å independent of the metal atom. These crystals exhibit excitations which lie at 17,000–18,000 cm^{-1}, 21,000–22,000 cm^{-1} and 24,000–25,000 cm^{-1} for the Pt, Pd and Ni derivatives respectively. It has been established[12] that the molecular orbital arising from the metal p_z orbital is stabilized by interaction with the π orbitals of the ligands lying above and below. In Figure 11 we exhibit the pressure shifts for these (essentially) $d \rightarrow p$ excitations[13] for the heptoxime. For the Ni derivative there is a red shift of $\sim$ 2000 cm^{-1} which is clearly leveling at high pressure. For the Pd derivative the red shift is nearly 8000 cm^{-1} before leveling commences. For the Pt compound the red shift is about 11,000 cm^{-1} in 80 kbar, but at higher pressures a distinct blue shift is observed. The first effect of compression is to stabilize further the metal p_z orbital (or its molecular equivalent). Since the original spacing, and presumably the compression is fixed by the ligand independent of the metal, at sufficiently high pressure the large Pt orbitals interfere and cause repulsion. This study illustrates how, for fixed geometry, ion size can modify the nature of intermolecular interaction. There are also intensity effects of interest. We discuss here the Pt derivatives

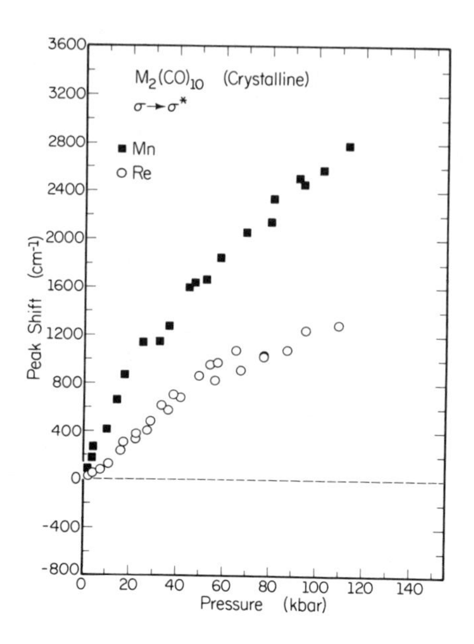

Figure 6. Peak energy vs. pressure: $\sigma \rightarrow \sigma$ excitation in crystalline $Mn_2(CO)_{10}$ and $Re_2(CO)_{10}$.

which provides the most dramatic case. In Figure 12 we show the normalized area under the peak as a function of pressure. This quantity drops by a factor of five in 60 kbar rises by $\sim$ a factor two by 90–100 kbar, and then drops again. The changes are reproducible and reversible. There is no sign of a phase transition in either the electronic or infrared spectrum. Ohasi et al.[14] have suggested that the intensities at ambient pressure are due in part to intensity borrowing from a series of charge transfer (CT) peaks at higher energy. Intensity borrowing between vibrational excitations (Fermi resonance) is a well established phenomenon and has been studied as a function of pressure[15–17]. However, intensity borrowing between electronic excitations has not been well analyzed. The change in the degree of borrowing would depend on the change in position of the high energy tail of the $d-p$ excitation relative to the low energy tail of the charge transfer peak. Since the d to p excitation peak does not change in shape with pressure we can use the difference in energy of the peak position relative to the tail of the CT peak at, say, an extinction coefficient of one. The change in this difference with pressure is plotted in Figure 13. It mirrors the behavior of the intensity in a most convincing way.

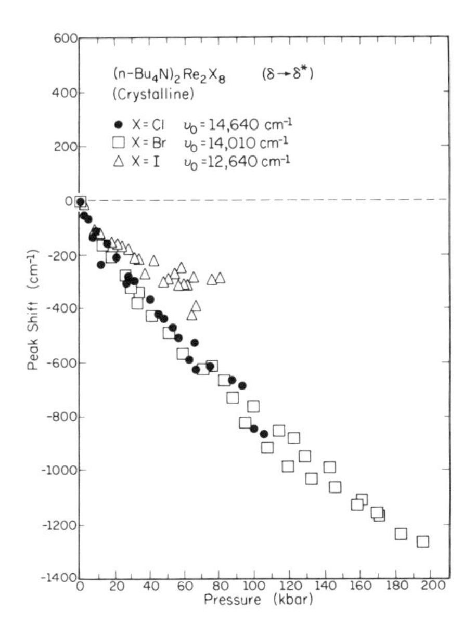

Figure 7. Peak energy vs. pressure: $\delta \rightarrow \delta^*$ excitation in crystalline compounds of $(Re_2X_8)^{-2}$; X = Cl, Br, I.

II. TESTING OF MODELS OR THEORIES

(A) First Example

One of the earliest aspects of pressure tuning which started nearly three decades ago was the measurement of the shift of the bottom of the conduction band relative to the top of the valence band with pressure. Perhaps the most studied materials have been Si, Ge and the III-V compounds like GaAs and GaP, because of both scientific and technological interest. Their band structure at one atmosphere has been well characterized. There is a triply degenerate maximum in the valence band at the center of the Brillouin zone (the 000 or Γ point). There are three minima in the bottom of the valence band. For silicon the lowest minimum is in the 100 or X direction. For GaAs, the material of interest here, the lowest minimum is at 000. The energy of the (direct) transition increases with pressure at a rate of 12.5 meV/kbar[18–20]. While this information is important, it would be useful for many calculations to know the shift of the *center* of the conduction band vis a vis the top of the valence band. The model used is the rigid band model wherein it is assumed that the conduction band moves up or down in energy with no change in width. This cannot be true over any large range of compression, but it would be helpful to establish whether it applies over a modest range. This can now be done using a layered material consisting of, say, 60 Å of GaAs, 60 Å of

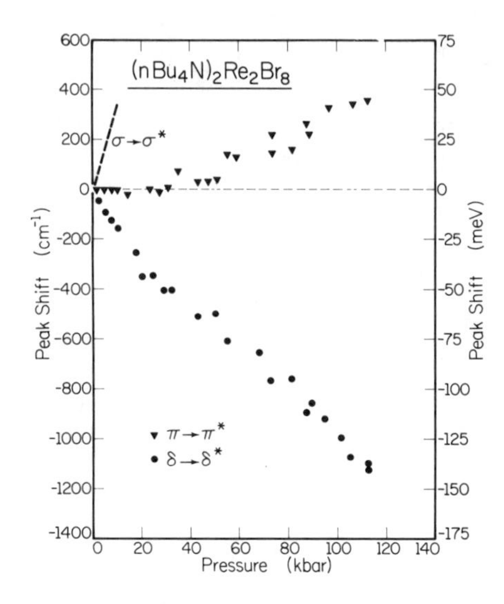

Figure 8. Peak energy vs. pressure: $\pi \longrightarrow \pi^*$ excitation in a crystalline compound of $(Re_2Br_8)^{-2}$ in comparison to $\sigma \longrightarrow \sigma^*$ excitation.

GaAlAs etc. The two materials are lattice matched at one atmosphere, but the absorption edge of GaAs is less than that of GaAlAs so a quantum well is formed (Figure 14)[21]. The degeneracy of the valence band maximum is partially removed, forming states with carries of different mobility labelled light and heavy holes. In the conduction well there are three discrete states separated by over 0.5 eV in energy. Figure 15[22] shows the absorption spectra as a function of pressure. The excitations from the two valence states to the three states of the conduction well area are clearly defined. At high energy one sees the 100 (X) minimum shifting slightly to lower energy with pressure as it does in Si.

In Figure 16 we exhibit the shifts with pressure. The essential point is that they are parallel to $\pm$ 0.01 meV/kbar so that the rigid band model is valid over this range of compression ($\sim$ 3%). It is of interest that the shift is 11 meV/kbar, distinctly less than the value for bulk GaAs. Although the materials are lattice matched at one atmosphere and the pressure is hydrostatic, the GaAlAs is less compressible than GaAs and transmits shear to the GaAs which reduces the change in excitation energy with pressure.

(B) Second Example

About 30 years ago Marcus developed a theory of electron transfer between ions in solution[23–24]. This theory has proved to be very significant for many applications and seminal as a basis for further developments. There

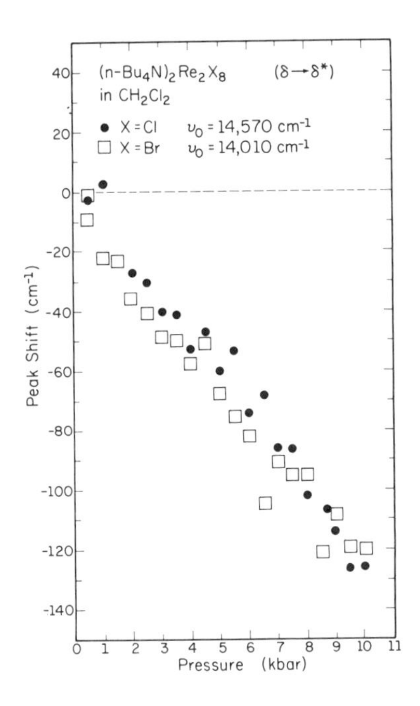

Figure 9. Peak energy vs. pressure: $\delta \rightarrow \delta^*$ excitation for $(Re_2Cl_8)^{-2}$ and $(Re_2Br_8)^{-2}$ in CH_3Cl_2.

is, however, one class of cases where we believe it has been misapplied. The theory, which is basically semiclassical, separates the energy involved in the transfer process into an inner energy E_i and an outer energy E_o. The former is associated with the change in bond lengths in the ions as the electron is excited and transfers. E_o contains the effect of rearrangement of the surroundings on the potential wells involved in the process. For his discussion of transfer between two ions Marcus assumed that this field came primarily from the dielectric continuum. In this case E_o contains a factor $\left(\frac{1}{n^2} - \frac{1}{D_{LF}}\right)$ where D_{LF} is the low frequency dielectric constant and n^2 represents the high frequency dielectric constant. (n is the refractive index.) This theory has had many successes.

Over the last decade or so the theory has been applied to electron transfer between Ru(II) and Ru(III) or Fe(II) and Fe(III) in molecular ions such as are shown in Figure 17[25-26]. Here the transfer takes place through the field supplied by the metal-ligand bonding electrons and π electron cloud of the aromatic bridge. The theory is applied by plotting $\left(\frac{1}{n^2} - \frac{1}{D_{LF}}\right)$ vs. peak energy and extrapolating to zero. Typically this gives values of E_o of 2500–5500 cm^{-1}.

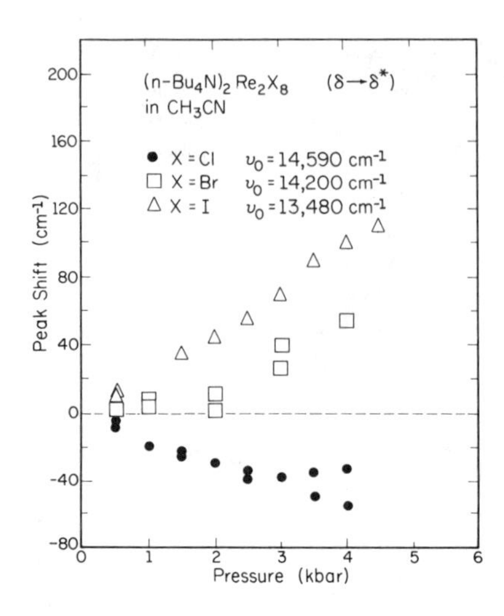

Figure 10. Peak energy vs. pressure: $\delta \rightarrow \delta^*$ excitation for $(Re_2Cl_8)^{-2}$, $(Re_2Br_8)^{-2}$ and $(Re_2I_8)^{-2}$ in CH_3CN.

There are several reasons for questioning this procedure. In the first place, the values of D_{LF} for solvents usable in the experiments is typically in the range 20–80, so that one is effectively plotting $\frac{1}{n^2}$, which could measure other properties such as polarizability. Secondly, the physical constants are such that $\left(\frac{1}{n^2} - \frac{1}{D_{LF}}\right)$ covers at most a range from 0.40–0.55 so the extrapolation is large. Thirdly, and most significantly, we have measured the peak location, shape and intensity as a function of concentration at ambient pressure[27]. The location may change as much as 600–700 cm^{-1} with concentration; the half width can change considerably, and the transition moment, as represented by the area under the peak divided by the concentration, can change significantly. The probable cause is change in the degree of ion pairing with concentration.

We decided on a more direct approach to the problem[28]. The freezing points of most liquids increase with pressure. We chose three solvents — CD_3CN, CD_3NO_2 and $C_6D_5NO_2$ which freeze at 25° below 10 kbar. It was desirable to have a modest freezing pressure so that one could use a cell with path length 2–3 mm and thus reasonably dilute solutions. We measured the peak location at pressures below and above the freezing point. Upon freezing the D_{LF} drops from values of the order 40–50 to 3.5–4.0, because one quenches out the orientational polarizability. One then would expect a large red shift of several thousand wavenumbers in the peak locations. Typical results are shown in Figure 18 and 19 and Table I. The observed shifts are either zero or at least very much smaller than predicted. While the results are still under analysis, we believe that for the class of materials under consideration the external field

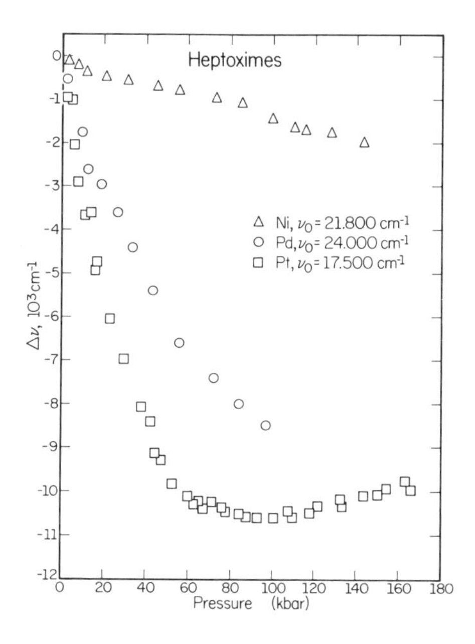

Figure 11. Peak energy vs. pressure: $d \rightarrow p$ excitation in Ni, Pd and Pt heptoximes.

which determines E_o comes from, in the first place the ligand π and ligand-metal bond electrons, secondly from the solvation shell, and then from the dielectric continuum. Thus the dielectric constant of the bulk medium is only of secondary importance for electron transfer in this class of compounds.

(C) Third Example

The final example involves energy transfer in phosphors. Over a decade ago we studied energy transfer between Tl^+ and Ag^+ as dilute dopants in alkali halides[29]. From measurements of the change in peak location and overlap and in lattice parameter as a function of pressure we demonstrated that Förster-Dexter[30–31] theory hold quantitatively over a range of transfer efficiency from 7–30%. Here I would like to take up a qualitative application of energy transfer for a system which exhibits a variety of facets.

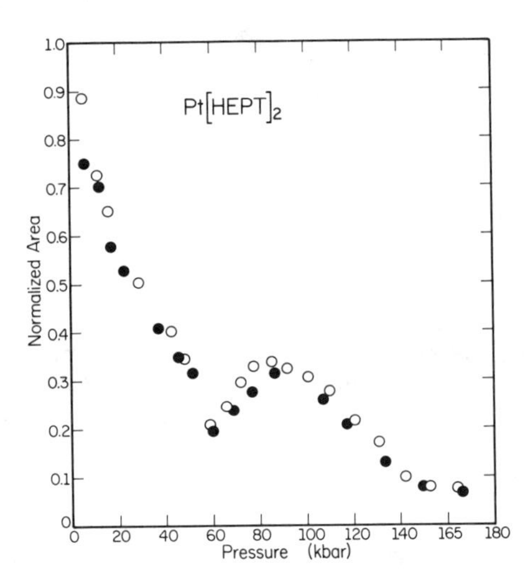

Figure 12. Change in peak intensity vs. pressure for $d \rightarrow p$ excitation in Pt heptoxime.

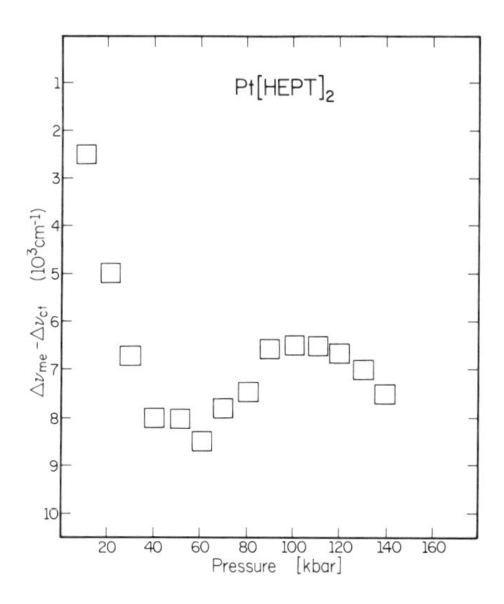

Figure 13. Difference in shift of metal $d \rightarrow p$ excitation peak and the tail of the charge transfer peak (ext. coeff. = 1) vs. pressure for Pt heptoxime.

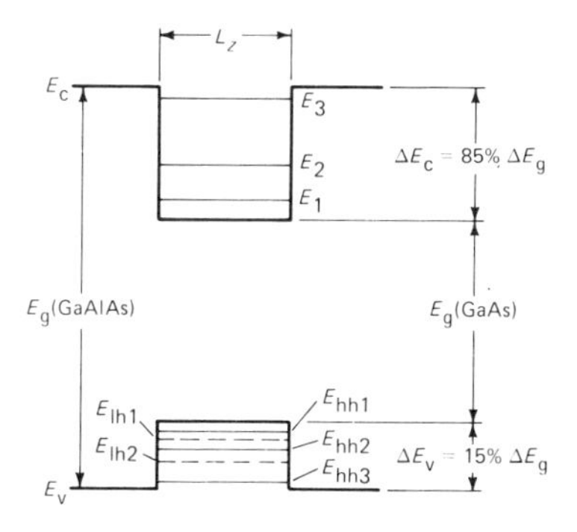

Figure 14. Quantum well for GaAs - GaAlAs superlattice.

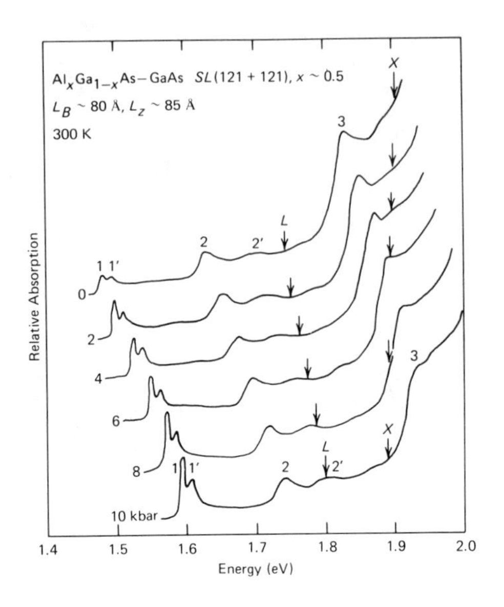

Figure 15. Absorption spectra vs. pressure: GaAs - $Ga_{0.5}Al_{0.5}As$ superlattice.

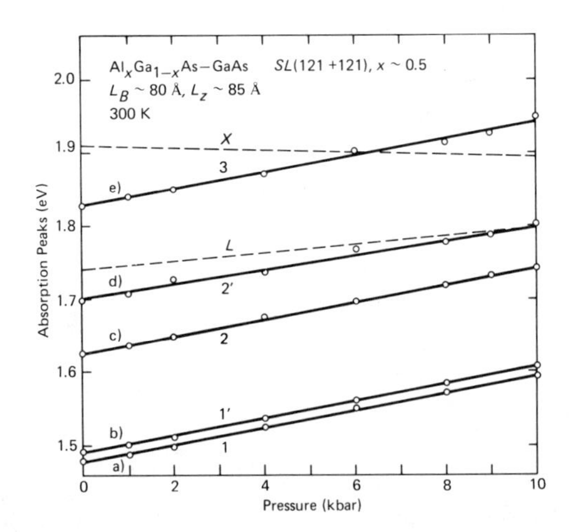

Figure 16. Shifts of quantum well excitations vs. pressure.

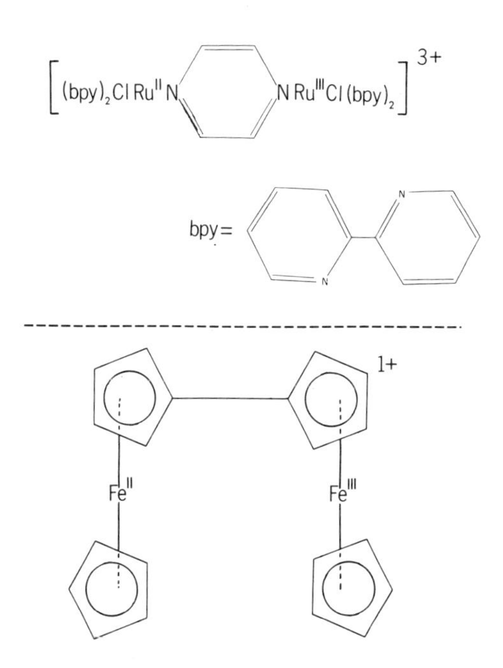

Figure 17. Mixed valence compounds used in study of intervalence transfer peak.

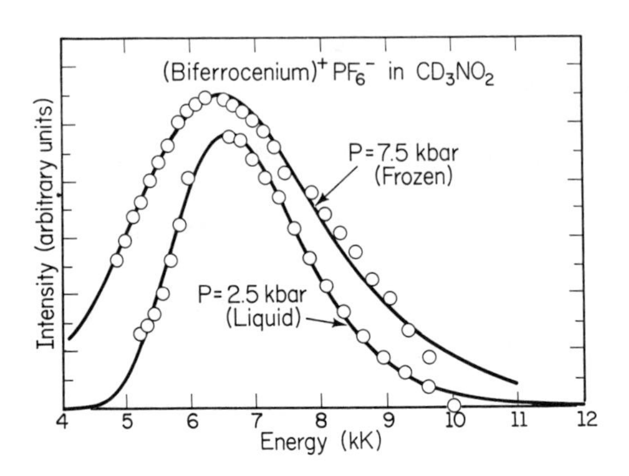

Figure 18. Near IR absorption spectrum of biferrocenium hexafluorophosphate in CD_3NO_2. (1kK = 1000 cm^{-1}.)

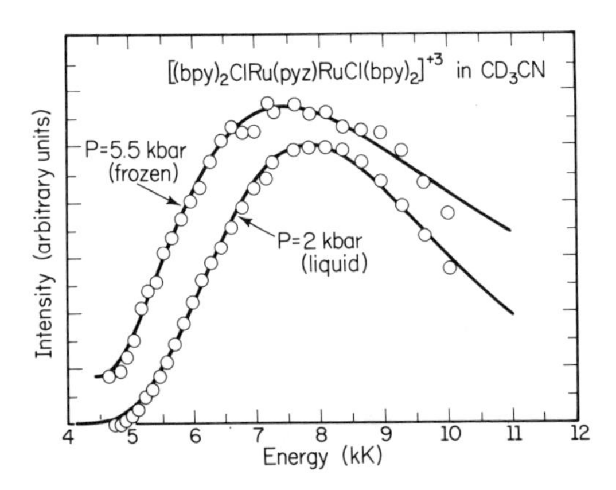

Figure 19. Near IR absoprtion spectrum of $[(bpy)_2ClRu(pyz)RuCl(bpy)_2]^{3+}$ in CD_3CN.

TABLE 1

Biferrocenium$^+PF_6^-$

Solvent	Conc. (mM)	Pressure (kbar)	IR Peak (kK)	D_{LF}
$C_6D_5NO_2$	4.9	1.0(L)	6.11	36.0
		2.5(S)	6.18	3.4
$C_6D_5NO_2$	6.2	1.0(L)	5.68	36.0
		2.5(S)	5.60	3.4
CD_3NO_2	4.9	2.5(L)	6.43	43.0
		7.5(S)	6.13	4.0
CD_3NO_2	7.1	1.5(L)	5.88	43.0
		7.0(S)	5.74	4.0

Biferrocenium$^+I_3^-$

Solvent	Conc. (mM)	Pressure (kbar)	IR Peak (kK)	D_{LF}
$C_6D_5NO_2$	4.0	1.0(L)	5.81	36.0
		2.5(S)	5.68	3.4
$C_6D_5NO_2$	0.8	1.0(L)	5.59	36.0
		2.5(S)	5.53	3.4
CD_3CN	4.9	1.5(L)	5.99	45.5
		5.5(S)	5.31	3.9

$[(bpy)_2ClRu(pyz)RuCl(bpy)_2)]^{+3}$

Solvent	Conc. (mM)	Pressure (kbar)	IR Peak (kK)	D_{LF}
CD_3CN	15	1.5(L)	7.81	45.5
		5.5(S)	7.48	3.9
$C_6D_5NO_2$	15	0.75(L)	7.85	36.0
		2.5(S)	8.10	3.4

L = Liquid: S = Solid

Rare earth ions have sharp emission peaks of scientific and practical interest. The problem is to get energy into the lowest excited state of the rare earth. Direct absorption is inefficient because the transition is forbidden on both orbital and spin considerations. The procedure is to surround the rare earth with ligands which have an intense absorption, and then to let some energy transfer to the rare earth center. We have studied three rare earths combined with four sets of ligands, all derivatives of acetyl acetonate[32]. Here we use Eu combined with dibenzyl methide (DBM) as our example.

The ligand has a broad intense absorption band in the UV into which one can excite effectively. The energy is rapidly transferred from the excited singlet S_1 to the lowest triplet T_1. The electron transfer is via T_1. One can visualize three regimes. T_1 may be above the *second* excited state of Eu($^5D_1$); then, because of the ΔE in the denominator of the transfer function, most of the transfer is to this state. Transfer from the second to the first excited state is slow because of poor overlap, so that this process is inefficient. If T_1 lies just above or just below the lowest Eu excited state $^5D_0$, the transfer to that state will be efficient, but so will be the back transfer to T_1 which can easily dispose of the excitation thermally. The ideal location for T_1 is near enough the lowest excited state of the Eu for effective transfer, but high enough to minimize back transfer.

We have studied the transfer both in the crystalline solid and in solution in PMMA. Figure 20 exhibits the energy levels. $S_1(0)$ and $T_1(0)$ lie lower in energy at one atmosphere in the crystal than in PMMA because the latter medium is less polarizable. The degree of tuning in 40 and 80 kbar is shown by the shaded areas. Triplet states in general tune less than singlet states because the necessity to keep the parallel spins apart decreases the deformability of the electron cloud. (Throughout we assume that the Eu levels are essentially unperturbed in energy by pressure.)

From Figure 20 we should expect, for the crystal, that at low pressures one should get a significant increase in transfer efficiency with pressure, then it should drop off rapidly as back transfer takes over. For the polymer essentially all transfer is via the second excited state $^5D_1$. One may get some increase in emission as this transfer improves and then some drop off as back donation takes over, but it will be strongly suppressed by the inefficient $^5D_1$ to $^5D_0$ transfer. Superimposed is an additional effect. With pressure there is a monotonic increase in emission intensity by a factor of 1.5 – 2.0 in 100 kbar. This is because of an increase in the radiative rate, k_r due to increased $d-f$ orbital mixing with pressure. In Figure 21 for the crystal we see an increase in efficiency by a factor of 7–8 in the first 25 kbar, then a drastic drop off by a factor of several hundred as back donation takes over. In the polymer, when the increase due to changing k_r is deleted, the effects are very modest.

From Figure 20 we can visualize several ways of getting energy into $^5D_0$. The first one is direct absorption in $^5D_0$. This is inefficient, but when we do it we observe only the modest linear increase due to changing k_r. We can normalize to these data as a fiducial marker. When we excite in the ligand we should expect a large increase as T_1 approaches $^5D_0$, then a drop off due to increased back donation. Thirdly, we can excite in $^5D_1$. Since direct transfer to $^5D_0$ is poor, we can expect the major transfer to be via T_1. Then exciting in

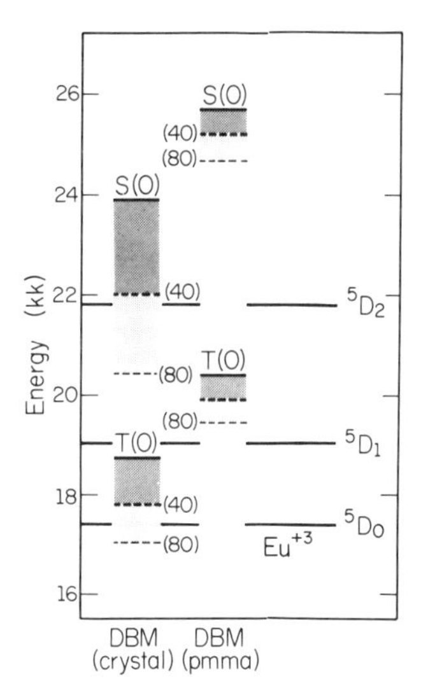

Figure 20. Pressure tuning of Dibenzylmethide (DMB) absorption peaks relative Eu^{+3}.

$^5D_1$ should, relatively, give the same result as exciting in T_1 directly. For the polymer, the relative efficiency should be independent of the exciting mode, as one is always limited by the weak transfer from $^5D_1$ to $^5D_0$. From Figure 22 we see that our expectations are met precisely.

Finally, we examine the back donation process in more detail. From Figure 23 we see that as the triplet state lowers in energy the barrier to back donation, E^0_A, decreases. We can measure the change in energy of the triplet state from the shift of the phosphorescence peak. The barrier height against back donation, ΔE_A, can be measured from the temperature coefficient of the efficiency or of the lifetime. (We obtained the same results from both methods.) If the simple picture holds the changes in these two energy quantities should be equal. As can be seen from Figure 24, where the dotted line is at 45°, the agreement is remarkable.

These are only a few of the wide variety of pressure tuning experiments which have had significant impact, and these few have been presented only in outline, eliding many details. I hope, however, that this picture, painted with a very broad brush, is sufficiently clear and well focused to portray the power and versatility of pressure tuning spectroscopy in modern science.

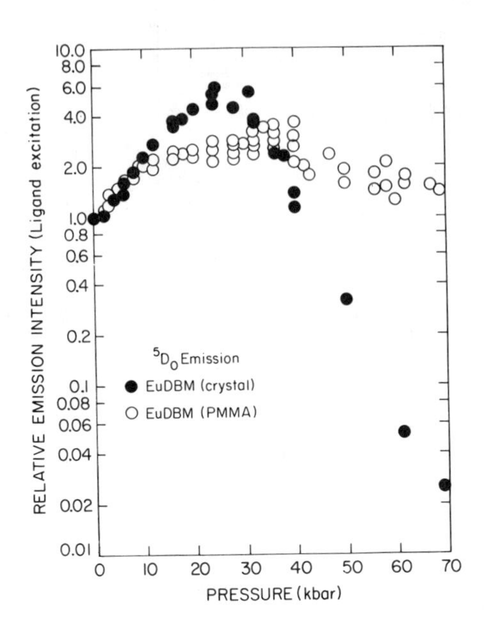

Figure 21. Effect of pressure on the $^5D_0$ emission efficiency of $Eu(DBM)_3$.

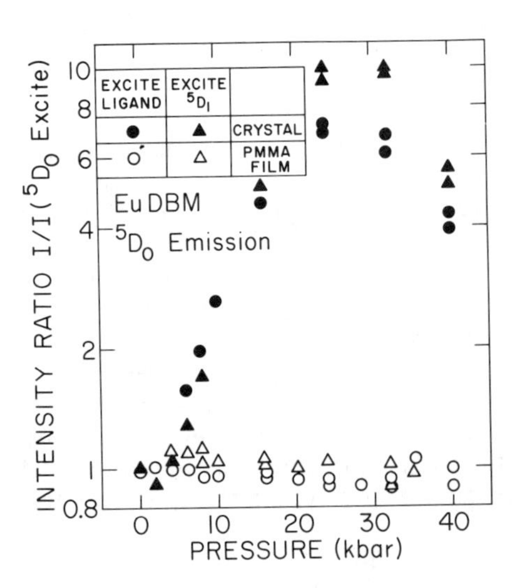

Figure 22. Relative effect of pressure on the $Eu(DBM)_3$ $^5D_0$ emission intensity to $^5D_0$ of different excitation modes.

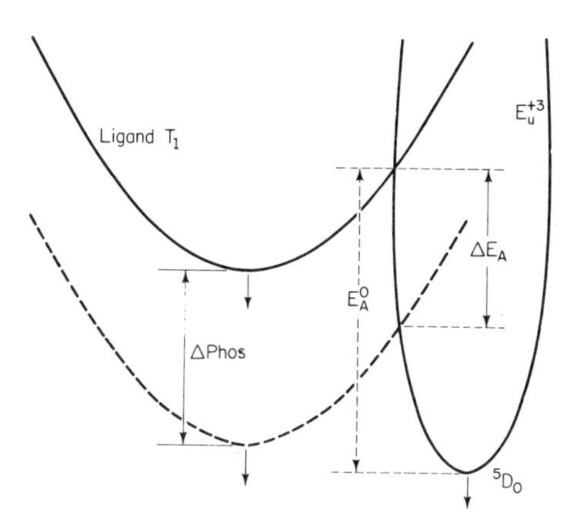

Figure 23. Schematic configuration coordinate diagram for back donation $^5D_0 \rightarrow$ ligand T_1.

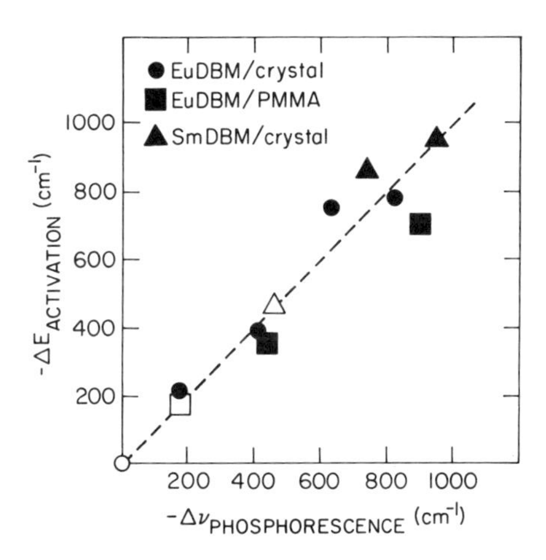

Figure 24. Correlation between shift of ligand T_1 phosphorescence peak and change of energy barrier for back donation.

ACKNOWLEDGEMENTS

The author would like to acknowledge the vital contributions of students and colleagues in this work. Support from the Materials Science Division, Department of Energy under contract DE-AC02-76ER01198 is gratefully acknowledged.

REFERENCES

1. H.G. Drickamer, in *Solid State Physics* Vol. 17, Seitz and Turnbull eds. (Academic Press, NY, 1965), p. 1.
2. G. Weber and H.G. Drickamer, Quart. Rev. of Biophysics 16, 89 (1982).
3. H.G. Drickamer and C.W. Frank, *Electronic Transitions and the High Pressure Chemistry and Physics of Solids* (Chapman and Hall, London, 1973).
4. H.G. Drickamer, Ann. Rev. Phys. Chem. 33, 25 (1982).
5. H.G. Drickamer, Int. Rev. Phys. Chem. 2, 171 (1982).
6. H.G. Drickamer, G.B. Schuster and D.J. Mitchell, in *Radiationless Transitions*, S.H. Lin, ed. (Academic Press, NY, 1980), p. 289.
7. H.G. Drickamer, Accts. Chem. Res. 19, 329 (1986).
8. D.J. Mitchell, H.G. Drickamer and G.B. Schuster, J. Am. Chem. Soc. 99, 7489 (1977).
9. T.L. Carroll, J.R. Shapley and H.G. Drickamer, Chem. Phys. Lett. 119, 340 (1985).
10. T.L. Carroll, J.R. Shapley and H.G. Drickamer, J. Chem. Phys. 85, 6787 (1986).
11. F.A. Cotton and R.A. Walton, *Multiple Bonds Between Metal Atoms*, (Wiley, NY, 1982), p. 439.
12. B.G. Anex and F.K. Krist, J. Am. Chem. Soc. 89, 6114 (1967).
13. M. Tkacz and H.G. Drickamer, J. Chem. Phys. 85, 1184 (1986).
14. Y. Ohashi, I. Hanazaki and S. Nagakura, Inorg. Chem. 9, 2551 (1970).
15. W. Schindler, T.W. Zerda and J. Jonas, J. Chem. Phys. 81, 4306 (1984).
16. T.W. Zerda, M. Bradley and J. Jonas, J. Chem. Phys. 117, 566 (1985).
17. M. Bradley, T.W. Zerda and J. Jonas, J. Chem. Phys. 82, 4007 (1985).
18. J. Feinleib, S. Groves, W. Paul and R. Zallen, Phys. Rev. 131, 2070 (1963).
19. A. Jayaraman, M.E. Sikovski, J.C. Irvin and G.H. Yates, J. Appl. Phys. 38, 4451 (1967).
20. B. Welber, M. Cardona, C.K. Kim and S. Rodriguez, Phys. Rev. *B*12, 5729 (1975).
21. N. Holonyak, Jr. and Karl Hess, in *Synthetic Modulated Structures*, L. Chang and B. Giessen eds., (Academic Press, NY, 1985), p. 257.
22. S.W. Kirchoefer, N. Holonyak, Jr., K. Hess, D.A. Gulino, H.G. Drickamer, J.J. Coleman and P.D. Dapkus, App. Phys. Lett. 40, 821 (1982).
23. R.A. Marcus, J. Chem. Phys. 24, 966, 979 (1956).
24. R.A. Marcus, Ann. Rev. Phys. Chem. 15, 155 (1964) and references therein.
25. T.J. Meyer, Accts. Chem. Res. 11, 94 (1978).

26. M.J. Powers and T.J. Meyer, J. Am. Chem. Soc. 100, 4393 (1978).
27. M.O. Lowery, D.N. Hendrickson, W.S. Hammack and H.G. Drickamer, J. Am. Chem. Soc. (submitted).
28. W.S. Hammack, H.G. Drickamer, M.D. Lowery and D.N. Hendrickson, Chem. Phys. Lett. 132, 231 (1986).
29. K. Bieg and H.G. Drickamer, J. Chem. Phys. 66, 1437 (1977).
30. Th. Förster, Ann. Physik 2, 55 (1948).
31. D.L. Dexter, J. Chem. Phys. 21, 836 (1953).
32. A.V. Hayes and H.G. Drickamer, J. Chem. Phys. 76, 114 (1982).

UNCONVENTIONAL SUPERCONDUCTORS: FROM 1 mK TO 90 K TO 10^{10} K

David Pines
Department of Physics
University of Illinois at Urbana-Champaign
1110 West Green Street
Urbana, IL 61801

ABSTRACT

Unconventional superfluids and superconductors are fermion systems in which the phonon-induced interaction between particles does not play an essential role, and in which the pairs of BCS theory may be found in higher angular momentum states. These are reviewed in the historical order in which their superfluidity was discovered: nuclei, neutron stars, liquid ^{3}He, and heavy electron metals. Some key experimental results on the copper oxide superconductors are reviewed, and the possibility that these are part of the heavy electron family, with a spin-fluctuation-induced interaction between quasi-particles being responsible for their superconductivity, is examined.

I. INTRODUCTION

It is a great pleasure to be speaking at this symposium which celebrates our colleague and friend, Hans Frauenfelder. Hans, who is at the cutting edge in the fields in which he works, has also found time to follow the latest developments in those frontier fields of science in which he is not working. For that reason I have chosen today to discuss unconventional superconductors, of which the new high temperature superconductors are the most recent, and perhaps the most exciting, example.

We live in a remarkable time for science. We have witnessed, in the span of a single year, the discovery by Georg Bednorz and Alex Mueller of a new class of superconductors based on the copper oxides, and the discovery of a supernova in a nearby galaxy at a time when the full power of the astronomical community can be turned to its observation. The work of Bednorz and Mueller has unleashed an extraordinary effort devoted to high temperature superconductivity, in which we are simultaneously trying to understand the physical properties of the new superconductors and the mechanism responsible for their superconductivity, searching for new, and higher temperature, superconducting materials, and beginning to develop technological applications based on the new superconductors. In my talk today I shall try to put the discovery by Bednorz and Mueller in historical and scientific perspective, a perspective which very nicely spans Hans' thirty-five year career here in Urbana. In so doing, I will introduce you to the "superfluid zoo," the appearance of superconductivity and superfluidity in the universe, which is found at temperatures as low as 1 mK and as high as 10^{10} K.

I shall begin by reviewing the pairing mechanism for "ordinary" low temperature metallic superconductivity, the phonon-induced interaction between conduction electrons, and the development here in Urbana of the microscopic theory of Bardeen, Cooper and Schrieffer. I will discuss briefly the search for higher temperature superconductors and for other kinds of pairing mechanisms in metals carried out in the decades between 1960 and 1980, which led to the proposal that there exists a maximum temperature for superconductivity in metals. This limit, which I shall call the "phonon barrier," represents an informed estimate of the highest superconducting temperature one might achieve using the phonon-induced interaction between electrons as the pairing mechanism for superconductivity. In the mid-portion of my talk I will consider *unconventional* superfluids and superconductors, fermion systems in which the phonon-induced interaction does not play an essential role, and in which the pairs of the BCS theory may be found in higher angular momentum states, in the historical order in which their superfluidity was discovered: nuclei, neutron stars, liquid ^{3}He, and heavy electron metals. I shall pay particular attention to the heavy electron systems because I suspect that few of you are aware that prior to 1987 superconducting metals had been discovered (at low temperatures to be sure), in which phonons did *not* provide the mechanism for superconductivity. In the final part I will discuss the search now underway for the physical mechanism responsible for the high temperature superconducting oxides (that part of my talk will be soon out-of-date) and speculate on further developments.

II. "CONVENTIONAL" SUPERCONDUCTORS AND THE PHONON BARRIER

The Coulomb interaction between electrons in metals is repulsive and has a characteristic energy scale $\sim 10^4$ K; the challenge which faced theorists was how such an interaction could give rise to superconductivity in metals at temperatures $\lesssim 10$ K? The answer was, in retrospect, quite simple: it does not. The key experimental clue came in the discovery of the isotope effect by Serin, Maxwell, and their collaborators in 1950: metals identical in every respect except for their ionic mass displayed superconducting transition temperatures which were different, and were inversely proportional to the square root of the mass of their constituent ions. Thus ion motion, described by phonons (the quanta of lattice vibrations) must play a role; the relevant energy scale, instead of being 10^4 K could be the much lower Debye temperature, $\theta_D \sim 200$ K, which characterizes that ion motion. Fröhlich and Bardeen independently proposed theories of superconductivity based on the electron-phonon interaction; indeed, Fröhlich had done so prior to the discovery of the isotope effect. In their work, however, it was assumed that the key physical effect was the resulting modification of the energy of the single electrons, the so-called self-energy, and this approach was soon found not to work. Fröhlich then suggested that perhaps the phonon-induced interaction between electrons played an essential role; in the language of quantum mechanics this is a process in which one electron emits a virtual phonon which is absorbed by a second electron. He showed that if one ignored the repulsive Coulomb interaction, and if the two interacting electrons possessed energies which differed by less than the characteristic phonon energy, θ_D, that phonon-induced interaction would be attractive.

But could the Coulomb interaction be ignored? While David Bohm and I had shown that the long range of the Coulomb interaction did not pose essential problems (it mainly gave rise to plasmons, the high energy collective oscillations of the electron system), the remaining short-range screened interaction between electrons in metals could modify their single particle motion in a significant way, and appeared to be significantly larger than the phonon-induced interaction. John Bardeen and I took up this problem during the years 1952-1954, when I was here in Urbana as a research assistant professor (I came to Urbana in the summer of 1952, at the same time that Hans Frauenfelder did). John and I adapted the collective coordinate approach which Bohm and I had used for electrons to the combined motion of electrons and phonons in metals. The result of the self-consistent description we developed is given in Equation (1). There we see that for a given momentum transfer $\vec{q}$, and relative energy difference, ω, the effective electron interaction,

$$V_{\text{eff}}(q,\omega) = \frac{4\pi e^2}{q^2\varepsilon(q,0)} + \frac{4\pi^2\omega_q^2}{q^2\varepsilon(q,0)[\omega^2 - \omega_q^2]} \tag{1}$$

can be represented as the sum of two terms: the direct Coulomb interaction between electrons, which is screened by the static electronic dielectric function, $\varepsilon(q,0)$; and a screened frequency dependent phonon-induced interaction. Equation (1) is appropriate for interacting electrons near the Fermi surface, and is valid in the "jellium" approximation, in which effects associated with the periodicity of the ion array in metals are neglected. For electrons of identical energies the two contributions cancel, while the net electron interaction is attractive so long as the electron energy difference, ω, is less than the energy ω_q, of the phonon which is being exchanged.

This interaction can be viewed in classical terms as a process in which the electrons polarize the medium in which they move; thus electron 1 acts to polarize the overall density fluctuations of the system (electrons *and* ions), while electron 2 is the analyser which samples that polarization field. We see that the medium provides the message; the presence of a second component (the ions) changes the effective electron interaction from repulsion to attraction. Quite generally, for all such polarization processes, if the characteristic energy difference of the medium is less than the characteristic energy differences of the electrons the resulting interaction will be attractive.

Leon Cooper, who succeeded me as a postdoc working with John Bardeen, was therefore led to study a "toy" problem based on this interaction; that of a pair of electrons outside the Fermi sea which interact with an attractive interaction $|V_{\text{eff}}|$, as long as their energy difference was less than a characteristic phonon energy, θ_D, and which had no interaction otherwise. He found that such electrons would form bound pairs; more generally, when one takes into account the presence of electrons below and above the Fermi surface, one finds that below some characteristic temperature, the Fermi surface is unstable against the formation of these pairs; this is the famous Cooper pairing instability which set the stage for the development, less than a year later, of the microscopic theory of superconductivity by Bardeen, Cooper, and Robert Schrieffer, who was then a graduate student working with Bardeen.

The BCS theory was truly remarkable: it not only provided a complete account of existing superconducting phenomena and successfully predicted or led to the prediction of many new results, but it also had a substantial impact on theories in nuclear physics, astrophysics, and particle physics, as well as the helium liquids. Key ideas of BCS theory, which provide the language we shall need for considering unconventional superfluids, include:

- Pairs of electrons of opposite spin and momentum in a singlet s state make up the *condensate*, a macroscopically occupied single quantum state, which is responsible for the superfluid behavior (flow without resistance, perfect diamagnetism, etc.) of the system.
- Excitation of single particles from the condensate costs an energy, Δ (the BCS energy gap) which increases rapidly to a constant Δ_0, as the temperature decreases below the superconducting transition temperature, T_c. As a result, single particle properties such as the specific heat and transport display exponential behavior at low temperatures.
- For the representation of the Bardeen-Pines effective electron interaction used by Cooper and BCS, T_c is given by

$$T_c = 1.14\,\theta_D \exp[-1/|N(0)V_{\text{eff}}|] \tag{2}$$

  where $N(0)$ is the density of states of electrons at the Fermi surface.
- The size of the pair wave function is the coherence length, $\xi \sim v_F/\pi\Delta$, where v_F is the electron Fermi velocity, and is $\gtrsim 10^3$ Å, quite large compared to the interelectron spacing, $r_0 \sim 1$Å. Thus the superconducting transition is fundamentally different from that which would occur for closely bound pairs ($\xi \lesssim r_0$); in the latter case, which had earlier been considered by Schafroth, electrons pair rather than forming a Fermi liquid, and one gets a Bose condensation, analogous to that found in liquid ^{4}He, of these pairs at a temperature $\sim T_F$ which depends mainly on the pair density and mass and is only weakly influenced by particle interaction.
- In BCS theory, the energy gap and T_c are intimately related; one finds $\Delta \sim 1.75\ k_B\ T_c$, while for the Schafroth pairs, T_c and Δ are essentially independent, one measuring the pair density, the other pair binding.

In the two decades following BCS, theorists explored the possibility of other kinds of condensate pairings (electron pairs in higher relative angular momentum states) and other kinds of pairing interactions in metals [e.g. an attractive electron interaction associated with the virtual exchange of spin waves or other purely electronic excitations ("excitons")] which might give rise to superconductivity, while experimentalists sought both higher superconducting temperatures and evidence for pairing induced by other than phonons. No evidence for the latter was found, while in the hundreds of new superconductors discovered by the late Bernd Mathias, Ted Geballe, John Hulm, and others, no transition temperatures higher than ~ 23 K were found. This led Bill McMillan, Phil Anderson, and Marvin Cohen to suggest that there is a maximum temperature, ~ 30 K, for a phonon-induced interaction between electrons to give rise to superconductivity. This "phonon barrier" is associated in part with the fact that for very strong electron-phonon interactions, one gets a lattice instability, or possibly a metal insulator transition, rather than a superconducting transition. During this same period theorists and experimentalists alike considered the possibility of superfluid behavior in other fermion systems, a search which turned out to be far more successful.

III. UNCONVENTIONAL SUPERCONDUCTORS

Nuclei

Following BCS theory it was quickly recognized that for any fermion system an attractive interaction between fermions in some part of (q, ω) space can lead to a pairing instability associated with some relative angular momentum state, $\vec{\ell}$ and hence to superconductivity (in the case of charged systems) or superfluidity (for neutral systems). The first successful proposal arose from a series of lectures on BCS theory which I gave in Copenhagen in the summer of 1957; Aage Bohr, Ben Mottelson and I realized that although atomic nuclei are comparatively small finite systems the basic long-range attraction between a nucleons would likely give rise to pairing phenomena which would be experimentally detectable. We found that in fact the pairing of nucleons outside closed shells plays a key role in nuclear behavior, from determining the nuclear inertial moments to influencing reaction rates significantly. The pairing energy, Δ, is $\sim 12/A^{1/2}$ MeV, where A is the total number of nucleons, while the corresponding coherence length is large compared to the average nuclear separation, so that one is clearly in the BCS, not the Schafroth, limit.

Neutron Stars

Our proposal led Arkady Migdal to suggest that since neutron stars are made up primarily of neutrons, if discovered these objects would also display superfluid behavior, while Vitale Ginzburg and David Kirzhnits pointed out that since neutron stars rotate, one would expect to find a vortex structure for the neutron superfluid, analogous to that observed when one rotates a vessel containing superfluid ^{4}He. The discovery of pulsars in1967, and their subsequent identification as rotating neutron stars, then raised the question of whether these *celestial superfluids* could be observed. The answer came quite soon; the comparatively young ($\sim 10^4$ years old) Vela pulsars which, like all other pulsars, had been seen to spin down in quite regular fashion, was observed in March, 1969, to undergo a remarkable *glitch*, in which the rotational period, ~ 89 ms, of the star suddenly decreased. The fractional changes in the period was $\sim 2 \times 10^{-6}$, which does not seem like much until one realizes that a comparable change in the earth's rotational period would require a sudden shrinkage of earth's surface by ~ 1 meter. Moreover, $\dot{P}$, the time derivative of the period, which underwent a still larger fractional change, $\sim 2 \times 10^{-3}$, was seen to relax on a macroscopic time scale $\sim$ months. This behavior led our Urbana group of newly minted theoretical astrophysicists

(Gordon Baym, Chris Pethick, and me) working together in Aspen with Mal Ruderman, to suggest that the physical origin of this extraordinarily long (for an object twelve kilometers in radius) time scale behavior was the delayed response of the core neutron superfluid to the sudden spin-up of the solid outer crust observed in the glitch. Our two-component (crust plus core neutron superfluid) theory provided a quantitative explanation of both the post-glitch behavior of the Vela pulsar, and that observed following the sudden spin-up of the Crab pulsar in September, 1969.

During the next decade, four more pulsar glitches with qualitatively similar post-glitch behavior were observed; however, the detailed post-glitch observation in both the Vela and Crab pulsars showed that this simple two-component theory did not work. Moreover to explain the physical origin of the glitches, Phil Anderson and Naoki Itoh invoked the *crustal* neutron superfluid, which Chris Pethick and I had suggested would likely be *intrinsically* pinned to crustal nuclei, since the coherence length in the crustal neutron superfluid would be comparable to the size of a nucleus there. Anderson and Itoh argued that a pulsar glitch represented the catastrophic unpinning of vortices in the crustal neutron superfluid, a phenomenon (induced by pulsar spindown) which is analogous to the flux jumps observed in type II superconductors (where a vortex structure enables a magnetic field to penetrate in much the same way as it permits the neutron superfluid to follow the crustal rotation). This proposal led Ali Alpar, Jacob Shaham, Anderson and me to propose that these same pinned vortices were the physical origin of the observed post-glitch behavior. Thus we argued that the core neutron superfluid would couple rapidly to the crust, while the longtime scale postflitch behavior represents the glitch-induced creep of the pinned crustal superfluid vortices, analogous to the flux creep observed in Type II superconductors.

It is thus the pinned crustal neutron superfluid, occupying a shell some 1400 m thick, at densities between 10^{13} g cm^{-3} and 2.4×10^{14} g cm^{-3}, which is responsible for both glitches and post-glitch behavior. since the crustal neutron superfluid can only change its rotation rate through vortex motion, it is the glitches which enable this superfluid to follow, over long time periods, the pulsar spin-down, while the rotational period behavior following a glitch reflects creep of vortex lines which have been redistributed in the course of a glitch. Typical energy gaps in this neutron superfluid range from a maximum of the order of 10^{10} K ($\sim$ 1 MeV) to a minimum of order 10^{8} K near the crust-core boundary. As a result one expects regions of both weak pinning (where $\xi \sim R_N$) and superweak pinning (where $\xi \gg R_N$) in the crust. I should add that not only is this celestial superfluid the highest temperature superfluid encountered in the universe, it is also the most distant (neutron stars are typically $\sim$ 1 kiloparsec away from us) and the most abundant (recall that one twentieth of a solar mass is $\sim 10^{31}$ grams, substantially in excess of the mass of existing terrestial superfluids and superconductors).

Rather to our surprise, it proved possible to develop a detailed quantitative description of vortex creep in neutron stars, which was in excellent agreement with the observational data for all nine glitches for which such data could be obtained. Thus thanks to the extraordinarily high density of neutron stars, and the remarkable accuracy with which astronomers can follow their rotational periods ($\sim$ nanoseconds), it is possible to use neutron stars a kiloparsec away as a celestial low temperature laboratory to follow changes in vortex structure over distances $\sim$ few hundred meters and deduce energy gaps, transition temperatures, and pinning energies in neutron matter, information unobtainable in any terrestrial laboratory.

Superfluid ^{3}He

The next unconventional superfluid to be discovered was liquid ^{3}He. Following BCS, there was little doubt in the theoretical community that at sufficiently low temperatures ^{3}He would become superfluid, since the long-range part of the bare ^{3}He atom-atom interaction, which comes from their van der Waals interaction, is attractive. The key questions revolved about the temperature at which the transition would take place, and the angular momentum of the pairing state responsible for the expected superfluidity. Here, in advance of the discovery, the theorists did not do especially well, being in general too optimistic about T_c, and backing *d*-state pairing as the state best able to take advantage of the van der Waals attraction.

The superfluid transition was discovered in 1972, by Doug Osheroff, David Lee, and Bob Richardson at Cornell. The transition temperature varies from $\sim$ 0.001 K at SVP to $\sim$ 0.003 K at 34 atm, while theoretical work by Tony Leggett, Phil Anderson, and Bill Brinkman, based on the experimental results of the Cornell group, and of John Wheatley and his collaborators at La Jolla, soon established that the pairs in the condensate were in an $q = 1$ state and, further that two distinct superfluid phases exist, corresponding to different condensate wave functions.

The physical origin of the superfluid transition is now understood. It turns out that the long-range attraction between the ^{3}He atoms plays little or no role in determining the pairing instability; what counts is the difference between the effective interaction of anti-parallel spin quasiparticles and that of parallel spin quasiparticles, the latter being weaker because the Pauli principle reduces somewhat the extent to which the latter sample their mutual short-range repulsion. This difference leads to a substantial enhancement of the spin susceptibility and low frequency spin fluctuations (the so-called paramagnons), and to *p* state pairing.

Liquid ^{3}He is therefore an example of a Pauli principle superfluid: the quasiparticle interaction is strongly repulsive in the density channel (hence *s* wave pairing is ruled out), attractive in the spin channel. The Debye temperature of the BCS theory of metallic superconductivity is replaced by the spin fluctuation temperature,

$$T_{SF} = E_F(1 + F_0^a) \tag{3}$$

where F_0^a is the dimensionless strength of the spin channel self-consistent field responsible for the susceptibility enhancement [$\chi_p \sim N(0)/(1 + F_0^a)$, where $N(0)$ is the density of states per unit energy.] An approximation expression for the transition temperature is

$$T_c \cong (T_{SF}/7) \exp\left(-\frac{1}{g}\right) \tag{4}$$

where the coupling constant g is a function primarily of F_0^a. At SVP, $T_{SF} \sim$ 500 mK, and $T_c \sim (T_{SF}/500) \sim (T_F/1600)$. For a one-component Pauli principle superfluid it is straightforward to show that with $F_0^a \sim -0.87$, one gets the maximum effect of the net attraction in the spin channel. The resulting trnasition temperature corresponds then to a "Pauli principle barrier" to superfluidity, and is

$$(T_c)_{\max} \sim E_F/20. \tag{5}$$

Heavy Electron Superconductors

Heavy electron systems are intermetallic compounds containing elements with unfilled f-electron shells, such as U or Ce, which at room temperature and above behave like a weakly interacting collection of f-electron moments and conduction electrons with ordinary masses, while at low temperatures the conduction electron specific heat becomes typically some hundred times larger than that found in most metals. During the period 1979-85 experiment showed that these highly correlated low temperature states display remarkable behavior whether the system remains normal down to the lowest temperature measured, becomes antiferromagnetic, or becomes superconducting. While in ordinary metallic superconductors a dilute concentration of magnetic impurities destroys superconductivity, in heavy electron systems superconductivity and antiferromagnetism can coexist; a transition to either ordered state may be followed by a second transition to a phase containing both states. Thus in both UPt_3 and URu_2Si_2 one finds on lowering the temperature that an antiferromagnetic transition is followed by a transition to the superconducting state, while in $U_{0.97}Th_{0.03}Be_{13}$ the order of the transition is reversed.

At high temperature heavy electron systems behave like a collection of weakly interacting f-electron moments and conduction electrons, while at very low temperatures, so far as thermal and transport processes are concerned, they behave like a system of strongly interacting itinerant electrons which scatter against impurities, against low frequency f-electron spin fluctuations, and against one another.

A physical picture of the transition betweenthese two regimes is that as the temperature is lowered the local moments and conduction electrons become more and more strongly coupled. The magnetic behavior is quenched while the effective mass of the itinerant electrons becomes substantially enhanced. As a consequence of this interaction, the f-electrons are no longer confined to the magnetic sites, but can hop into the conduction band, as in the Anderson model. The itinerant heavy electron states at low temperatures are therefore superpositions of localized f-electrons and conduction electrons. Their quite strong interaction reflects not so much their direct Coulomb interaction, as it does an interaction induced by their coupling to spin fluctuations on the magnetic sites, and it provides a natural explanation for the large finite temperature corrections to the low-temperature form of the specific heat, the strong temperature dependence of the electrical resistivity and other transport coefficients, and the appearance of superconductivity.

The spin fluctuations, which play a role in heavy electron systems analoguous to that played by phonons in ordinary superconductors, are described by the spin-spin correlation function. The neutron scattering results for UPt_3 and U_2Zn_{17} may be fit by a model for the spin-spin correlation function in which fluctuations of the magnetic moment at the f-atom site are coupled to those at other sites by an effective exchange interaction. If one assumes that all the magnetic moment is associated with electron in f-orbits, this leads to an expression for the wavenumber- and frequency-dependent spin-spin correlation function of the form

$$\chi(\vec{q},\omega) = \frac{\chi_\mu(\omega,T)}{1 - J(\vec{q},\omega,T)\chi_\mu(\omega,T)} \tag{6}$$

where $\chi_\mu(\omega,T)$ describes the correlations of the spin at a single f-site, including the effects of interaction with the compensating electron cloud, and $J(q,\omega,T)$ is an effective exchange interaction which describes the coupling between spins at different sites.

Chris Pethick and I have proposed that an expression of this form provides a useful starting point for the examination of all aspects of heavy fermion behavior. From a microscopic point of view the induced spin-spin interaction is given by an expression of the form

$$J(\vec{q},\omega,T) = -\sum_{\vec{K}_n} \left|V_{\vec{q}+\vec{K}_n}\right|^2 \chi_c(\vec{q}+\vec{K}_n,\omega,T) \tag{7}$$

where V describes the coupling of a conduction electron-hole pair to the local spin fluctuations described by $\chi_\mu(\omega,T), K_n$ is a reciprocal lattice vector, and χ_c is the conduction electron-hole spin-spin response function. As a result of the coupling, J, the characteristic energies which enter into the low frequency limit of χ, (Eq. 4), become wavevector- and temperature-dependent, being given by

$$\theta_{\text{coh}}(\vec{q},T) \cong T_K\left[1 - J(\vec{q},0,T)\chi_\mu(0,T)\right], \tag{8}$$

where T_K, the Kondo temperature, is the characteristic energy for an isolated impurity. We argued that the presence of a second energy scale, lower than T_K, is a characteristic feature of all heavy fermion systems, and may be a necessary condition for observing heavy fermion behavior. Put another way, if $J\chi_\mu \ll 1$, one is likely in a weak coupling limit, and no heavy fermion behavior results. On the other hand if, as in U_2Zn_{17}, for some wavevector $\vec{q}$, and temperature T, $J\chi_\mu = 1$, then an antiferromagnetic phase transition occurs. Both normal and superconducting heavy fermion compounds would seem to lie in the strong coupling regime, $J\chi_\mu \sim 1$.

The coupling between the heavy quasiparticle pairs and the local spin fluctuations gives rise to an induced wavevector-, frequency- and temperature-dependent, heavy electron interaction,

$$U_{\text{ind}}(\vec{q},\omega,T) = -V_{\text{eff}}^2\chi(\vec{q},\omega,T) = -\frac{V_{\text{eff}}^2\chi_\mu(\omega,T)}{1 - J(\vec{q},\omega,T)\chi_\mu(\omega,T)}. \tag{9}$$

For frequencies low compared to the characteristic frequencies which enter into χ, that interaction will be attractive between like spins and repulsive between unlike spins; to the extent that χ exhibits antiferromagnetic correlations (and neutron scattering experiments suggest that this might quite generally be the case), $U_{\text{ind}}(q,\omega)$ will behave in similar fashion. This induced interaction is the physical origin of the pairing instability which gives rise to superconductivity. The proposed approach is quite reminiscent of the electron-phonon interaction problem, with the local moment spin fluctuation frequency-dependent susceptibility playing the role of a phonon progagator. In the present theory there is a considerable amount of feedback, and possible non-linear behavior, in that, for example, J depends on χ_c which in turn depends on J through electron-local moment fluctuation coupling.

Perhaps the clearest experimental indication that heavy electron superconductors are unconventional comes from the fact that no equilibrium or transport properties in the heavy fermion superconductors exhibit the exponential behavior expected for states with a non-zero energy gap everywhere on the Fermi surface; rather both specific heat and transport measurements display the power-law behavior characteristic of states with gaps which vanish at points or along lines on the Fermi surface. The exact nature of the condensate wave function has not yet been established, although considerable information about where the nodes lie on the Fermi surface has been obtained through measurements of transport coefficients such as acoustic attenuation.

IV. THE COPPER OXIDE SUPERCONDUCTORS

The discovery in 1986 of superconductivity at 30 K in the oxide system $La_{2-x}Ba_xCuO_{4-\delta}$ by Bednorz and Mueller, which was soon followed by the discovery at 90 K in the system $YBa_2Cu_3O_{7-\delta}$ by Chu, Wu and their collaborators at Houston and Alabama, was totally unexpected, coming as it did in an obscure, semi-metallic, not previous superconducting, corner of the periodic table, ceramic copper oxide materials with crystal structures derived from perovskites. Subsequent work on these materials has shown that Cu-O planes, in which a copper atom is surrounded by four oxygen atoms, provide the major common feature and that it is holes in these planes (believed to lie in an oxygen band), produced by departures from the anti-ferromagnetic insulating compositions, which give rise to metallic behavior and reduce the Néel temperature at which the materials undergo an antiferromagnetic transition. Experiments on single crystals have shown that the electrical resistivity and Hall effect of the doped materials are highly anisotropic, with a substantially lower resistivity, generally varying linearly with temperature being found in the Cu-O (ab) planes. The materials are Type II superconductors with very large Ginzburg-Landau parameters, K_{G-L}, and, again, a considerable degree of anisotropy; in polycrystalline samples the average coherence length is quite short, being ~ 12 Å.

Theoretical work on the copper oxide superconductors has tended to focus on the physical origin of the superconducting behavior, and especially on the nature of the elementary excitations in the Cu-O planes. Quite generally, their high transition temperatures, unusual normal state properties, and short coherence lengths have led most theorists to regard them as unconventional superconductors, in which the physical mechanism responsible for superconductivity is of mainly electronic origin. The absence of an isotope effect in $YBa_2Cu_3O_{6.9}$ thus provided experimental confirmation of a previously held set of prejudices. The field is much too new for well-developed theories; what one has instead is the development of alternate scenarios (a situation familiar in astrophysics) in which one attempts to develop physical pictures consistent with known physical laws to account for the key experimental facts. A central question is whether the excitations in these materials resemble the quasiparticles found in ordinary metals; thus do they move first (i.e. exhibit Fermi liquid behavior, albeit highly anisotropic) and pair later (as a consequence of an attractive interaction induced by their coupling to spin fluctuations, acoustic plasmons, other electronic excitation modes (excitons) etc.)? Or are we dealing with a quite new physical situation, as Anderson has proposed, in which strong Coulomb and spin correlations lead to quite new kinds of soliton-like excitations in the normal state which possess charge but obey Bose statistics? In this case the superconducting transition could correspond to a condensation of these preformed bosons, as in Schafroth's picture; this kind of theory may be classified in the "pair first, move later" category. It is only through careful comparison of theory with experiment that one will be able to settle this question.

Are the High Temperature Superconductors A Branch of the Heavy Electron Family?

Chris Pethick and I have taken the point of view that the normal states of the ceramic oxide superconductors is best described as an anisotropic Fermi liquid, and that since heavy electron systems provide us with a new mechanism for superconductivity and new pairing states in metals, it is useful to inquire whether the physical origin of superconductivity in the ceramic oxides is similar; thus does superconductivity in these systems arise from an attractive interaction between holes induced by their coupling to spin fluctuation excitations, and are these superconductors another banch of the heavy electron family? Models in which spin fluctuations induce superconductivity have been proposed by a number of authors, while the detection of antiferromagnetic ordering in La_2CuO_{4-y} provides support for this hypothesis. For polycrystalline $YBa_2Cu_3O_{7-\delta}$, Kevin Bedell and I have carried out a self-consistent isotropic analysis of the superconducting properties and concluded that itinerant carrier mass is in the range 3 to 5 m_e, and that one may have a substantial exchange enhancement of the carrier magnetic susceptibility. The family resemblance to heavy electron systems in therefore striking. Quite generally in the high T_c superconductors, the copper-oxide planes represent a promising source of spin fluctuations, while if the itinerant carriers belong to a distinct but nearby (in energy) band, the basic physics would be remarkably similar, with a spin-fluctuation induced interaction being responsible for both the enhanced carrier mass and superconductivity; it is also possible that the carriers and spin fluctuations are excitations which belong to the same band.

The reason, then, that one achieves high superconducting temperatures in the ceramic oxides is that the itinerant carrier density is so small that the spin-spin interaction induced by coupling to the itinerant carriers plays a small role; rather the magnetic behavior is associated with exchange interactions between copper spins or nearest neighbor and next nearest neighbor sites. Since this characteristic energy can be large compar d to room temperature, superconductivity at high temperature can easily result. Put another way, what spoils the chances for high T_c in the heavy electron systems is that the conduction electrons are sufficiently dense to screen that f-electron moments, so that the scale over which the heavy electron interactions can be attractive is one or two orders of magnitude smaller than the Kondo temperature; that same phenomenon is responsible for the fact that the Néel temperature in heavy electron systems are $\lesssim$ 20 K, rather than being comparable to or greater than room temperature. Thus heavy electrons have lower characteristic spin fluctuation energies, low transition temperatures, and very large carrier masses, while in the copper oxide superconductors one finds high characteristic spin fluctuation energies, high T_c, and modest spin fluctuation enhancement of carrier masses.

In common with the plasmon-exchange and exciton-exchange mechanisms, the spin-fluctuation mechanism provides a natural explanation for the observed absence of an isotope effect in $YBa_2Cu_3O_7$ and $EuBa_2Cu_3O_7$. Unlike the above mechanisms, it also provides a natural explanation for the extreme sensitivity of the copper-oxide superconductors to substitutions for the copper ions. Such substitutions, it may be argued, can change dramatically the nature of the spin fluctuation excitations in the copper-oxide layers, and easily destroy superconductivity, while it is difficult to see why these substitutions would affect the exchange of virtual plasmons or excitons between carriers, and hence affect any superconductivity arising from that exchange. As is the case for heavy electrons, spin fluctuation exchange can give rise to pairing in higher angular momentum states.

Much theoretical and experimental work will be required to test the spin fluctuation mechanism for ceramic oxide superconductivity. It took some three years of intensive experimental and theoretical investigations for the heavy electron community to arrive at a consensus on the physical picture we have set forth; it would not be surprising if a comparable period of time might be required to arrive at a comparable consensus on the new high T_c materials.

V. FUTURE DEVELOPMENTS

Let us now ask about the next 35 years of superconductivity and pairing (I see this as Hans' mid-career phase). My own prediction, cloudy crystal ball and all, is that we will have room-temperature superconductors in the next five years. The phonon barrier has been broken, and if one is clever and lucky enough, there is no reason one cannot find an electronic mechanism that will increase the transition temperatures by a factor of three or more. Assuming there is room-temperature superconductivity, there is a further very interesting question: does one encounter pairing phenomena in biomolecules? If you look about a little bit you find an extraordinary amount of the right ingredients: oxygen, spin fluctuations, highly polarizable chains, resonating valence bonds (those were, after all, invented by chemists), excitons, and even acoustical plasmons. So one does not lack for mechanisms and one does not lack for theories who will write down theories based on these ideas. The key question is: did Nature order that pairing phenomenon? That is a more subtle question. [Yesterday I put it to Manfred Eigen and Martin Karplus; neither immediately said yes or no.] In conclusion, therefore let me list a set of homework assignments for Hans, in the hope that he will report back to us at his 100th birthday symposium: are pairing phenomena in biomolecules necessary for reaction rates, tunneling, prebiotic evolution, population genetics or neural networks? I know you will join me in wishing Hans every success in this endeavor.

ACKNOWLEDGEMENTS

The section on heavy electron superconductors is taken in part from manuscripts written in collaboration with Chris Pethick (a Ginzburg Festschrift, and a review article prepared for *Science* in collaboration with Z. Fisk, D. Hess, J. Smith, J. Thompson, and J. Willis), while that on copper oxide superconductors represents a June, 1987 perspective, and is taken in part from a report to the Berkeley meeting on Novel Superconductivity prepared in collaboration with Chris Pethick. I have profited greatly from discussions with Kevin Bedell, Bertram Batlogg, Don Ginsberg, Chris Pethick, Myron Salamon, Charlie Slichter, and Russ Walstedt in the course of preparing this manuscript. This work is supported in part by grants from the National Science Foundation, NSF DMR85-21041 and NSF DMR86-12860.

POSTSCRIPT

During the months since June, 1987 (these lines are being written in March, 1988) experimentalists working on the copper oxide superconductors have done a superb job of making single crystals, characterizing their behavior as a function of doping (for the $La_{2-x}M_xCuO_4$ materials) and oxygen concentration (for $YBa_2Cu_3O_{7-\delta}$), and measuring key physical properties. For low carrier concentrations ($x \lesssim 0.06; \delta \gtrsim 0.5$) the materials are antiferromagnets for which theoretical descriptions based on a quantum spin 1/2 Heisenberg model appear quite promising. At higher carrier concentrations, antiferromagnetism disappears, and superconductivity is found. Experiments on superconducting single cyrstals of the 1-2-3 materials ($YBa_2Cu_3O_{7-\delta}$) can be quantitatively described by anistropic Ginzburg-Landau theory, and yield results consistent with effective mass anisotropy ratios, $(m_\perp/m_\parallel) \sim 50\%$ where $m_\parallel$ is the quasiparticle mass in the Cu-O, a-b plane, and $m_\perp$ that along the c axis, perpendicular to the a-b plane. The corresponding coherence lengths are sohrt: $\xi_\parallel \sim 30$ Å, and $\xi_\perp \sim 4$ Å; the latter is close to the distance between adjacent Cu-O planes.

There is a widespread consensus that the physical origin of superconductivity is purely electronic (i.e. non-phonic) in origin. Two experiments which strongly suggest that the condensate wave function is highly anistropic are:

- Measurements by Don Ginsberg, Myron Salamon, and Sue Inderhees of Gaussian fluctations in the specific heat near the superconducting transition, which show that the transition is three dimensional and that the number of independnet components of the order parameter which describe the superconducting state is 7 ± 1 (for an isotropic s-state, $n = 2$).
- Nuclear magnetic relaxation measurements by Russ Walstedt and Bill Warren involving ^{63}Cu and ^{65}Cu nuclei in both the "plane" sites and "chain" sites of the 1-2-3 materials show that, unlike conventional BCS superconductors, the relaxation times of Cu nuclei at both sites drop dramatically below T_c, with however a different temperature dependence. With the site assignment obtained by Charlie Slichter and Charles Pennington from aligned single crystal experiments, it turns out that below T_c the drop is far sharper in the "plane" site Cu nuclei, with a temperature dependence which is roughly consistent with a large, "strong-coupling", energy gap, $\Delta \sim 4k_BT_c$; that for the chain site can be approximately fit with either a much smaller energy gap, or a power law behavior.

Detailed studies of antiferromagnetic spin fluctuations in both the 1-2-3 and 2-1-4 materials have been carried out by Paul Fleury and Ken Lyons; they find such fluctuations in the 1-2-3 system are present as well in superconducting samples.

The nature of the normal state of the superconducting phases is still not understood. Here I believe the results of NMR and NQR experiments on the 60 K ($YBa_2Cu_3O_{6.7}$) and 90 K ($YBa_2Cu_3O_{6.9}$) materials hold the key to any successful explanation of the normal state elementary excitations. With the Slichter-Pennington assignment, the experiments of Walstedt, Warren, and others show that the "chain" sites in both the 60 K and 90 K materials exhibit Korringa-like relaxation behavior ($(1/T_1) \sim T$), of quite similar magnitudes, of the kind seen for relaxation by conduction electrons in ordinary metals. Thus Chris Pethick and I argue that these experiments show that the normal state in these materials is sufficiently three-dimensional for quasiparticles in the CuO planes to have a sufficiently large amplitude at the chain Cu sites to give the usual Korringa result. The "plane" sites on the other hand exhibit a transition from Korringa behavior, albeit with markedly longer ($\sim 10^3$) relaxation times, in the 60 K materials to a much more rapid, but temperature independent, relaxation in the 90 K material. We conjecture that a qualitative explanation of the latter results is that the 90 K relaxation is associated with low frequency spin fluctuations, which are suppressed in the 60 K material by their more nearly antiferromagnetic vehavior, as seen in Raman scattering experiments. If, as seems plausible, the linear in-plane resistivity observed in the 90 K material is associated with scattering against spin flucutations, then the fact that Bertram Batlogg and his collaborators see an almost temperature independent resistivity in their 60 K samples would likewise reflect that change in the spin fluctuation spectrum.

ORDER AND CHAOS

Edgar Lüscher
Physik Department, Technische Universität München
D-8046 Garching, West Germany

1. GENERAL REMARKS

The greek $\chi'\alpha \circ \rho$ can be translated as empty, infinite space or primary matter without any form. The initial state of our earth described in hebrew with "tohu wa bohu" is formulated in Genesis 1.2: "The earth was without form and void, and darkness was upon the face of the deep ...". According to Friedrich Nietzsche chaos represents the basic character of the world. Faust sings to mephistopheles in the magnificent opera "Mefistofele" from *Arrigo Boito*: "Strano figlio del chaos." In the last issue 1986 of the American Scientist a cartoon from Sidney Harris demonstrated the entrance of fractals even into private rooms. The pictures on the walls, the tapestry, the upholster are covered with fractals.

We may start the history of the physics of chaos with Leonardo da Vinci. in his studies of the flow of water, the originals are in the possession of Queen Elizabeth II at the Windsor Castle, he has designed effectively the chaotic turbulence flow. Large swirls break into smaller ones and these again break up into smaller ones etc. The studies of bifurcational instabilities can be traced back to Henry Poincaré and his work[1] on chaotic solutions of the Hamilton equations of rotating planetary masses. But these papers have not been noticed from more than 70 years by most of the physicists.

The turning point and the start of modern "chaos physics" was the publication of E.N. Lorenz[2] form the MIT. He found that even sets of coupled, first order, nonlinear differential equations can lead to chaotic trajectories. By then a real explosion of publications in this field took place. A small selection of important contributions are mentioned in Reference 3.

In a first approximation we will call aperiodic behavior in time and/or space chaotic. A wide variety of natural phenomena exhibit complicated, unpredictable random behavior. At the present stage of endeavor the chaotic behavior of deterministic dynamical systems can be indistinguishable from a random process. Contrary to what was believed in earlier time, the deterministic equations of classical mechanics do not imply a regular, ordered universe. The use of very fast computers is the key to the recent progress.

The novel approach which combines real and numerical "experiments" with mathematical analysis has given rise to the new interdisciplinary field of nonlinear dynamics, according to H. Haken a partial set of synergetics. The applications are stretched over a wide field, ranging from feedback control of mechanical and electrical systems to biology and even to economic and arms-strategic problems.

2. EXAMPLES

In this section some typical "experiments" are reviewed in which deterministic chaos can be detected.

Example 1: Pendulum

The equation of motion of a periodically driven simple pendulum where θ denotes the deviation angle, is:

$$\ddot{\theta} + \gamma\dot{\theta} + g\sin\theta = F\cos\omega t$$

with

$$\gamma = \text{damping factor}$$
$$g = 9.81$$
$$\omega = \text{frequency of the driving force.}$$

If the driving force F exceeds a critical value F_c, the function $\theta(t)$ becomes aperiodic.

A more illustrative example represents the double pendulum[4], Figure 1.

At the end of the second pendulum a light diode is mounted. The trajectories of a certain time interval are shown in Figure 2.

Example 2: Julia Sets

A simple example of a non-linear dynamical system is the logistic map[5] which can be described by a single difference equation

$$x_{i+1} = ax_i(1 - x_i).$$

The value of the variable x_{i+1} is determined at time-step $i+1$ from the value at time-step i. The time evolution of the variable x exhibits a transition from order to chaos with increasing value of the parameter a which is a measure of non-linearity. Such non-linear difference equations of this type arise naturally in the study of evolution mechanism for example by

Manfred Eigen[6] or Hermann Haken[7] in his laser thoery. For $a > 3.2$, there are two fixed points between which the value for x oscillates. With increasing values of the parameter a, there are more and more fixed points.

The graphic representations of Julia sets[8] boundary can be interpreted with the concept of logistic maps. Julia sets are fractals and the motion of iterates on these sets is chaotic. The generation of these sets are obtained by the mathematical feedback mechanism represented in Figure 3.

A very important property of Julia sets is their self-similarity. Beautiful pictures by Mandelbrot[9], Richter and Pleitgen[10] of such sets are well known, even through the popular press. At the UNESCO meeting on stochastic phenomena 1986 Manfred Peschel made an amusing poem on Julia sets:

Figure 1. Experimental double pendulum.

O Julia, ich dachte you where quite in order,
but too much chaos gibts leider at the border,
Trotz dieser Schwächen lieben wir dich sehr,
und deine Schönheit lockt uns mehr und mehr.

Example 3: Fractals[11]

In the February 1986 issue of "Physics Today" Leo Kadanoff asked the question: "Why all the fuss about fractals?" The concept of fractals plays a role in many areas of natural science and is of useful practical importance. In the material sciences, new structures, such as sponges, rough surfaces, porous materials, percolation clusters and random aggregates, can be described using fractal dimension. The most commonly used concept was introduced 1919 by Hausdorff: Let us consider a set of points in a D-dimensional space. The

Figure 2. Trajectories of an end-point of the second pendulum; light paths of a photodiode.

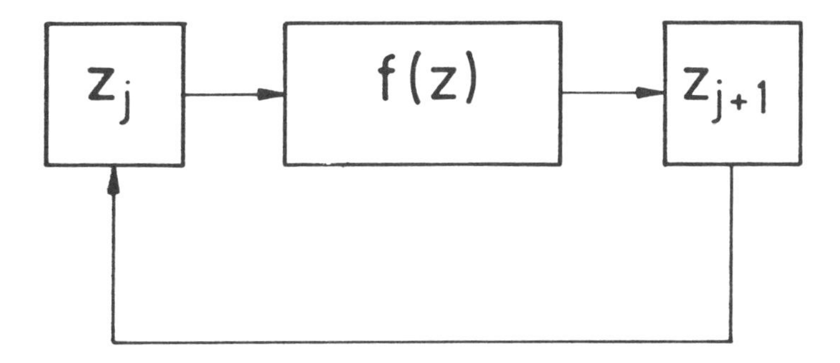

Figure 3. Mathematical feedback for the generation of Julia sets.

number of D-dimensional spheres $N(L)$ with L as diameter is needed to cover this set of points. If for $L \to 0$ this number increases like

$$N(L) \sim L^{-d_H},$$

there d_H is called the Hausdorff dimension. Some dimensions for typical self-similar sets are given in Table 1.

Table 1

Set	Hausdorff Dimension
Straight line	1
Cantor set	0.6309
Koch's snowflake	1.2618
Arrowhead curve	1.5849
Sierpinski sponge	2.7268

The field of fractals was born in the last century, when mathematicians asked the question whether all continuous functions were differentiable. It was demonstrated that functions existed which were continuous everywhere but not differentiable anywhere. Koch's snowflake is a typical example.

In Figure 4 the origin of the Rhine River network in the Swiss Canton Grison is plotted in a 1:300,000 scale and in Figure 5 the source of the Rabiuso section, marked in Figure 4, on a 1:25,000 scale. This river network can be considered as fractal curves with a Hausdorff dimension $d_H \approx 1.3$.

An example of fractal growth is the electrolytic deposition of copper[12] shown in Figure 6.

Simple reproducible dendritic pattern can be generated in high ciscous fluids like grease[13]. A chop of grease is placed between two parallel plates which are pressed together in order to create a circular spot. Dendritic structures from this spot (Figure 7) are obtained by pulling the plates with a well defined force.

For the computer simulation we consider the Navier-Stokes equation in an xy-plane neglecting inertial and gravitational forces. The boundary condition for the pressure is given through the surface tension. The calculated structures for four different pull-forces are given in Figure 8.

Despite the elegance and aesthetic value of the experiments and simulations Leo Kadanoff's statement is very true: "The physics of fractals is, in many ways, a subject waiting to be born."

Example 4: Medical Sciences

The electrocardiogram of a healthy heart shows a periodical pattern, Figure 9a. In contrast Figure 9b demonstrates a disorganized spread of impulses due to an irregular and chaotic arrhythmia, called ventricular fibrillation which is fatal if not immediately terminated. The most common cause of this ventricular fibrillation is an acute myocardial infarction.

In constrast to the heart the encophalogram of a normal adult is chaotic and it becomes periodic for example during a grand mal seizure (idiopathic epilepsy) as shown in Figure 10[14]. The sharp spikes occur with a rate of 25 to 30 per second.

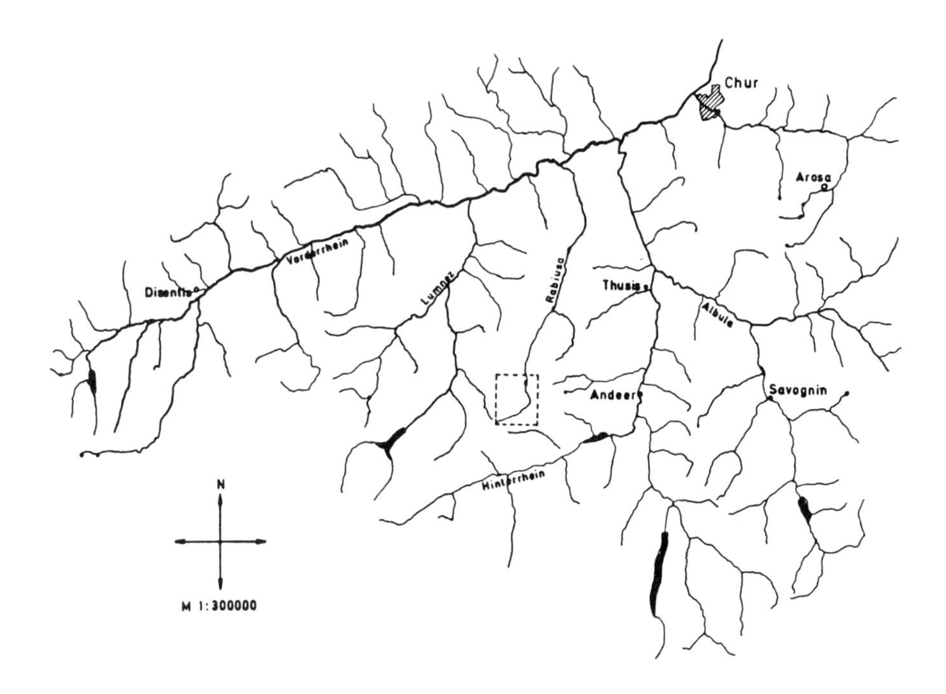

Figure 4. The sources of the Rhine River, scale 1:300,000.

Example 5: Biochemistry

Benno Hess and his school observed many biochemical reactions with quasiperiodic and chaotic oscillations, for example by monitoring the fluorescence of NADH in glycolyzing baker's yeast under sinusoidal glucose input flux[15] as shown in Figure 11.

Example 6: Aerodynamic

For airplanes, it is essential to know the onset of the stall which occurs when the angle of attack is increased too much. In the classical aerodynamics the critical angle of attack was defined with the onset of turbulence on the upper wing surface. In modern aircraft design, the lift can still be increased if the turbulences have not reached the chaotic regime. In Figure 12, the lift coefficient c_A in function of the angle of attack α of a Duck-Wing aircraft model is represented. Stall occurs when the ordered turbulence street regime turns off to the chaotic phase (top of Figure 12).

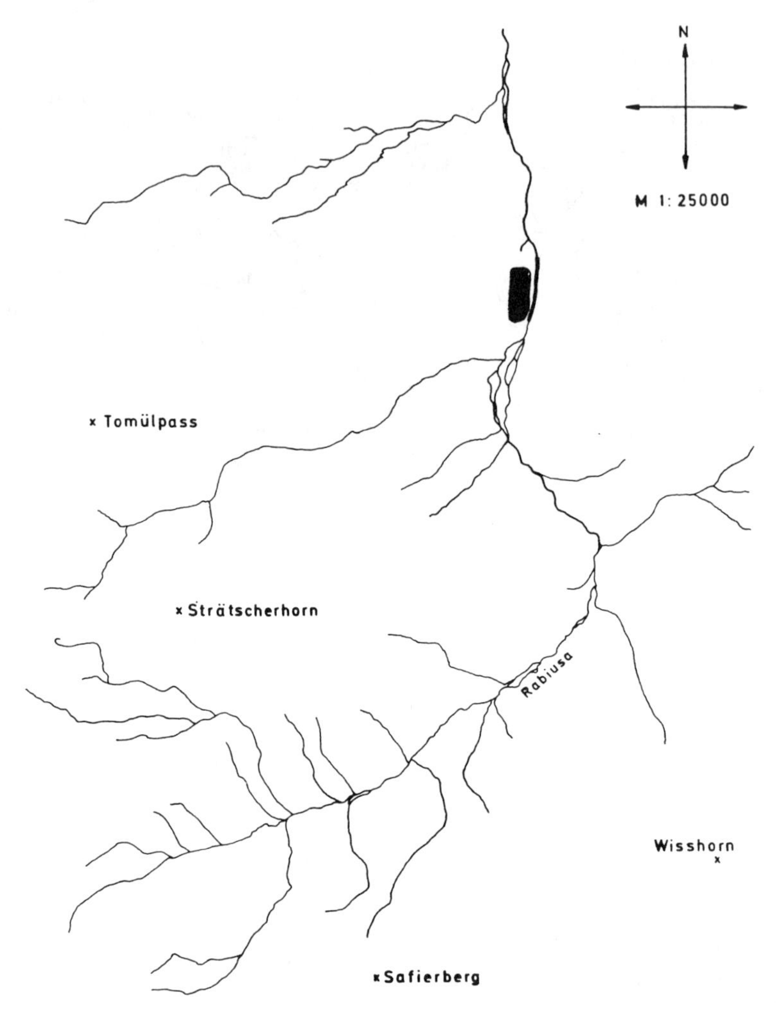

Figure 5. Detail of a source section: Rabiusa-creek, scale 1:25,000.

Example 7: Wind-instruments

The generation of musical sounds in wind instruments is due to periodic oscillations of turbulences. Let us consider the simple case of a labial organon pipe with a rigid lip (Figure 13). The coupline of this periodic turbulence oscillation leads to stationary states of three-dimensional waves. If the Reynold number reaches a critical value, these oscillations become aperiodic and result in non-musical sounds.

The number of examples could easily be enlarged to infinity because in nature the chaotic events outnumber by far the periodic ones.

Figure 6. Dendritic growth from the electrolytic deposition of copper.

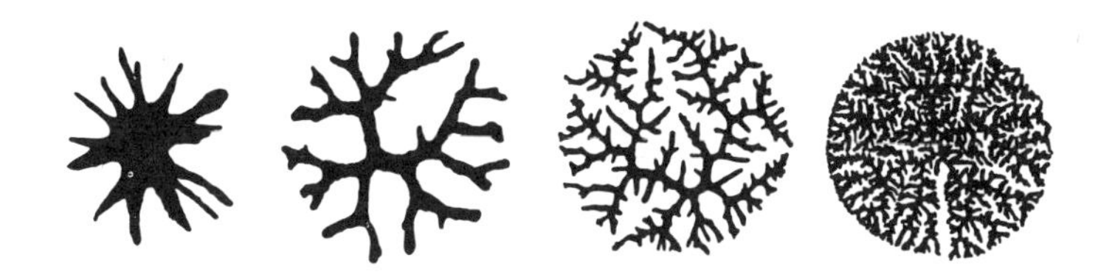

Figure 7. Dendritic structures obtained from a chop of grease between two plates.

3. DESCRIPTION OF CHAOTIC STATES

A very illustrative way for the demonstration of the onset of a chaotic regime are studies of ideal fluid flow past a cylinder, Figure 14. The time evolution of the fluid velocity is measured at a fixed point M with the Reynolds

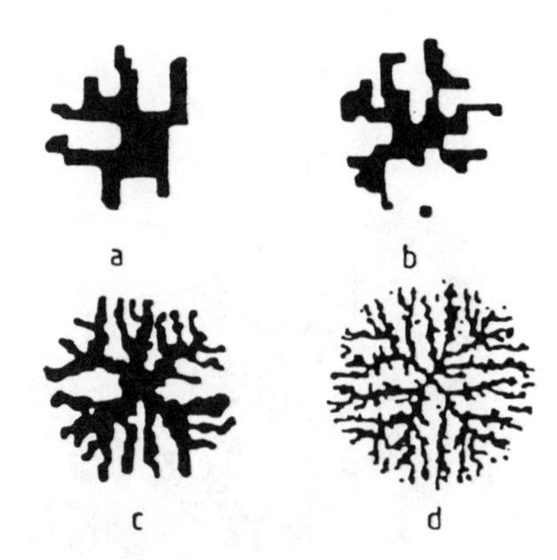

Figure 8. Calculated dendritic structures from a chop of grease.

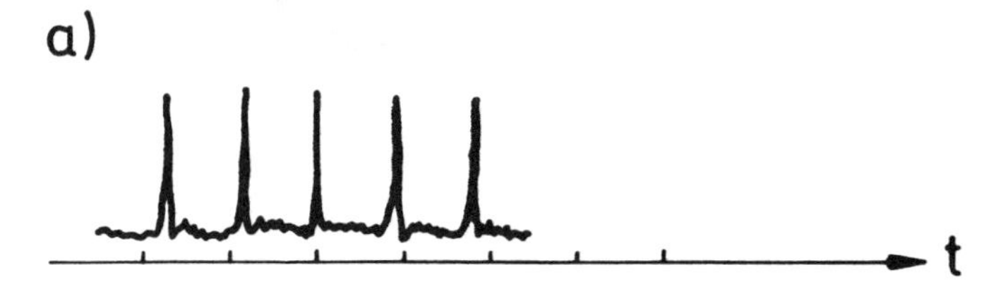

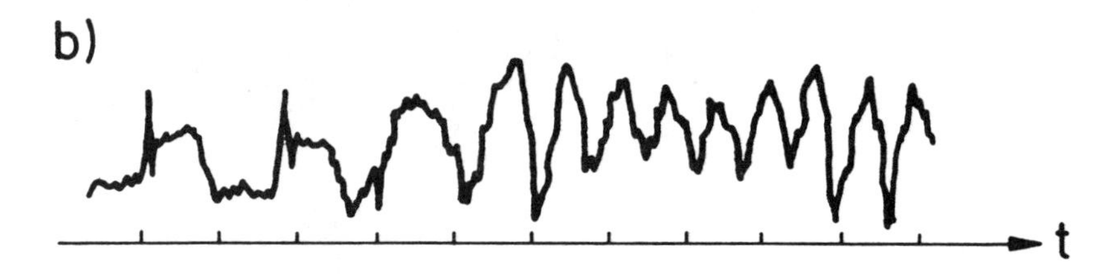

Figure 9. (a) Electrocardiogram of a healthy heart. (b) Electrocardiogram during ventricular fibrillation.

number as a parameter. It is convenient to consider the Fourier transform of the function $v_M(t)$. In Figure 15 the flow pattern,

$$p(\omega) = T \left[\int_0^T dt \exp(2\pi i \omega t) v_M(t) \right]^2$$

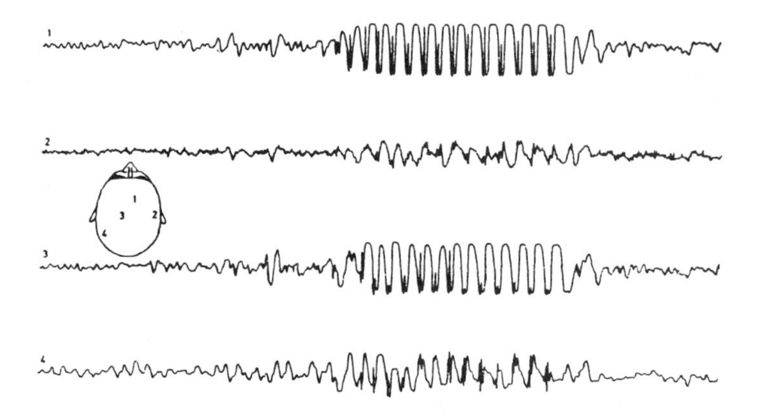

Figure 10. Electroencephalogram before, during and after a grand mal seizure at 4 different derivation points (1,2,3,4) on the skull.

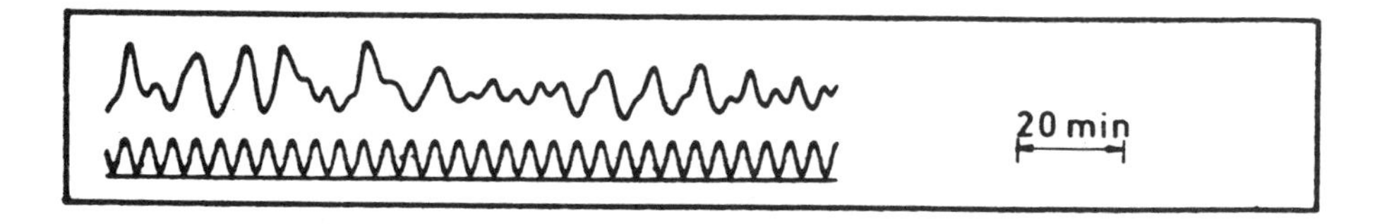

Figure 11. Chaotic fluorescence of NADH in glycolyzing yeast. The lower curve represents the periodic input flux of glucose[15].

for various Reynolds numbers R_e are plotted.

For small Reynolds numbers the oscillations of the turbulences are periodical. For $R_e \approx 10^4$ there can still be detected some periodicity although a continuous background in the power spectrum can be recognized. At very large Reynolds numbers, the full chaotic regime has developed. The transition occurs somewhere between $R_e = 10^4$ and $R_e = 10^6$. In 1932 Sir Horace Lamb, the famous author of "Hydrodynamics," published in 1879 under the title "Treatise on the mathematical theory of the motion of fluids," made the following statement: "I am an old man, and when I die and go to heaven, there are two matters on which I hope for enlightenment: one is QED and the other is the turbulent motion of fluids. About the former I am rather optimisitic."

If the power spectrum $p(\omega)$ has broad spectral features, the regime is called chaotic. This is one of the several possible definitions of chaos.

In the field of non-linear dynamics it is necessary to apply discrete mathe-

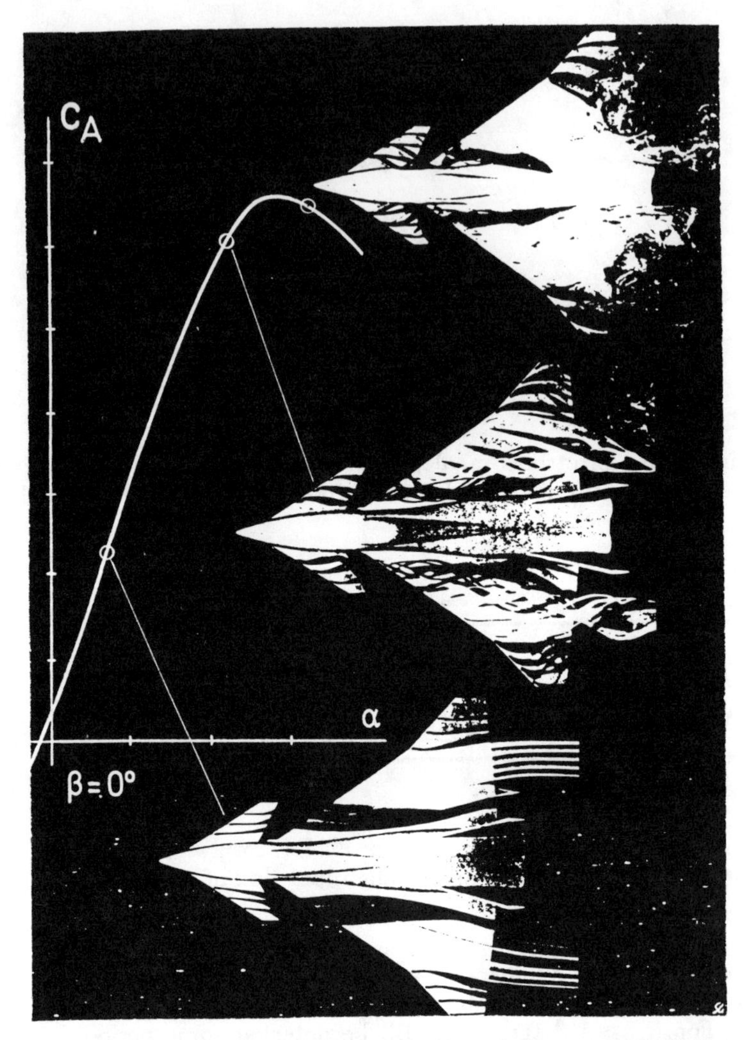

Figure 12. Laminar (bottom), ordered turbulences (middle) and chaotic flow (top) of an aircraft model (Courtesy B. Sacher, MBB Munich).

matics[16]. If the differential equations cannot be solved analytically, which mostly occurs, it is necessary to approximate the differential equations by different equations in order to compute it at all. The famous example of a logistic map, found in every book or introductory paper on chaos, is written as

$$x_{i+1} = R(x_i) = rx_i(1 - x_i).$$

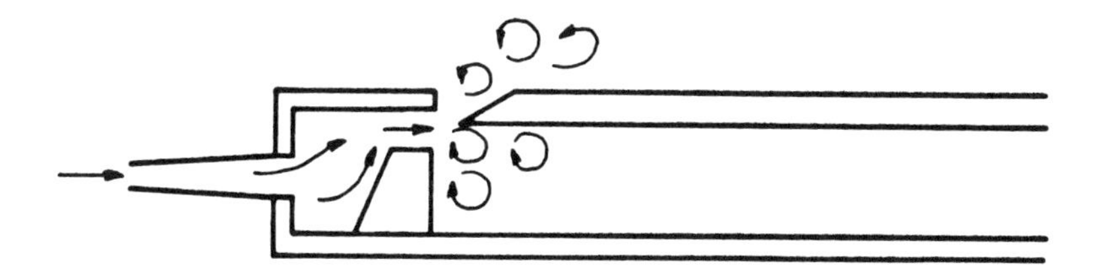

Figure 13. Scheme organon pipe with turbulences.

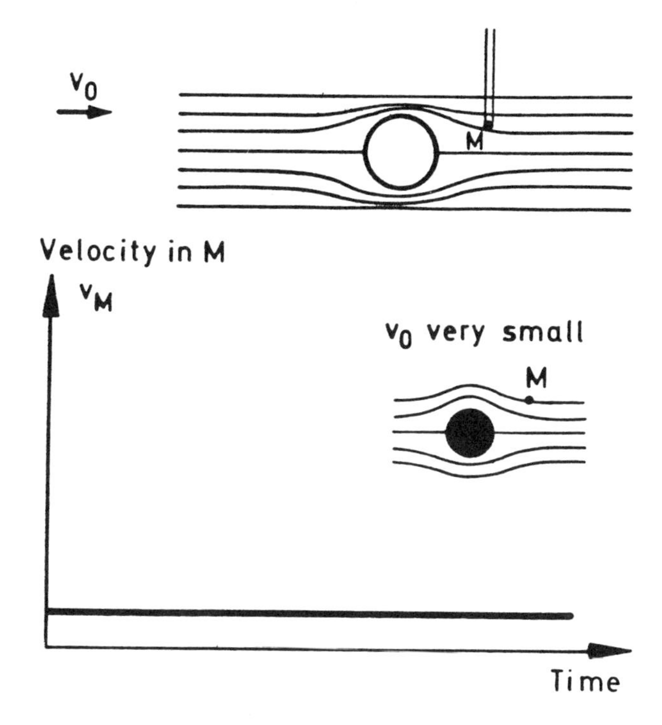

Figure 14. Velocity vs. time diagram and the corresponding lines of flow for very low speed; very small Reynolds number. The position of the flow-speed meter is marked with M.

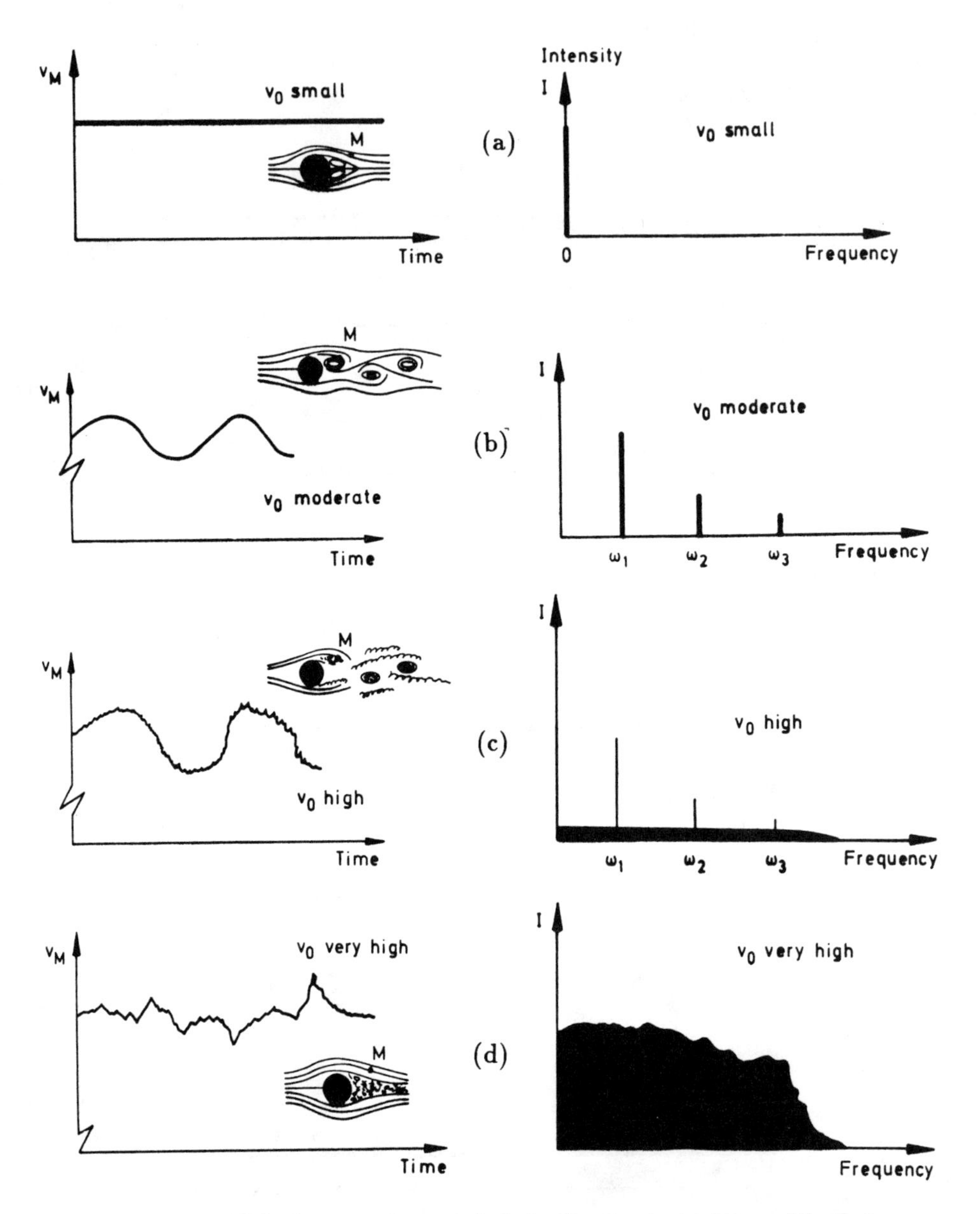

Figure 15. Velocity vs. time and their Fourier transform with their corresponding lines of flow for: (a) small Reynold number; (b) moderate Reynold number; (c) high Reynold number; (d) very high Reynold number.

Grossmann[17] and Feigenbaum[18] have studied the long term behavior of this equation specially the dependence on the growth rate parameter r. We can think of r as being akin to the Reynolds number R_e. They demonstrated that for large r the behavior becomes chaotic.

The transition from periodic to chaotic behavior can be illustrated in a nowaday well-known $x_i(r)$ graph introduced by Feigenbaum and is called bifurcation diagrams or Feigenbaum graphs. These graphs display the underlying structure of chaos, an evolution from regular to chaotic behavior. In the chaotic region the values of x_i cover continuous intervals. The discovery and explanation of regular structure domains in the chaotic region may have important practical applications. An analytical description of a chaotic evolution is in principle not possible, but we can hope that statistical theory may predict a probability for x_i for any particular i and r value. A simple example for such a regular domain in the chaotic region is an overcrowded swimming pool or ice-rink, where the mean free path for each individual in order to be able to swim or skate is practically zero. If a supervisor gives the strict order that everybody in the pool or on the rink has to swim/skate on a trajectory parallel to the boundary of the pool/rink, then everybody has the possibility to move, of course on one trajectory only with one speed.

A further possibility to characterize a chaotic regime is a measure for the average loss of information after a certain number N of steps of iterations.

In a linear dimension we start with a length of l_o on a logistic map:

$$x_{i+1} = R(x_i) \qquad i = 0, 1, \ldots N$$

where

$$l_o = (x_o + l_o) - x_0.$$

After N iterations we obtain a "length" of

$$l_N = l_o \exp[N \cdot \lambda(l_o)]$$

λ is called the Liapunov exponent and measures the exponential separation, giving an idea how chaotic a certain process is[19]. In Figure 16 the Liapunov exponent is plotted against r for the well-known logistic equation

$$x_{i+1} = r x_i (1 - x_i)$$

for $N = 10^4$ iteration steps.

The chaotic domain is characterized by all positive values

$$\lambda > 0.$$

There exists a view of more different techniques[20] for the description of chaotic regimes like Kolmogorov entropy[21] which measures also the average rate of loss of information. The Kolmogorov entropy is proportional to the sum of all positive Liapunov exponents.

Despite all the beautiful work being done by many excellent scholars there does not yet exist a general theory of chaotic regimes.

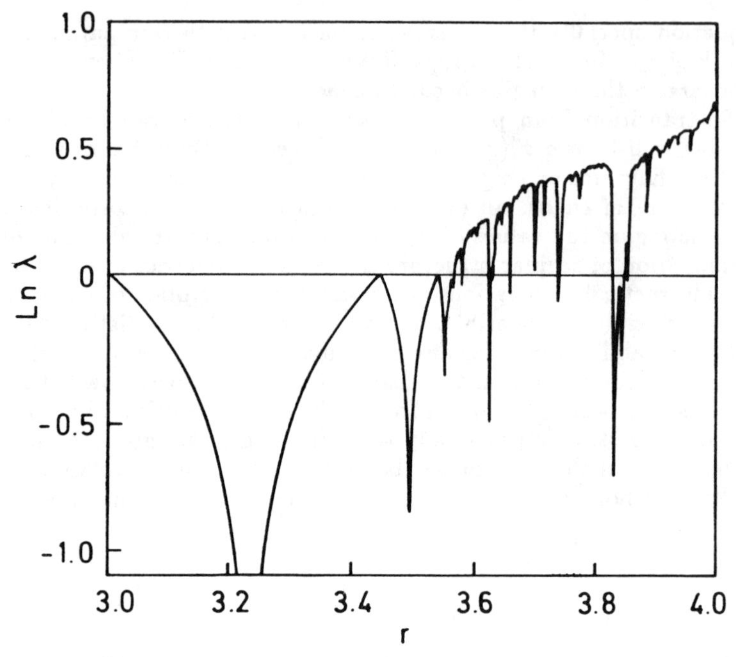

Figure 16. Liapunov exponent as a function of the growth parameter r of the logistic equation $x_{i+1} = rx_i(1 - x_i)$.

A different but related question is the role of determinism in classical physics. Already in 1955 Max Born mentioned the indeterministic character of classical mechanics due to the growth of unavoidable experimental errors. At that time not many physicists took this statement serious. The "quantum-billiard" is an illustration of this statement. A number of billiard balls on a very long table has a distance of 2 meters from each other; they are fixed in such a position that the moving ball is reflected in order to hit the next ball according to the scheme in Figure 17.

The problem is now the following: How many collisions of the moving ball A with the fixed balls $B_1, B_2, \ldots$ are possible until the uncertainty due to the quantum effects has reached such a value which makes further collisions unlike? After only eight collisions the uncertainty factor has grown by 10^{16}. The exponential growth of error leads to chaos. This is the main reason why accurate weather-forecasts for a longer period are in principle not possible. The deterministic character in classical physics is limited by the exponential growth of errors.

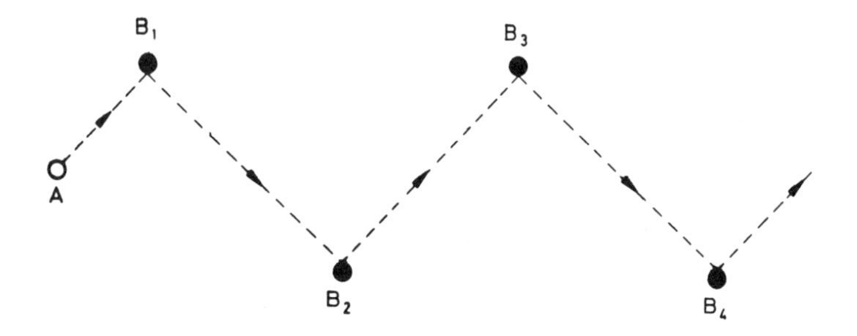

Figure 17. Quantum billiard ball setting: A, moving ball, and B_i, fixed balls.

4. EXPERIMENTS ILLUSTRATING ORDER-DISORDER TRANSITIONS

In this section some simple experiments and computer simulations will be discussed.

1st Example:

A well known class-room experiment[22] are flames which demonstrate the presence of standing acoustical waves in a tube filled with illuminating gas. If the intensity of the sound field is very high - over 110 dB - the onset of a chaotic regime can be observed as shown in Figure 18.

The same effect can be demonstrated with a standing wave field over Glycerol, Figure 19.

Figure 18. Rubens flame tube.

Figure 19. Kundt tube filled with glycerol. The sound frequency is 100 sec^{-1} and the number 409.9 is a measure for the sound amplitude.

2nd Example: Spatial-chaotic Structures

A mixture of glycerol and water placed on an insulating plate is exposed to a radial hgih voltage field (25 kV). A slow formation of radial dendrites can be observed[23] as shown in Figure 20.

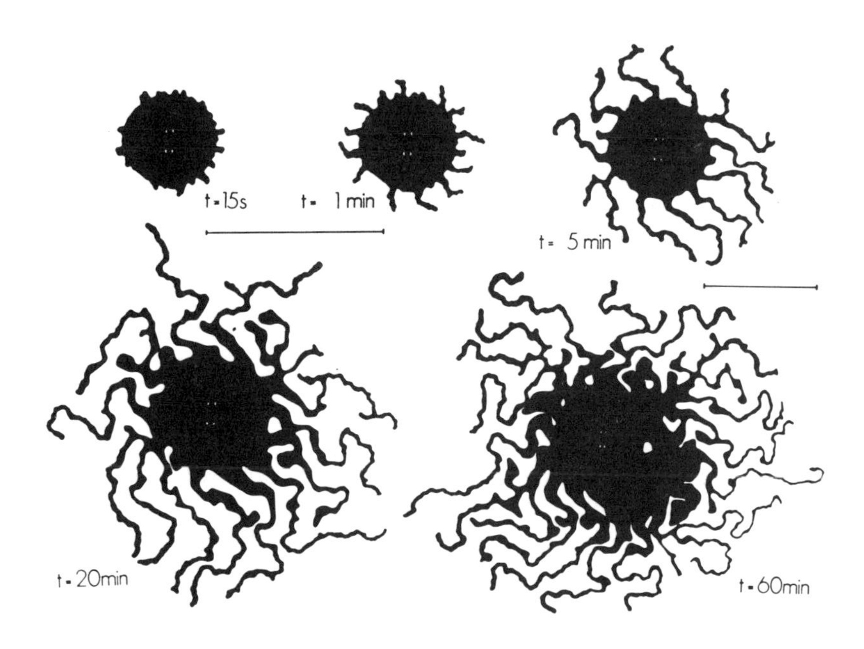

Figure 20. Time development of the dendritic growth in an electric field of a mixture of glycerol and water. The applied voltage was 25 kV.

In Figure 21 the fractal dimension in function of the raster width L is plotted with time as the parameter.

3rd Example: Metallic Particles in Electric Fields

A mixture of small steel balls or iron powder and castor oil contained in an acrylic box is placed between two capacitor plates. Applying a high voltage between the two plates, the iron particles start to move forming a non-periodic structure[24], Figure 22a. For the computer simulation it was assumed that the maximal growth of the electrically charged iron-strings takes place at the highest electrical field strength of the order 6 kV/cm. Long range, isotropic monopol interaction is simulated by using the diffusion model of Niemeier et al.[25]. Furthermore, a short-range anisotropic dipole-dipole interaction for the partial sticking probability was assumed.

The results of this simulation are shown in Figure 22b.

4th Example: Pattern Formation of Powder on a Vibrating Disc

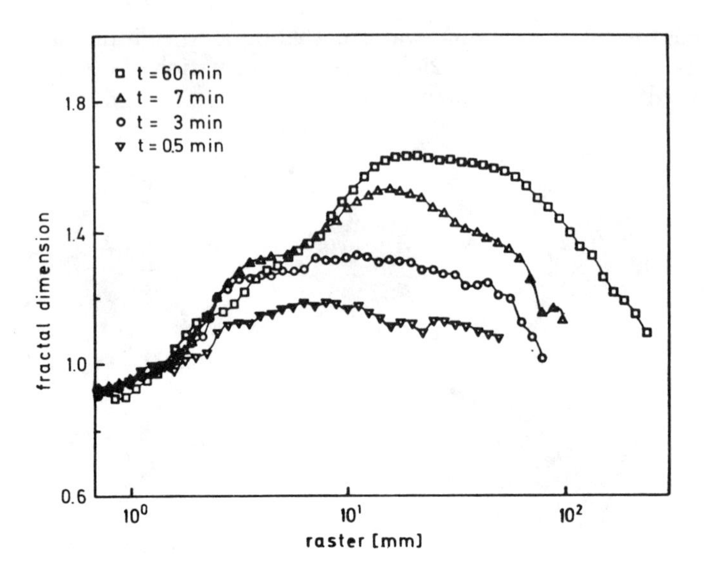

Figure 21. Fractal dimensions of the dendrites plotted against the scaling length with time as parameter.

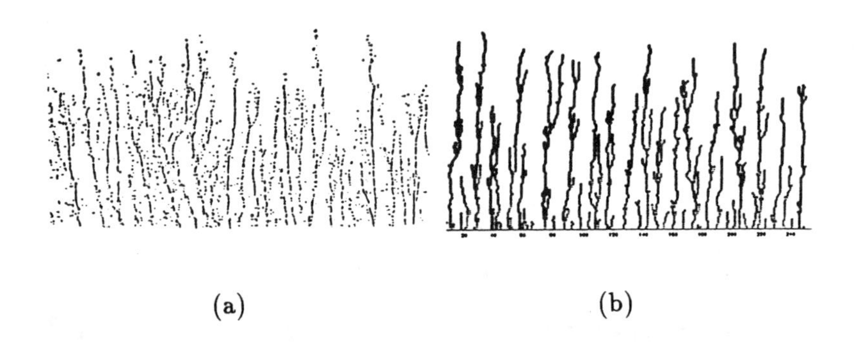

(a) (b)

Figure 22. Structuring of iron particles in an electric field of 5.7 kV/cm. (a) real experiment; (b) computer simulation.

Fine grain powder, placed on a periodic vibrating disc, shows dissipative patterns[26]. In Figure 23 the time-development of an initially cone-shaped powder "mountain" with an projected y-x-area of 2 cm^2 is demonstrated.

The correlation between maximum vibrating disc acceleration and the (x-y)-projections of the "hills" is plotted in Figure 24.

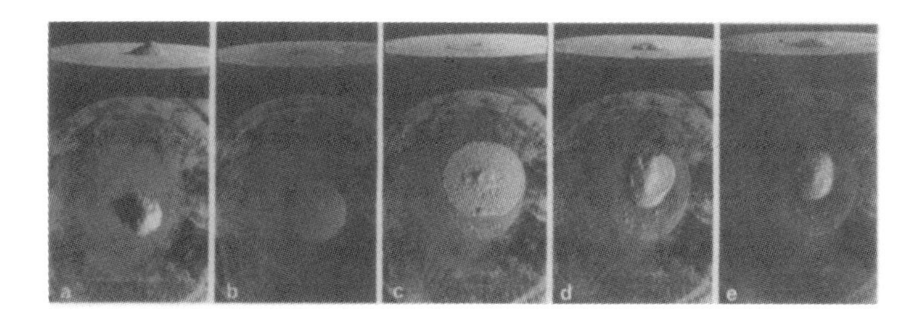

Figure 23. Powder structuring sequence of an initially cone shaped distribution. Vibrating frequency: 85/sec. Acceleration of the vibrating plate in units of g: (a): 0; (b): 3; (c): 8.25; (d): 8.25 after 10 sec; (e): 9.25.

The acceleration was increased every 30 sec and the vibrating frequency was 85/sec. A very sharp "phase transition" can be observed at a = 8g, g = 9.81 m/sec^2, where a flat area turns to a "hill."

For a computer simulation a hopping powder-particles model was developed. A diffusion-like term was introduced as a product of the hopping rate and the square of the hopping width. In Figure 25 four steps of a computer simulation of this pattern formation starting out with a "flat-mesa" are illustrated.

The list of examples could easily be expanded to a very large extend, because in the last ten years considerable progress has been made using combination of empirical approaches, numerical simulations and analytical approximations. It is for example possible to reconstruct differential equations using chaotic oscillations[27]. New methods for studying chaotic behavior permit a better understanding of the unpredictable. R.V. Jensen[28] compares chaotic dynamical systems with football games; even with the largest computer it is not possible to predict the outcome with certainty. The players themselves provide the fastest analogous computation of the evolution of this dynamical system. Roulette adicts may even learn from "chaos-physicists" how to improve their odds at the roulette table in Monte Carlo[29].

The growing literature and the desirability of including more non-linear phenomena in freshman or even in high-school physics may also reveal a very important ethic effect: only a very small number of problems of our world can exactly be solved by scientific methods. Restricted to science itself we have to ask the question, if the present description of nature with out macroscopic and microscopic models and theories is really adequate - an exciting program of research for many years to come.

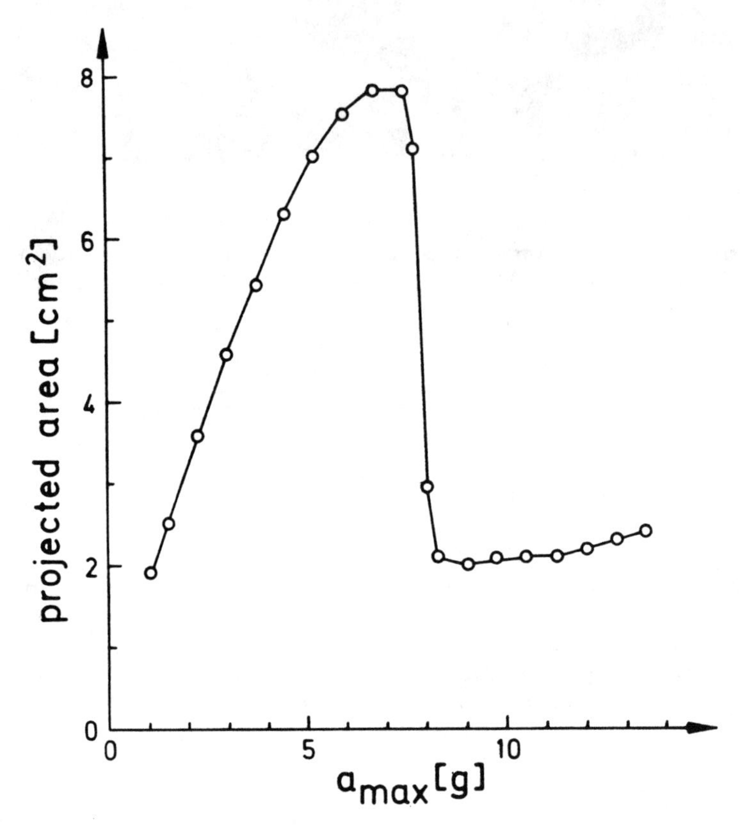

Figure 24. Projected area of the hilly landscape, initially cone shaped, as a function of the maximum acceleration $a_{\max}$. At $a_{\max} \approx 8$ g occurs a kind of phase transition.

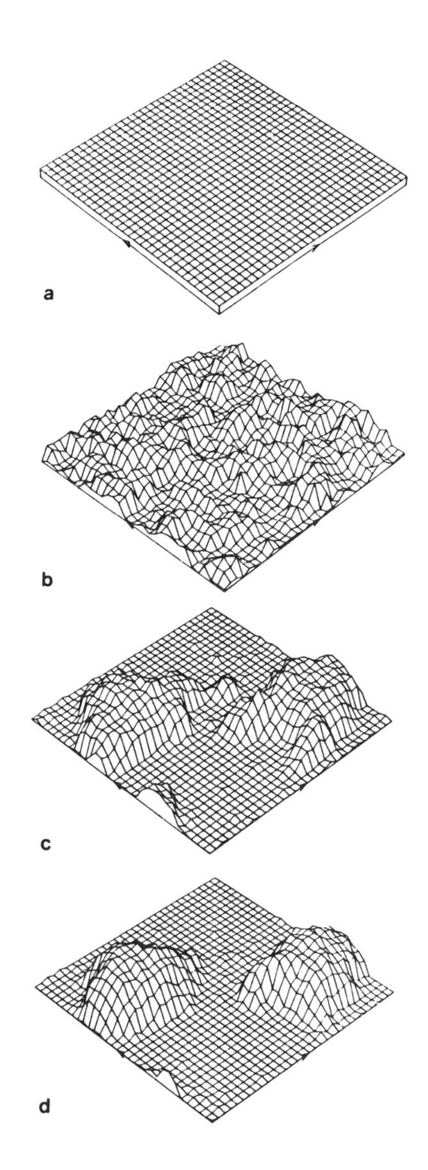

Figure 25. Computer simulation of the pattern formation (hilly landscape) from an initially flat distribution.

REFERENCES

1. H. Poincaré, Les métodoes nouvelles de la mécanique celeste, Paris, 1892.
2. E.N. Lorenz, J. Atmos. Sci. 20, 130 (1963).
3. D. Ruelle and F. Takens, Commun. Math. Phys. 20, 167 (1971); S. Grossmann and S. Thomae, Z. Naturf. 32*A*, 1353 (1977); J.P. Eckmann, Rev. Mod. Phys. 53, 643 (1981); H. Haken, Chaos and Order in Nature (Berlin, 1981); B.B. Mandelbrodt, The fractal geometry of nature (S.F., 1982); J.P. Eckmann, D. Ruelle, Rev. Mod. Phys. 57, 617 (1985).
4. R.M. May, Nature 261, 459 (1976).
5. P.R. Richter and H.J. Scholz, in *Stochastic Phenomena and Chaotic Behavior in Complex Systems*, P. Schuster ed. (Berlin, 1984).
6. M. Eigen and P. Schuster, Naturwiss. 65, 341 (1978).
7. H. Haken, Handbook of Physics, XXV/2c (Springer, Berlin 1970).
8. G. Julia, J. Math. Pures et Applic. 4, 47 (1918).
9. B.B. Mandelbrot, Ann. N.Y. Acad. Sci. 357, 249 (1980).
10. P.H. Richter and H.O. Peitgen, Morphologie komplexer Grenzen (Universität Bremen, 1984).
11. S.H. Lin, Sol. Stat. Phys. 39, 207 (1986).
12. W. Kropf, A. Hübler and E. Lüscher, to be published.
13. G. Feuerecker, A. Hübler and E. Lüscher, Biol. Cyb. 56, 57 (1987).
14. E. Bleuler, Lehrbuch der Psychiatrie (Springer, Berlin 1969).
15. M. Markus, S.C. Müller, B. Hess and Ber. Bunsenges. Phys. Chem. 89, 651 (1985).
16. K.A. Ross and C.R.B. Wright, Discrete Mathematics (Prentice Hall 1985).
17. S. Grossmann and S. Thomae, Z. Naturf. 32*A*, 1353 (1977).
18. M.J. Feigenbaum, J. Statistic. Phys. 19, 25 (1987).
19. J.D. Farmer, E. Ott and J.A. Yorke, Physica 7*D*, 153 (1983).
20. J.P. Eckmann and D. Ruelle, Rev. Mod. Phys. 57, 617 (1985); H.G. Schuster, Deterministic chaos (Physik-Verlag, Weinheim, 1984); E. Ott, Rev. Mod. Phys. 53, 655 (1981).
21. A.N. Kolmogorov, Dokl. Akad. Nauk. USSR 98, 527 (1959).
22. R. Resnick and D. Halliday, Physics, Part 1, (J. Wiley, NY, 1966), p. 504.
23. K. Klotz, A. Hübler and E. Lüscher, Helv. Phys. Act. 59, 124 (1986).
24. A. Rosenberger, A. Hübler and E. Lüscher, Biol. Cybern. 56, 151 (1987).
25. L. Niemeyer, L. Pietronero and H.J. Wiesmann, Phys. Rev. Lett. 52, 1003 (1984).
26. F. Dinkelacker, A. Hübler and E. Lüscher, Biol. Cybern. 56, 51 (1987).
27. T. Kautzky, A. Hübler and E. Lüscher, Helv. Phys. Acta 60, 222 (1987).
28. R.V. Jensen, Am. Sci. 75, 168 (1987).
29. T.A. Bass, The Endemonic Pie: or Why Would Anyone Play Roulette Without a Computer in His Shoe? (Houghton Mifflin, 1985).

SPONTANEOUS MIRROR SYMMETRY BREAKING IN NATURE AND THE ORIGIN OF LIFE

V.I. Goldanskii and V.V. Kuz'min
Institute of Chemical Physics
Academy of Sciences of the USSR
Ulitsa Kossygina 4, Moscow 117334, USSR

ABSTRACT

Theoretical and experimental studies of the problem of mirror symmetry breaking in Nature in the course of the origin of life are discussed.

The interplay of three main factors – statistical fluctuations, advantage factor (AF) and racemizing processes is taken into account.

The detailed analysis of existing data leads to the following scheme of main stages of the formation of life:

1. Prebiotic monomer formation, i.e. the formation and accumulation of various components of future biopolymers in racemic form ("warm"-terrestrial, and "cold"-extraterrestrial scenarios are discussed).
2. Chiral purity formation, i.e., the cooperative, bifurcational - type chiral polarization of the prebiotic medium. The sign of such polarization at the Earth (*D*-sugars and *L*-amino acids) seems to be accidental rather than determined by the violation of parity in weak interactions.
3. Formation of oligomeric tracers of chiral specific activity, i.e. the formation of protobiological structures (apparently, starting with RNA) up to several hundreds monomer units by matrix polymerization in chirally pure media.
4. Self-enzymatric block synthesis of biological macromolecules (RNA, DNA, proteins) and more complex systems.

The possibilities of the destruction of biosphere due to long-term racemizimg inpacts are also discussed.

INTRODUCTION

More than one hundred years have passed since Louis Pasteur discovered the "demarcation line" between the living and non-living: mirror dissymetry of organisms[1]. It would seem that any field of research, including this particular one, must have been exhausted during such a long period of time. Nevertheless, the fact that living organisms use only one of the two mirror isomers of such molecules as amino acids and sugars but do not use the othe isomer at all (nucleic acids comprise only *D*-isomers of sugars, while enzymes comprise only *L*-isomers of amino acids) is still an intriguing riddle.

This phenomenon – breaking of the mirror symmetry of the molecular basis of life – is, apparently, the first and one of the most dramatic examples of symmetry breaking in natural science. Interest to this seemingly purely biological problem was expressed to a greater extent by chemists and physicists than by biologists. P. Curie[2], Lord Kelvin[3], P. Jordan[4], E. Fischer[5], J. Bernal[6], and many others paid tribute to it.

Before entering into a discussion of the present-day state of the problem of mirror symmetry breaking in living nature, to be more exact, on its molecular level, we should like to remind the reader of some basic concepts which will be used in our further exposition.

Objects having no center and plane of inversion are called chiral (or mirror antipodal). If such an object is reflected in a mirror, the result will be an object incompatible in space with the initial one, as the left and right hands. By the way, the very term "chirality" has its origin in the Greek $\chi\varepsilon\iota\rho$, meaning "hand." The term was brought into scientific use by Lord Kelvin. To chiral objects there belong, in particular, molecules comprising so-called asymmetrical carbon atom: amino acids, sugars, etc. These molecules possess the property of chirality, if all the four substituents (ligands) associated with the central carbon atom are different (Fig. 1). Mirror isomers (enantiomers) of such molecules are usually called "left" and "right" isomers; in the biophysical literature they are denoted by the letters L (*Laevo* - left) and D (*Dextro* - right). We shall use these symbols in our paper. (It should be noted that in the chemical literature so-called R, S-nomenclature, adopted by the IUPAC, is used more often for denoting the mirror isomers). The mirror isomers are remarkable in that substances formed by L- or R-isomers alone (for instance, solutions, crystals) possess identical physical properties: the same internal energy, solubility, melting point, boiling point, etc. Their only physical difference consists in that these substances rotate the plane of polarization of the light passing through them in opposite directions, i.e., they possess optical activity. The chemical properties of the mirror isomers in their interaction with achiral molecules are also identical. It should be pointed out that, in addition to carbon, phosphorus and nitrogen atoms can also behave as an asymmetrical center. Molecules are known, whose chirality is caused not by the presence of some asymmetrical center (atom), but by the spatial structure, for instance, biphenyls (Fig. 2), having a "propeller" structure[7]. Sugars and amino acids, however, are of greatest interest as regards the problem of breaking the mirror symmetry of the molecular basis of the living.

Chiral purity, in contrast to the chirality of an individual small molecule (such as an amino acid molecule or a sugar molecule), is a property of an object (e.g. of a protein macromolecule or of a nucleic acid molecule) or of a medium consisting of monomers of molecules-isomers, formed by chiral molecules. This property resides in that such an object contains molecules of one type of chirality only (either L or D). In this case an object contains equal quantities of mirror antipodes, it is racemic.

In non-living nature the left and right molecules are present in equal quantities, whereas in living nature a sharp unbalance between the right and left isomers is observed: as stated above, the molecular basis of the living is chirally pure. Thus, chiral purity, alongside of the genetic code, common to all the living on the Earth, is a distinctive and key property of the living.

Naturally, not a single scenario of the origin of life can be "written" without solving the problem about the origin of chiral purity (or, as may be stated in other words, about mirror symmetry breaking). In this connection, for more than one hundred years attempts have been made to answer that question. Such as unabated interest to the problem is explainable also by the fact

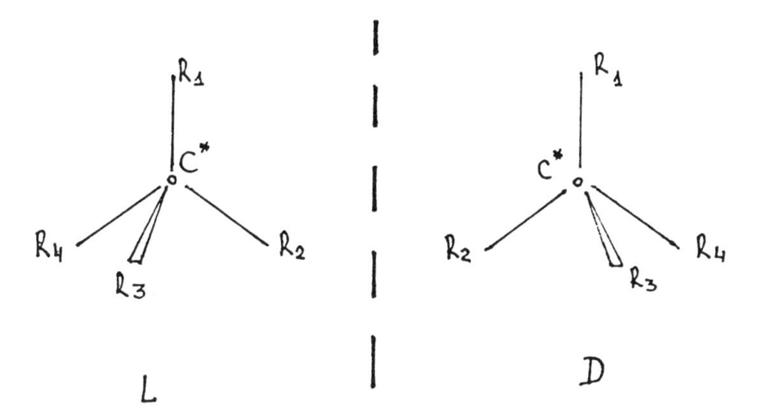

Figure 1. Mirror isomers (L and D) of molecule with asymmetrical carbon atom (C*).

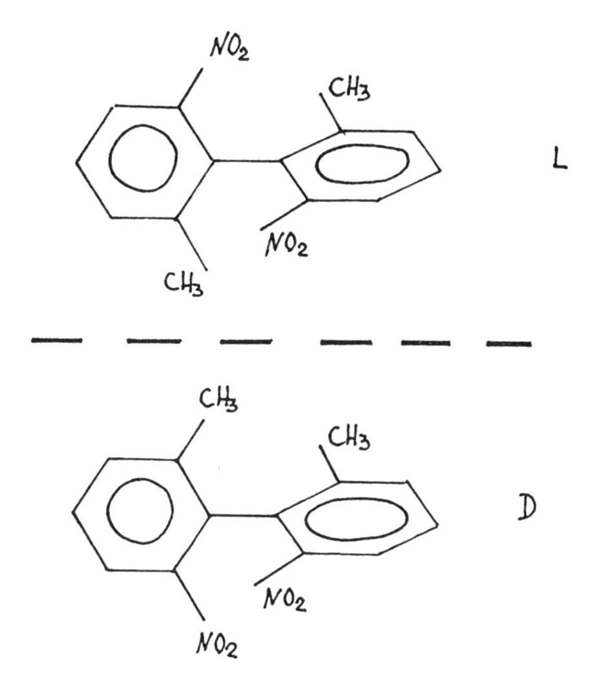

Figure 2. Mirror isomers of binaphthyl molecule (chirality is caused by "propeller" structure of molecule).

that the chiral purity of biopolymers is a necessary condition for their normal functioning. For this very reason the problem concerning the origin of this property becomes one of the key ones in the understanding of processes leading to the formation of self-replicating biological systems.

Until recently one could distinguish two main approaches to solving the problem of the origin of chiral purity. One of them is based on the idea of biogenic origin of this property. The essence of this approach consists in that one type of isomers was selected in the course of the vital activity of first organisms[8]. It is assumed that chirally pure primary organisms, built up of the same isomers as the contemporary ones, had a certain evolutionary advantage over racemic organisms built up of a racemic mixture of isomers, and the first have won in the course of struggle for existence. Such an approach, however, involves principal difficulties discussed in Section 2.

The second approach is based on the idea of an abiogenic origin of the chiral purity at the prebiological, chemical stage of evolution, due to physico-chemical processes occurring in the "primordial soup."

The choice between these two approaches can be made proceeding from the solution of an "auxiliary" problem, whether the property of self-replication of molecular structures (of the type of proto-nucleic acids) can originate in a racemic medium. If the answer is positive, room is left for the version of biogenic origin of the chiral purity; if the answer is negative, the mirror symmetry breaking occurred in an abiogenic manner. An analysis of this question is a central point in Section 2.

Within the framework of the abiogenic approach two trends have been developed for solving the question about the origin of the chiral purity. One of these trends relates the mirror symmetry breaking of the organic medium with the action of some asymmetrical factor ensuring an advantage to one type of mirror isomers; a gradual "accumulation" of this advantage led to the dominance of the isomer of this type. Such a scenario of mirror symmetry breaking may be called a scenario of successive evolution or simply an "evolutionary scenario." This approach is discussed in Section 5.

A totally different way of the origin of the chiral purity of the organic medium is considered within the framework of a scenario of spontaneous mirror symmetry breaking at the stage of chemical evolution. Its essence is that the process of deracemization of the prebiosphere is associated with self-organization of chirality in the "primordial soup" rather than with the external assymetrical action. The attaining of chiral purity by the initially racemic medium occurs as a kind of a nonequilbrium "phase transition." It should be noted that behind this scenario is not only a theoretical analysis of the kinetics of processes in nonequilibrium chiral systems (Section 4), but also the cooperative character of the behavior of even low-viscosity solutions of mirror isomers, observed experimentally[9–12]. The scenario of spontaneous mirror symmetry breaking is discussed in Section 6. In the same Section we shall analyze the question about the possibility of intensification of weak asymmetric actions by systems with spontaneous mirror symmetry breaking, which is a subject of intensive discussions nowadays. Interest to this problem is associated, first of all, with the hypothesis that deracemization of the prebiosphere is connected with parity nonconservation in weak interactions of elementary particles.

Considering the process of deracemization of the prebiosphere as a kind of a "phase transition", it is possible to assess the physical conditions necessary for the attaining of chiral purity and, in partiuclar, the expectation time for the commencement of irreversible mirror symmetry breaking. Section 7 is devoted to this question.

In recent years a renewed interest in "cosmic" scenarios of the origin of life on the Earth has been inspired again. On the one hand, this is due to the discovery of sufficiently complex organic compounds in space (up to amino acids), this being indicative of the synthesis of such molecules proceeding under extreme conditions from the view point of terrestrial chemistry[13,14]. On the other hand, impressive geological discoveries made in the last years definitely point to the existence of a sufficiently developed biosphere on the Earth as long ago as 3.5 – 3.8 billion years[15,16]. Since the origin of life is possible only after the formation of solid Earth's crust, which took place 4 billion years ago, the duration of the prebiological stage of evolution cannot exceed 0.2 – 0.5 billion years. Such severe time limitations for the process of bipoiesis (the process of the origin of life), covering the period from the formation of "primordial soup" comprising all the necessary molecular "bricks" of the living to protocells and primitive organisms, but some researchers in doubt as to the truth of the "standard" model of the origin of life – the theory of Oparin-Haldane[17–18]. It should be noted, however, that the hypothesis of extraterrestrial origin of life dating back to Arrhenius' "panspermia" as such does not clear away the basic question of this problem – the question about the processes capable of leading to the appearance of the key properties of the living and, first of all, to the appearance of chiral purity of the organic medium. The hypothesis of the "cold prehistory of life" must be analyzed from this standpoint. The possibilities of mirror symmetry breaking in such a way are considered in Section 8, where the attention is focused on the "stabilization" of chirality under the conditions of low temperatures (when chemical reactions proceed by the tunneling mechanism).

Thus, the present review is devoted, mainly, to the physical aspects of the problem of mirror symmetry breaking in the course of prebiological evolution of the organic medium. We believe that invoking a physical ideas and methods for solving this problem not only promotes a more thorough understanding of the very problem of the origin of the chiral purity of the biosphere, but also makes it possible to analyze the processes responsible for the course of biopoiesis.

2. CHIRAL PURITY AND SELF-REPLICATION

The biogenic scenario of the origin of chiral purity assumes implicitly that the functioning of the information carrier – of the nucleic acid molecule – does not depend in principle on the chiral composition of its primary sequence. Let us consider this question in more detail. As is known, the DNA molecule is a double-stranded structure, whose primary sequence is formed by nucleotides (phosphate group + sugar molecule + nitrogen base) (Fig. 3). The strands are interconnected in accordance with the rule of complementarity: each of the four nitrogen bases – adenine (A), guanine (G), thymine (T), and cytosine

(C) – forms bonds only with the base complementary to it, namely A-T, G-C. As stated above, only *D*-isomers of sugar molecules are used in nucleic acids. Is there a connection between the chirality of sugars and the property of complementarity of the double helix of the nucleic acid? The answer to this question was given in Refs. 19 and 20.

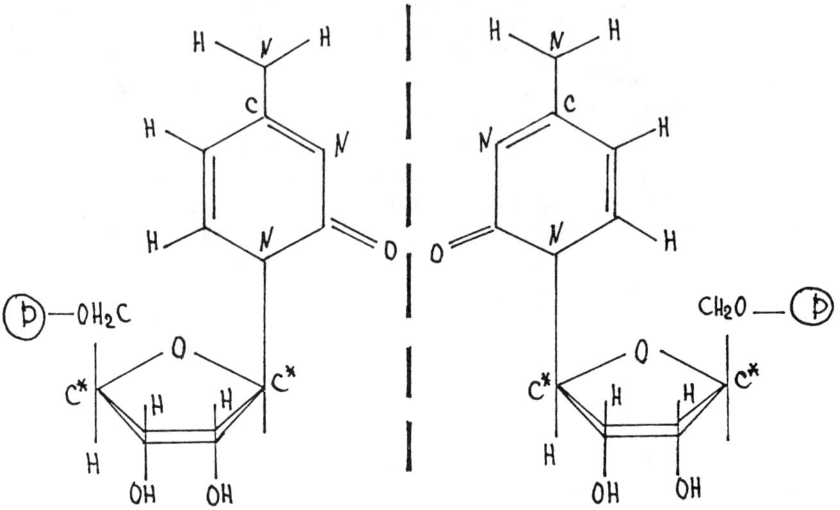

Figure 3. Mirror isomers of nucleotides (sugars possess opposite chirality).

It seems almost apparent that replacement of a unitary naturally occurring *D*-nucleotide (i.e. of a nucleotide containing *D*-isomer of the sugar molecule) in the double-stranded structure (for instance, DNA) by its mirror *L*-isomer will lead to the formation of a structure with a greater energy. This raises the question whether it is possible to preserve the complementarity of such a chirally deficient pair in the double-stranded structure, and, moreover, is such a defect local or does it destroy a large domain of the polynucleotide chain? A qualitative answer to these questions can be obtained from a molecular model constructed to simulate this situation. Corey-Pauling-Koltun spaced molecular models are most suitable for the purpose.

For the sake of comparison two fragments of a double-stranded structure have been constructed, each fragment comprising five complementary paris A-T. The nucleotide strands of each fragment contained only one type of nucleotides (poly-A strands and poly-T strands respectively). These gragments simulate the polynucleotide chain characteristic of DNA (or RNA, if T-base is replaced by U-base and deoxyribose by ribose).

In the first fragment both strands were chirally pure. It is natural that such a structure is regular (Fig. 4) and forms a fragment of so-called B-form of the usual double helix (Fig. 5). The second fragment differs from the first

Figure 4. Homochiral double-stranded A-T–structure.

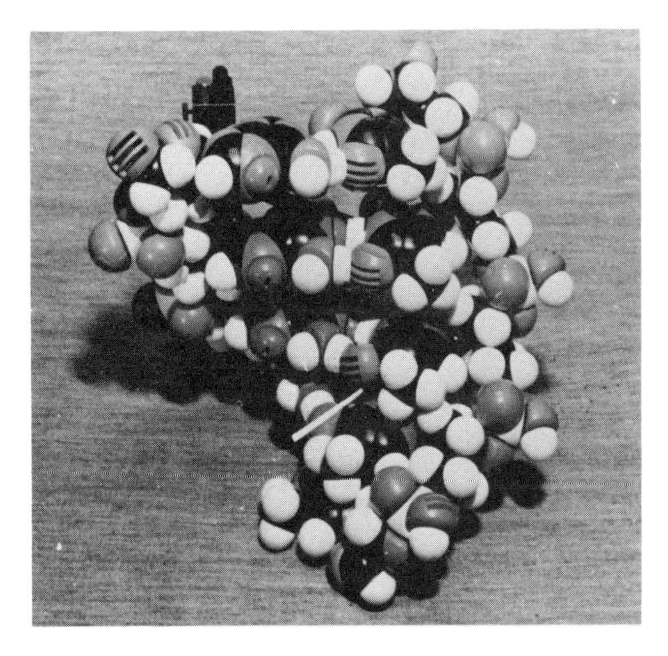

Figure 5. Fragment of double helix: poly-A – poly-T.

in that the T-nucleotide in the third unit is replaced by its mirror isomer (i.e. *D*-sugar is replaced by the unnatural *L*-sugar). This nucleotide was inserted in such a manner that the integrity of the poly-T strand should be preserved (Fig. 6). This structure simulates a chiral defect in the polynucleotide chain.

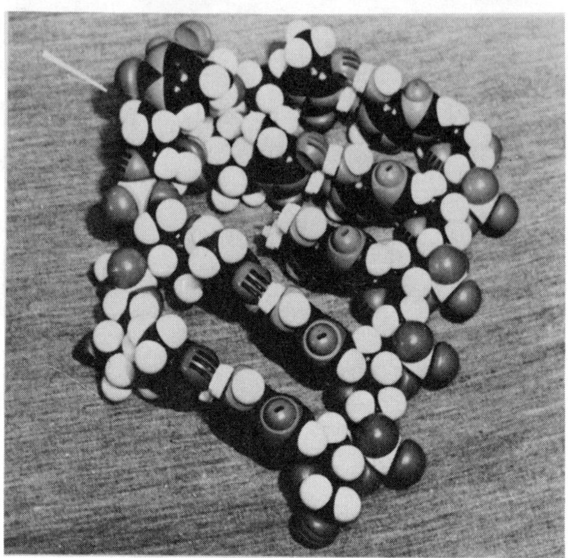

Figure 6. Double-stranded structure with chiral defect.

The model of the B-form containing a chiral defect is shown in Fig. 7. In such a structure the position of the pyrimidine ring of the "defective" T-nucleotide proves to be turned with respect to the normal position of the nitrogen base in the double helix through an angle $\sim 100°$! This rules out the possibility of the formation of hydrogen bonds between the bases A and T of the defective pair and, hence, removes the main cause of the origin of the A-T (and, similarly, of G-C, or A-U for RNA) complementarity, since in this case the combination of the nucleotide bases of the deficient pair may be arbitrary. Such a situation takes place also in the case of another occurring form of the double helix, so-called A-form containing a chiral defect.

It is easy to understand what will happen upon insertion into a homochiral structure of a unit consisting of a pair of nucleotides, the chirality of both partners of which is opposite to the chirality of other nucleotides. In this case the formation of hydrogen bonds between the bases of the inserted foreign pair will not take place either.

Thus, the chiral purity of the nucleotides is a necessary condition of complementarity.

In the absence of enzymatic mechanisms of replication (characteristics of comtemporary organisms) this problem becomes quite topical. In particular, the formation of a chiral defect can take place during the process of matrix oligomerization. Such a process of abiogenic formation of polynucleotides is characteristic of the prebiological evolution conditions[21] and is of considerable

Figure 7. Fragment of double helix with chiral defect.

interest, especially after it has been found that RNA (highly probable ancestor of DNA) possesses autocatalytic activity and the ability to self-replicate in the absence of a special enzymatic apparatus[22,23].

However, after an unambiguous connection between complementarity and the chiral purity of the polynucleotide chains has been established, the following question arises. Suppose that an individual chirally pure polynucleotide chain (proto-RNA) has assembled in a random manner in a racemic medium. Will this chain be able to self-replicate in such a medium by way of matrix oligomerization? Experiments for studying the influence of the chiral composition of the medium on the process of matrix oligomerization were carried out in 1984[24–26]. In those experiments it was found that the process of assembling of a complementary chain on a chirally pure matrix was strongly inhibited if the solution of monomers was racemic. This result can eaisly be understood, if we take into consideration the connection of the chirality of the polynucleotides with their complementarity. Moreover, experiments of such type make it possible to construct a qualitative model, relating the process of replication with the chiral composition of the medium[27,28].

The essence of the experiments was as follows. Synthesized polynucleotide chains (matrices) were placed into a solution of nucleotide-monomers, which in one case was chirally pure and in the other case, racemic. After a definite period of time the result of the process of matrix oligomerization of the nucleotides on the matrices was studied, i.e. the dependence of the length of chirally pure "replicas" on the chiral composition of the monomeric medium was studied.

The distributions of the chirally pure polynucleotides, accumulated during the experiment, depending on the polymer length n, given in Refs. 24 and 25 (Fig. 8), may be interpreted as the distributions of the probabilities of formation of such polymers in the medium with the definite value of chiral polarization $\eta = (L - D)/(L + D)$[29], where D and L are the quantities of the right and left monomers in the medium; for the racemate $\eta = 0$, while for the chirally pure medium $|\eta| = 1$. The form of these distributions, at least starting with a certain n', may be approximated by an exponential distribution

$$P(n, \eta) \sim \exp[-\alpha(\eta)n]$$

with the exponent α diminishing as η increases. The data reported in Refs. 24 and 25 demonstrate that the value of the coefficient α for the racemic medium $\alpha(\eta = 0) = \alpha_0$ exceeds the value α for the chirally pure medium $\alpha(\eta = 1) = \alpha_1$.

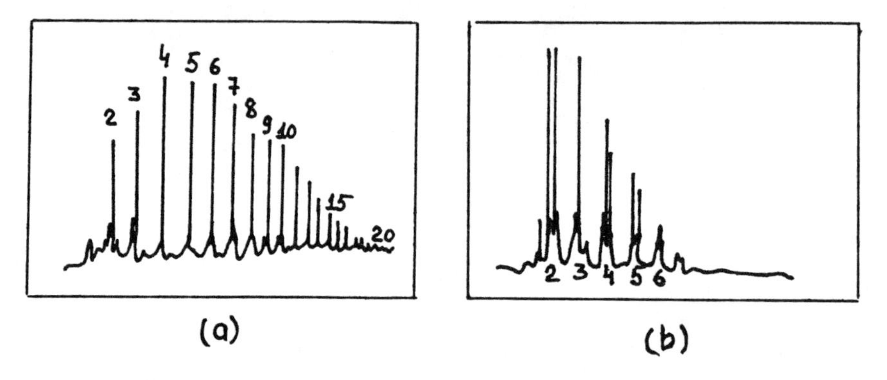

Fig. 8. Distribution of replicas synthesized on chirally pure matrix: (a) replication in chirally pure solution of nucleotides; (b) replication in racemic solution of nucleotides[24].

Such an interpretation of the results of the experiments on the matrix oligomerization allows one to construct a simple model, on the basis of which it will be possible to analyze the question concerning the ability of the medium with chiral polarization η to ensure self-replication of polynucleotides having a definite length n.

Suppose that some quantity of chirally pure nucleotides having a length n has originated in such a medium in a random manner. We have already seen that a prerequisite condition of the complementarity of each unit of the initial matrix and of the replica is the same chirality of the nucleotides, entering into these units in pairs (for instance, A-T, G-C pairs). Then the condition indispensable for the formation of replica of any essential length at any η is the chiral purity of the initial polynucleotide matrix (at $\eta = 1$ this is specified by the initial condition itself; at $\eta \neq 1$ this is a consequence of the exponential inhibition demonstrated in Refs. 24 and 25). For the sake of simplicity, in what follows we shall not distinguish the replica from the initial matrix (this attitude is justifiable, since at a subsequent stage the replica itself becomes a matrix). We shall introduce a constant $K_n^+(\eta)$ for the rate of replication of polymers having a length n in a medium with chiral polarization η and a constant $K_n^-(\eta)$ for the rate of their demise. We understand the process of demise of a chirally pure polymer to consist not only of its destruciton, but also of any processes which result in the loss of the property of self-replication by a polymer, e.g., due to the appearance of a monomeric unit with an antipodal chirality in the polymeric chain. We shall write down a dynamic equation for the number x_n of chirally pure polymers having length n as

$$\frac{dx_n}{dt} = [K_n^+(\eta) - K_n^-(\eta)]x_n \quad (1)$$

We shall call the replication process stable if $dx_n/dt > 0$ (x_n grows with time) and unstable if $dx_n/dt < 0$. Equation (1) is linear, and the replication rate constant $K_n^+(\eta)$ differs from the value $P(n,\eta)$ only in the dimensional factor s^{-1}. Consequently, it is possible to choose the simplest interpolation of $K_n^+(\eta)$, using the linear dependence of the exponent α on η:

$$K_n^+(\eta) = A_n \exp[(\alpha_0 - \alpha_1)n\eta] \quad (2)$$

where $A_n = A\exp(-\alpha_0 n)$ is the constant for the rate of replication of a polymer of a length n in a racemic medium. For $K_n^-(\eta)$ the following simplest approximation may be adopted: $K_n^-(\eta) = K^- =$ constant, which corresponds to an assumption that the characteristic time of polymer decomposition ($\tau_d = 1/K^-$) is dependent neither on the polymer length nor on the chiral polarization of the medium η. Since $K_n^-(\eta)$, generally speaking, must increase both as n grows and as η diminishes, the actual dependence of $K_n^-(\eta)$ on these values can only strengthen the results obtained below. From (1) and (2) it follows that the process of replication of chirally pure polymers of length n in a medium with chiral polarization η is stable if

$$K_n^+ = A_n \exp[(\alpha_0 - \alpha_1)n\eta] > K^-$$

i.e.

$$\eta > \frac{ln(K^-/A) + \alpha_0 n}{(\alpha_0 - \alpha_1)n} \equiv \eta_n^{cr} \quad (3)$$

It should be noted that if for some values $\alpha_0, \alpha_1, K^-, A$ and $n, \eta_n^{cr} > 1$, this means that the process of replication is unstable in the medium with any chiral polarization, since $|\eta| \leq 1$ by the very definition of the value η. We should also like to point to a weak (logarithmic) dependence of η_n^{cr} on variations of the demise rate constant K^- and a considerably stronger dependence on the polymer length n.

Let us estimate the value $\Delta\eta_n^{cr} = (1 - \eta_n^{cr})$, describing the deviation η_n^{cr} from the polarization of the chirally pure medium. From Equation (3) we obtain

$$\Delta\eta_n^{cr} = \frac{(1/n)ln(A/K^-) - \alpha_1}{\alpha_0 - \alpha_1}. \quad (4)$$

Equation (4) demonstrates that as the polymer length increases, the "gap" $\Delta\eta_n^{cr}$ becomes all the narrower: for a stable replication of longer polymers, a much closer approximation to the chiral purity of the medium is required. Moreover, from Equation (4) it can be seen that there exists a maximum length $n_{\max}$ of polymers capable of stable replication. For all polymers having a length $n > n_{\max}$ the process of replication is unstable even in a chirally pure medium. One can find $n_{\max}$ by putting in Equation (4) $\eta_n^{cr} = 1$:

$$n_{\max} = (1/\alpha_1) ln(A/K^-).$$

The upper limit of $n_{\max}$ values can easily be obtained by using rough estimates of the parameters α_0, α_1 and A, which can be derived from the experimental data given in Refs. 24 and 25 for the nculeotides: $\alpha_0 \sim 3, \alpha_1 \sim 10^{-1}, A \sim 10^{-5}s^{-1}$ and by assuming that the characteristic time of decomposition of these polymers τ_d is commensurable with the time of existence of the Earth, i.e. by adopting an obviously underrated value $K^- \approx 10^{-17}s^{-1}$. This gives $n_{\max} \sim 3 \times 10^2$. Thus, polynucleotides consisting of more than 10^3 monomeric units are not capable of stable replication even in a chirally pure medium. Chirally pure polymers with $n > 300$, capable of replication, originated, apparently, already at further stages of prebiological evolution as a result of interaction of chirally pure polynucleotides with length $n \lesssim 300$ with chirally pure oligopeptides and/or polypeptides (and, possibly, with protoenzymes as well). It should be emphasized that the length of polymers capable of stable replication diminishes rapidly as η decreases, and in a racemic medium it amounts only to $n_{\max} \sim 10^1$ (with the fact that K^- is underrated being taken into account).

These qualitative estimates, in combination with the established stringent connection between the chiral purity of polynucleotides and their property of complementarity, suggest a conclusion that the origin of even simplest self-replicating systems could occur only in a medium with strong mirror symmetry breaking, practically, in a chirally pure medium only. Consequently, it was just at the prebiological stages of evolution that the physico-chemical processes leading to global breaking of the mirror symmetry of the organic medium had to be necessarily realized.

These results allow us to come to a conclusion that the biogenic scenario of the orginization of chiral purity of the biosphere could not, even in principle, be realized in the course of evolution, since without the chiral purity of the medium, the apparatus of self-replication, which is the basis of the process of self-reproduction of any organisms, cannot appear: life cannot appear in a racemic medium.

Moreover, scenarios in which chiral purity of the prebiosphere is associated with the stage of selection on the level (to be exact, at the stage) of formation and selection of the precursors of the most important biopolymers, prove to be unacceptable either. As shown above, the process of their formation and accumulation cannot proceed in a racemic medium.

Thus, an abiogenic scenario remains the only real scenario of the origin of the chiral purity. All the subsequent sections of the present review are devoted to a detailed consideration of the various aspects of this approach to the problem of mirror symmetry breaking.

3. RACEMIZATION, HUND'S PARADOX, AND PROBLEM OF DERACEMIZATION

For a long time, it was believed that any chemical transformations leading to the formation of chiral molecules and any transformations of the chiral molecules themselves could lead only to their racemic mixture. Most widespread arguments can be summarized as follows. Consider the reaction of formation of a chiral molecule from an achiral one, namely, the reaction of substitution of one of two similar ligands Y at a tetrahedral atom of carbon by a ligand R (Fig. 9). It is quite obvious that the ligands Y are on a par and in such a reaction the substitution of either ligand is equiprobable. Consequently, the result will be a racemic mixture of L and D isomers.

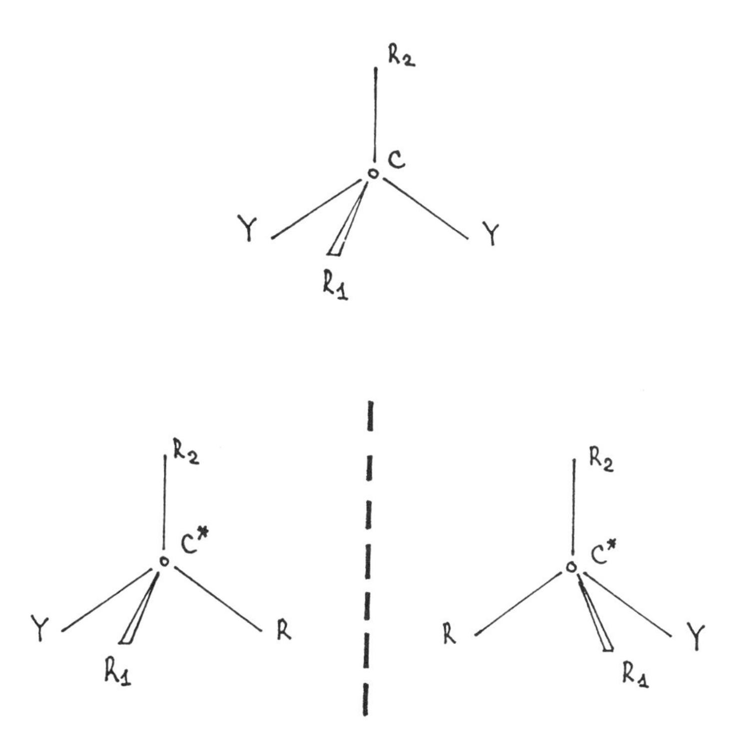

Fig. 9. Reaction of formation of chiral molecule from achiral precursor.

It is also easy to see that under symmetrical conditions (in an isotropic space) to the minimum of the thermodynamic potential $\mathcal{F}$ of a system of mirror isomers with their concentrations x_L and x_D

$$\mathcal{F} = N_K T(x_L ln x_L + x_D ln x_D) + U(x_L + x_D)$$

there corresponse a racemic mixture, since in this case the entropy of mixing is maximal (U is the internal energy of the molecules-isomers). Therefore, even if at the initial moment of time one of th eisomers was in excess, sooner or later, as a result of chemical transformations, the system will come to the racemic state, corresponding to the thermodynamic equilibrium. The system is said to racemize.

The process of racemization (or inversion of antipodes) may be represented in the form of the following reaction:

$$L \underset{k_r}{\overset{k_r}{\rightleftarrows}} D \tag{5}$$

(k_r being the racemization rate constant). This scheme describes not only the process of direct inversion of the antipodes, but also other processes, effectively reducible to Scheme (5). At temperatures on the order of room temperature (and higher) racemization occurs due to the superbarrier transition, i.e. it is conditioned by thermal excitations. The process of racemization can also be induced, e.g. by radiations – so-called photo- and radioracemization. It should be noted that the characteristic time of racemization is strongly dependent on external conditions (temperature, pH of the medium, presence of the ions of heavy metals in the medium, etc.)[30–34]. Some experimental data on racemization are listed in Table 1. It should be pointed out that as far back as 1927 Hund[35] paid attention to the fact that racemization was, so to speak, inherent to chiral molecules. The fact is that mirror isomers represent, from the standpoint of quantum mechanics, an energy-degenerated state (Fig. 10), and, consequently, they are not eigenstates of the molecule. That is $|L\rangle$ and $|D\rangle$ states possess chirality, but they do not possess a definite parity. The eigenstates of the molecule capable of existing in two mirror-isomeric forms are symmetrical and antisymmetrical combinations of the $|L\rangle$ and $|D\rangle$ states

$$|+\rangle = (1/\sqrt{2})(|L\rangle + |D\rangle)$$
$$|-\rangle = (1/\sqrt{2})(|L\rangle - |D\rangle) \tag{6}$$

which possess a definite parity rather than chirality. Consequently, the molecule "oscillates" between L and D states by tunnelling with the characteristic frequency

$$\Omega = \tau^{-1} \exp[-(2VM)^{1/2} Q] \tag{7}$$

where M is the mass of the molecule, V is the height of the energy barrier separating the L and D states, Q is its width, τ is the characteristic time of intramolecular motions. Consequently, if at the initial moment of time the molecule was, say, in the L state, then after the time $T > \Omega^{-1}$ it will be in the state D. Thus, during the times $t > T$ the probability of finding the molecule either in the L or in the D state will be the same and equal to 1/2. This means that the system of isomers at times $t \gg T$ is racemized.

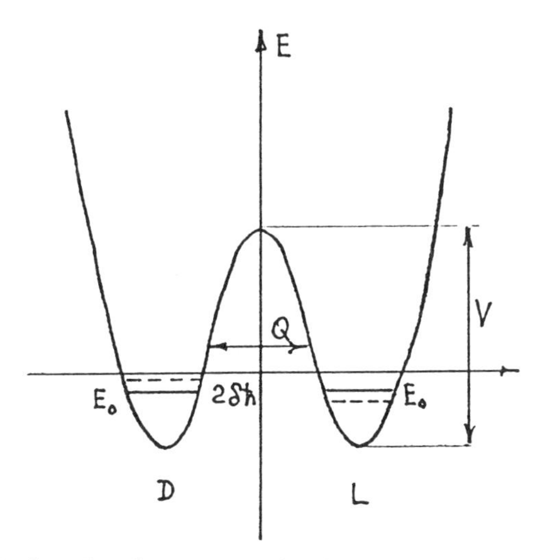

Fig. 10. Chiral molecule as a two-level system. x–collective coordinate of molecule, V–height of the barrier separating L and D state, Q–width of the barrier.

Table 1
Characteristic Times of Racemization of Amino Acids

Amino Acid	Temperature (°C)	pH of Medium	Racemization Time (Years)
Alanine	0	7	1.4×10^{6}
	25	7	1.2×10^{4}
Aspargic Acid	0	7	4.3×10^{5}
	25	7	3.5×10^{3}
Valine	135	7	2.5×10^{-1}
	135	0.9	1.33
Isoleucine	0	7	6.0×10^{6}
	25	7	4.8×10^{4}
	100		1.6
	200		5.0×10^{-5}
Phenylalanine	0	7	1.6×10^{5}
	25	7	2.0×10^{3}

Racemization processes oppose mirror symmetry breaking by tending to return the system to the racemic state. Consequently, racemization could occur only if there exist processes capable of effectively opposing the tendency to racemization.

4. MAIN TYPES OF PROCESSES IN CHIRAL SYSTEMS

Suppose there is a system containing chiral (L and D) molecules (with concentrations x_L and x_D) and achiral molecules $A, B, \ldots$ (with concentrations $x_A, x_B, \ldots$), which may undergo various chemical transformations. The kinetics of the reactions leading to changes in the concentrations of chiral molecules is described by equations

$$\frac{dx_L}{dt} = f_L(x_i, \{k_i\})$$
$$\frac{dx_D}{dt} = f_D(x_i, \{k_i\})$$

where $x_i = (x_L, x_D, x_A, \ldots)$, and $\{k_i\}$ is a set of the reaction rate constants. Usually models of chiral systems are written down and studied in variables x_L and x_D. However, as shown in Refs. 29 and 36, the dynamics of chiral systems can be studied much more conveniently in the variables of chiral polarization

$$\eta = (x_L - x_D)/(x_L + x_D)$$

and dimensionless concentration of the antipodes in the system

$$\theta = \nu(k_i)(x_L + x_D)$$

where $\nu(k_i)$ is the coefficient defined by the reaction rate constants.

Chiral polarization η plays for chiral systems the role of the parameter of order[29]: to the completely ordered, chirally pure state there corresponds $|\eta| = 1$, while to the racemic state there corresponds $\eta = 0$. Investigation of the behavior of chiral systems in the variables (η, θ) and (x_L, x_D) is completely equivalent, since the transformation $(x_L, x_D) \rightarrow (\eta, \theta)$ changes neither the number nor the type of singular points of the system.

Depending on the character of chemical transformations in the system, i.e. depending on the processes in the chiral subsystem and on the interaction of the chiral and achiral subsystems, one can specify the main types of dynamic processes in chiral systems. Altogether, there are three such processes (see Table 2).

(a) Racemizing

The process belonging to this type cause relaxation of chiral ordering to racemic state. From Table 2 it is seen that these are, in the first place, reactions of zero order with respect to chiral reagents and, in the second place, the process of racemization proper and processes stemming from the non-absolute stereoselectivity of chemical reactions with the participation of chiral molecules. The general form of dynamic equation for the processes of racemizing type (blocks III, V, VI, VII, XI in Table 2) can be written down as:

$$\left(\frac{d\eta}{dt}\right)_r = -f(\eta).$$

Table 2
MAIN TYPES OF DYNAMIC PROCESSES IN CHIRAL SYSTEMS

Reactions		Final State ($\eta^{(s)}$)		Deterministic equations in the (η,θ) space	
Blocks	Scheme	Symm. Case	Assymm. Case $g \neq 0$	Symmetric Case	Asymmetric Case $g \neq 0$
1	2	3	4	5	6
I	synthesis (r) $A \xrightarrow{k^L} L$ $A \xrightarrow{k^D} D$	$\eta^{(s)} = g$	$\eta^{(s)} = g$	$\dot{\eta} = -\frac{kA}{\theta}$ $\dot{\theta} = 2$	$\dot{\eta} = \frac{kA}{\theta}(g - \eta)$ $\dot{\theta} = 2kA$
II	autocatalysis (n) $A + L \xrightarrow{k^L} 2L$ $A + D \xrightarrow{k^D} 2D$	$\eta^{(s)} = \eta_0$	$\eta^{(s)} = \lvert 1 \rvert$	$\dot{\eta} = 0$ $\dot{\theta} = k\theta A$	$\dot{\eta} = kAg(1 - \eta^2)$ $\dot{\theta} = kA\theta(1 + g\eta)$
III	"erroneus" autocatalysis (r) $A + L \xrightarrow{k_m} L + D$ $A + D \xrightarrow{k^D} L + D$	$\eta^{(s)} = 0$		$\dot{\eta} = -2k_m A\eta$ $\dot{\theta} = k_m A\theta$	
IV	destruction (n) $L \xrightarrow{k^L} A$ $D \xrightarrow{k^D} A$	$\eta^{(s)} = \eta_0$	$\eta^{(s)} = \lvert 1 \rvert$	$\dot{\eta} = 0$ $\dot{\theta} = -k\theta$	$\dot{\eta} = kg(1 - \eta^2)$ $\dot{\theta} = -k\theta(1 + g\eta)$
V	racemization (r) $L \underset{k_r}{\overset{k_r}{\rightleftarrows}} D$	$\eta^{(s)} = 0$		$\dot{\eta} = -2k_r\eta$ $\dot{\theta} = 0$	
VI	pair destruction (r) $L + L \xrightarrow{k^L} A + B$ $D + D \xrightarrow{k^L} A + B$	$\eta^{(s)} = 0$	$\eta^{(s)} = g$	$\dot{\eta} = -\frac{k}{2}\theta(\eta - \eta^3)$ $\dot{\theta} = -\frac{k}{2}\theta^2(1 + \eta^2)$	$\dot{\eta} = \frac{k}{2}\theta(g - \eta)(1 - \eta^2)$ $\dot{\theta} = \frac{k}{2}\theta^2(1 + \eta^2) + kg\theta^2\eta$
VII	pair racemization (r) $L + L \xrightarrow{k^L} L + D$ $D + D \xrightarrow{k^D} L + D$	$\eta^{(s)} = 0$	$\eta^{(s)} = \frac{g}{2}$	$\dot{\eta} = -2k\theta\eta$ $\dot{\theta} = 0$	$\dot{\eta} = k\theta(g\eta^2 - 2\eta + g)$ $\dot{\theta} = 0$
VIII	"annihilation" (dr) (antagonism) $L + D \xrightarrow{k_a} A + B$	$\eta^{(s)} = \lvert 1 \rvert$	$\eta^{(s)} = \lvert 1 \rvert$	$\dot{\eta} = \frac{k_a}{2}\theta(\eta - \eta^3)$ $\dot{\theta} = -\frac{k_a}{2}\theta^2(1 - \eta^2)$	
IX	cross-inversion (n) $L + D \xrightarrow{k^L} L + L$ $L + D \xrightarrow{k^D} D + D$	$\eta^{(s)} = \eta_0$	$\eta^{(s)} = \lvert 1 \rvert$	$\dot{\eta} = 0$ $\dot{\theta} = 0$	$\dot{\eta} = 2kg\theta(1 - \eta^2)$ $\dot{\theta} = 0$
X	trimolec. autocatalysis (superconcurrence) (dr) $A + 2L \xrightarrow{k^L} 3L$ $A + 2D \xrightarrow{k^D} 3D$	$\eta^{(s)} = \lvert 1 \rvert$	$\eta^{(s)} = \lvert 1 \rvert$	$\dot{\eta} = \frac{k}{2}A\theta(\eta - \eta^3)$ $\dot{\theta} = \frac{k}{2}A\theta^2(1 + \eta^2)$	$\dot{\eta} = k\theta(g + \eta)(\eta - \eta^3)$ $\dot{\theta} = kA\theta^2(1 + \eta^2) + kgA\theta^2\eta$.
XI	trimol. "erron." autocat. (r) $A + 2L \xrightarrow{k_e} 2L + D$ $A + 2D \xrightarrow{k_e} 2D + L$	$\eta^{(s)} = 0$		$\dot{\eta} = \frac{k_e}{2}A\theta(\eta + \eta^3) - k_e A\theta\eta$ $\dot{\theta} = \frac{k_e}{2}A\theta^2(1 + \eta^2)$	

Symbolism: $\eta = \frac{L - D}{L + D}$; $\theta = L + D$; $\dot{x} = \frac{dx}{dt}$

$g = \left|\frac{k^L - k^D}{k^L + k^D}\right|$; $k = k^L + k^D$; A,B, = achiral substrate and product.

Types of process: r - racemizing
n - neutral
dr - deracemizing

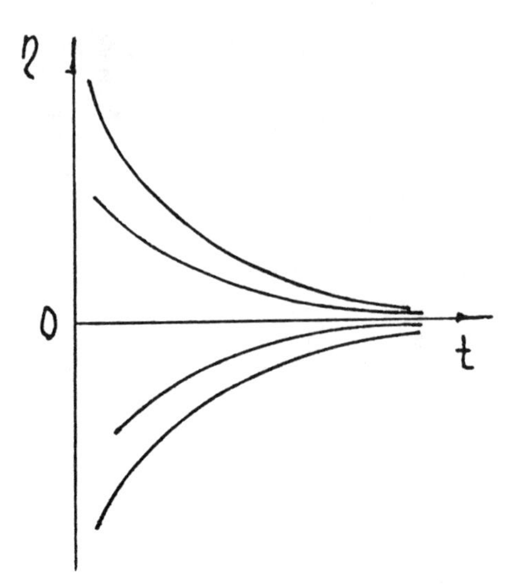

Fig. 11. $\eta(t)$ plot for process of racemizing type.

The characteristic form of the dependence $\eta(t)$ is shown in Fig. 11.

(b) Neutral

The characteristic feature of the processes of this type is that they do not change the chiral ordering of the system, leaving the initial value of its chiral polarization unchanged:

$$\left(\frac{d\eta}{dt}\right)_n = 0$$

From Table 2 it follows that to the processes of neutral type there belong processes of autocatalytic synthesis of mirror isomers (II), processes of their destruction (IV) (i.e., all the irreversible reactions of the first order in terms of concentrations of the antipodes). The form of the dependence $\eta(t)$ for the processes of neutral type is shown in Fig. 12.

(c) Deracemizing

This type of processes leads to an increase in the chiral ordering of the system (Fig. 13). The characteristic feature of the processes of this type is the presence of positive feedback in terms of chiral polarization, arising due to the "interaction" of mirror isomers. As is seens from Table 2 (VIII, X) to the processes of this type there belong "annihilation" of the antipodes in thier interaction ($L + D \rightarrow A$) and "superautocatalysis" ($A + L + L \rightarrow 3L, A + D + D \rightarrow 3D$). The general form of dynamic equation for the chiral polarization in the case of the processes of deracemizimg type

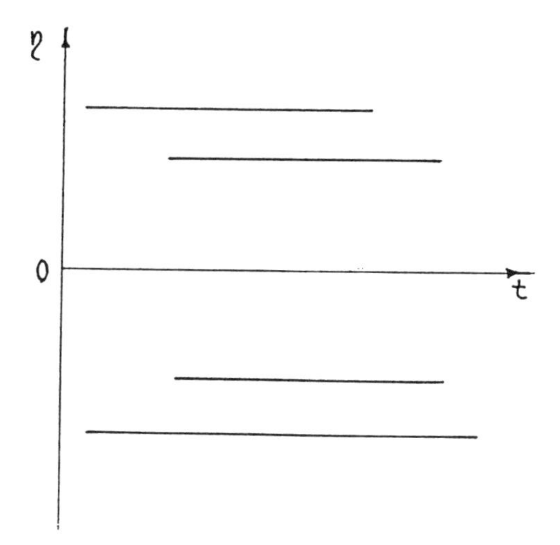

Fig. 12. $\eta(t)$ plot for process of neutral type.

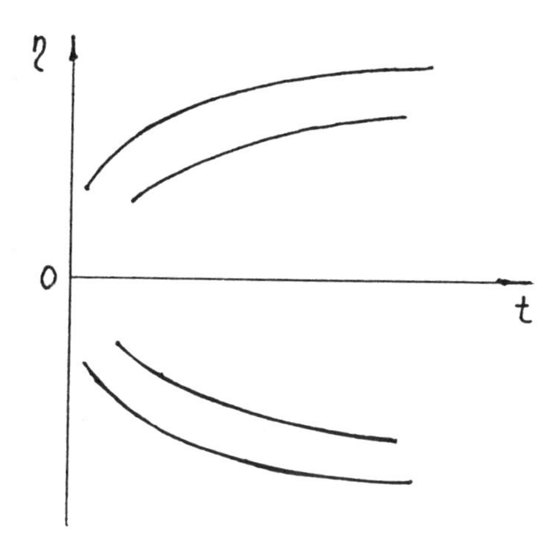

Fig. 13. $\eta(t)$ plot for process of deracemizing type.

$$\left(\frac{d\eta}{dt}\right)_{dr} = \alpha(\theta)\eta - \beta(\theta)\eta^3, \quad (\alpha, \beta) > 0 \tag{9}$$

is analogous to the well-known equations of the theory of nonequilibrium phase transitions (cf, e.g., Refs.37 and 38). Such a form of the equation is characteristic of systems with a critical behavior: at $(\alpha/\beta) < 0$ a racemic state is stable in the system; upon attaining critical conditions – $(\alpha/\beta) = 0$ – the racemic state loses stability and at $(\alpha/\beta) > 0$ there arise stable chirally polarized conditions – spontaneous mirror symmetry breaking takes place (Fig. 14).

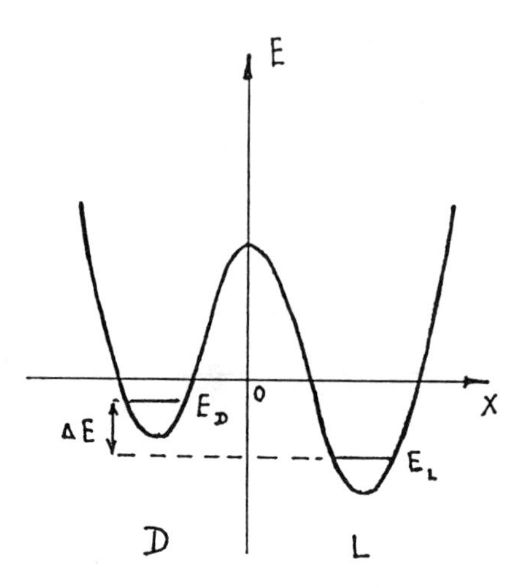

Fig. 14. Chiral potential with parity nonconservation taken into account. $E_{L,D}$ — distribution of levels L and D states of molecule.

Table 2 should be briefly commented upon. In this Table "reaction blocks" are presented, by means of which sufficiently complicated process schemes can be constructed very easily. There is no difficulty in writing down dynamic equations directly pointing to the character of behavior of chiral polarization in the system constructed. On the other hand, this Table may be used as a "template" in the analysis of any suggested scheme of reactions. This is the major advantage of the classification based on the use of the variables (η, θ). It should be noted that it allows one to single out the class of processes capable of effectively ensuring mirror symmetry breaking – of deracemizing processes – and indicates the minimal set of reactions-"blocks", realizing such a process (Table 3).

To delineate the main types of processes in chiral systems – the racemizing, neutral, and deracemizing ones – with a greater accuracy, we left out of account possible influence of asymmetrical effects – of the advantage factor – (AF) - on these processes. The presence of the advantage factor leads to the rate constants of mirror-conjugated reactions for L and D isomers) being no longer equal. A measure of AF may be introduced as a relative difference in the rate constants of mirror-conjugated reactions:

$$g = (k_i^L - k_i^D)/(k_i^L + k_i^D).$$

TABLE 3

MIRROR SYMMETRY BREAKING PROCESSES

Number, Type of Scenario	Involved Blocks	Set of Reactions	Final State	References
1 "evolution"	II (n)	$A + L \xrightarrow{k} 2L$; $A + D \xrightarrow{k} 2D$ Jordan-Kuhn scheme	$\eta^{(s)} = \eta_o$	[39], [78]
2 "evolution"	IV (n,g) g → CPL	$L \xrightarrow{k^L} A$; $D \xrightarrow{k^D} A$	$\eta^{(s)} = 1$	[44], [45]
3 "evolution"	I (r) IV (n,g) V (r) g → WNC	$A \xrightarrow{k} L$; $A \xrightarrow{k} D$ $L \xrightarrow{k^L} A$; $D \xrightarrow{k^D} A$ $L \overset{k^r}{\rightleftarrows} D$	$\eta^{(s)} \simeq \frac{g}{k_R}$; $K_R \equiv \frac{k_r}{k^L + k^D}$	[94], [153]
4 "evolution"	IV (n,g) V (r) g → β-radiolysis	$L \xrightarrow{k^L} A$; $D \xrightarrow{k^D} A$ $L \overset{k_r}{\rightleftarrows} D$	$\eta^{(s)} \simeq \frac{g}{k_R}$	[93]
5 "bifurcat"	II (n) VIII (dr)	$A + L \rightarrow 2L$; $A + D \rightarrow 2D$ $L + D \rightarrow A + B$ Frank Scheme	$\eta^{(s)} = \pm 1$	[126]
6 "bifurcation"	I (r,g) II (n) VI (r) VIII (dr) g → WNC	$A + B \xrightarrow{k^L} L$; $A + B \xrightarrow{k^D} D$ $L + A + B \rightleftarrows 2L$; $D + A + B \rightleftarrows 2D$ $L + D \rightarrow A + B$	$\eta^{(s)} = 0$ at $\rho < \rho_c$ $\eta^{(s)} = \pm\sqrt{1-\rho_c/\rho}$ at $\rho > \rho_c$ $\rho = f(k_i, A, B)$	[120], [130], [131]
7 "bifurcation"	X (dr) XI (r)	$A + 2L \xrightarrow{k} 3L$; $A + 2D \xrightarrow{k} 3D$ $A + 2L \overset{k_e}{\rightleftarrows} 2L + D$; $A + 2D \overset{k_e}{\rightleftarrows} 2D + L$	$\eta^{(s)} = \pm 1$	[122]
	All blocks I-XI (Table 2) and their sets	Various sets of reactions without and with FA are analysed	$\eta^{(s)}$ and phase portraits $(\eta = \eta^{(\theta)})$ are presented	[36], [103]

The taking of the advantage factor into account leads to the appearance of additional terms in the equations for chiral polarization. For the processes of racemizing type the contribution of the AF has the form

$$\left(\frac{d\eta}{dt}\right)_r^{(g)} = \alpha\eta$$

while for the processes of neutral and deracemizing types it has the form

$$\left(\frac{d\eta}{dt}\right)_{n,dr}^{(g)} = bg(1-g^2) \tag{10}$$

It we recall that in the absence of AF the equation for chiral polarization for the processes of neutral type has the form $d\eta/dt = 0$, an impression may be formed that in the presence of the AF the neutral process becomes the deracemizing one. But this is not so. The fact is that in nonliving nature there are no processes possessing an absolute stereoselectivity. This means that "erroneous" reactions occur alongside of the "correct" ones. For example, the process of neutral type (II)

$$A + L \xrightarrow{k^L} 2L$$
$$A + D \xrightarrow{k^D} 2D$$

in reality is always accompanied by the process of racemizing type

$$A + L \xrightarrow{k_m} L + D$$
$$A + D \xrightarrow{k_m} L + D$$

Such a process gives the following contribution to the equation for the chiral polarization: $-2k_m\eta$. As a result, we obtain the following equation for the chiral polarization:

$$\frac{d\eta}{dt} = g(1-\eta^2) - k_R\eta, \qquad (\tau = (k^L + k^D)At) \tag{11}$$

$k_R = \frac{2k_m}{k^L+k^D}$ is the effective rate constant of racemizing process. Consequently, the chiral polarization of the system tends not to 1, as it followed from Equation (9), but to the value g/k_R, as it follows from Equation (10). We shall estimate the order of magnitude of this value in Section 5, when discussing the hypothesis about the origin of chiral purity of the organic medium due to the effect of the factor of advantage to the "primordial soup."

It should be emphasized that the analysis of the dynamics of chiral systems in the variables (η, θ) gives an unambiguous answer about the type of processes in a given system, in contrast to the similar investigation in other variables.

We shall illustrate this point, using as an example a rather popular hypothesis about the autocatalytic origin of chiral purity, which was put forward by some authors[4,45,39,40].

Footnote: In Ref. 45, it was proposed to use Eigen-type selection model for the explanation of the origin of optical asymmetry:

$$\frac{dx_i}{dt} = (A_i Q_i - D_i)x_i - \phi_0 x_i + \sum_{i \neq j} \phi_{ij} x_j$$

A_i – the self-replication rate constants;
D_i – the destruction rate constants;
Q_i – the part of the exactly reproducing molecules;
ϕ_0 – the dilution rate;
ϕ_{ij} – rates of spontaneous production of substance i due to the erroneous copying of substance j.

The Eigen-type model's equations for mirror isomers can be rewritten in the following way:

$$\frac{d\eta}{dt} = g(1 - \eta^2) - k_R \eta$$

Here: $g = (W^L - W_D)/(W^L + W^D), (W^i = A_i Q_i - D_i$, with selective values put by Eigen[151]), and $k_R = \phi_{ij}/(W^L + W^D)$, the relative rate of erroneous copying.

It is easy to see that this model cooresponds to the process of neutral type with an AF (see Equation 11). Such systems can ensure "strong" breaking of the chiral symmetry of the final state only in ideal systems (i.e., in systems with an absolutely precise reproduction of chirality in duplication and in the absence of processes of racemization ($k_R = 0$)). Under real physical conditions ($k_R \neq 0$) in such systems , for the chiral purity to be ensured, the ratio of the advantage factor to the racemizing process (g/k_R), unrealistic for the physico-chemical systems, is required (the real estimates for g/k_R see below).

It is based on the following scheme of reactions:

$$A + L \xrightarrow{k} 2L; \quad A + D \xrightarrow{k} 2D$$

(the rate constants of the mirror-conjugated reactions coincide, for the AF is absent). The authors of the hypothesis proceeded from the fact that the dynamic equation for the difference of the antipodes $x_L - x_D$ had the form

$$\frac{d}{dt}(x_L - x_D) = kA(x_L - x_D)$$

and, basing of its solution

$$x_L(t) - x_D(t) = (x_L(0) - x_D(0))\exp(kAt)$$

they concluded that, insofar as the difference of the antipodes (which originated due to fluctiations at the initial moment of time) grows exponentially, such a process could ensure domination of one of the antipodes in the biosphere. However, this conclusion is erroneous. Indeed, the equation for the sum of the antipodes

$$\frac{d}{dt}(x_L + x_D) = kA(x_L + x_D)$$

has the same form as the equation for the difference; consequently, the sum of the antipodes grows exponentially too (with the same of the antipodes grows exponentially too (with the same exponential term). The chiral ordering of the system, described by the value of chiral polarization $\eta = (x_L - x_D)/(x_L + x_D)$, does not change, remaining equal to its initial value: $\eta(t) = \eta(0)$. Thus, in the analysis of the processes occurring in chiral systems, with a view to establishing their capability of strong mirror symmetry breaking, one should exercise great caution in making conclusions on the basis of equations for the concentrations of mirror isomers (from their form it is difficult to infer the applicability of a given scheme in the problem of the origin of chiral ordering in the course of evolution) and of equations for the difference of the antipodes, since they may lead to erroneous conclusions.

We shall now discuss two alternative hypotheses of the abiogenic orgin of chiral purity.

5. HYPOTHESIS OF THE ADVANTAGE FACTOR — SCENARIO OF SUCCESSIVE EVOLUTION

The central idea in this hypothesis is that the rate constants of those reactions in which molecules-mirror isomers participate may differ in the presence of an asymmetrical effect (AF) and these differences have played a decisive role in the mirror symmetry breaking of the prebiological organic medium. This hypothesis may be represented schematically in the following manner:

Primordial racemic soup	AF $\longrightarrow$	Development and enhancement of external asymmetry in physico-chemical transformations	$\longrightarrow$	Chiral purity of prebiosphere

The idea that dissymetry of living matter has some physical causes goes back to Louis Pasteur[41]. This idea was put forward at the dawn of the problem of the origin of the chiral purity of the biosphere and dominated in this problem until recently, since the researchers were convinced that in the absence of the AF physico-chemical processes had no resources for breaking the mirror symmetry (see Section 3).

At present most diverse assumptions concerning the nature of the AF which caused mirror symmetry breaking in the course of biopoiesis can be found in the literature (cf, e.g., references cited in Ref. 42). All the physical sources suggested in explanation of the right and left isomers being not on par may be subdivided into two classes.

First class: local AFs. To such factors there belong, for example, circularly polarized light[43–46], different combinations of static electric and magnetic fields[47–49], so-called "mechanical" AFs, for instance, a combination of the gravitation field and forces originating in rotation, the Coriolis force[50–52], and also "mixed" AFs: a combination of mechanical forces and electromagnetic fields[48]. The effect of lightning discharges[51], a combination of a magnetic field and linearly polarized light[53,54] may also be attributed to this class. More exotic "candidates" were also suggested, for instance, circularly polarized electromagnetic radiation generated by solar flares[55]. The term "local AFs" is used with a view to emphasizing that the AFs of such kind may exist or be absent in some region on the earth's surface and vary from region to region, or act during a definite period of time.

Second class: global AFs. The AF sources belonging to this class are caused by parity nonconservation in the weak interactions of elementary particles. Such AFs may manifest themselves either in the effect of polarized products of β-decay (so-called Vester-Ulbricht hypothesis[57]) or in the effect of weak neutral currents[58,59].

Let us consider how the difference in the rate constants of chemical processes is induced by the AF.

There are two possible mechanisms. One of them is associated with cancelling the energy degeneration of the molecules-mirror isomers (see Section 3) (Fig. 14). In this case the inner energy of the antipodes is already different ($E_L \neq E_D$) and due to this difference their reactivity somewhat differs. Such a mechanism leads to an internal difference of the molecules-isomers. It is realized in the case of the AF caused by weak neutral currents. If $\hat{H}_{WNC}$ is a Hamiltonian responsible for weak neutral currents, the energy difference of the isomers ΔE_{WNC} is defined as

$$\Delta E_{WNC} = \langle L|\hat{H}_{WNC}|L\rangle - \langle \hat{\sigma}^{-1}\hat{H}_{WNC}\hat{\sigma}|D\rangle$$

($\hat{\sigma}$ is the operator of inversion).

In the second mechanism the difference in the reactivities of the molecules-antipodes arises due to their different interaction with a chiral AF (for instance, with circularly polarized light, polarized radiations of β-decay, etc.). Under the effect of the AF of a definite chirality (say, left or right circularly polarized light), isomers of one type of chirality are excited more effectively than isomers of the opposite chirality and, consequently, undergo chemical transformations more easily, though their internal energy remains the same. The difference in the activation energy of the isomers is defined as follows:

$$\Delta E = \langle L|\hat{H}_{int}|L\rangle - \langle D|\hat{H}_{int}|D\rangle$$

($\hat{H}_{int}$ being the Hamiltonian of the interaction of the isomer molecule with the AF). If we define the measure of the factor of advantage as a relative difference of the rate constants of mirror-conjugated reactions

$$|g| = \left| \frac{k_i^L - k_i^D}{k_i^L + k_i^D} \right|$$

and use the Arrhenius equation which relates the reaction rate constant and the activation energy

$$k^L = A \exp\left(-\frac{E_L}{kT}\right); \quad k^D = A \exp\left(-\frac{E_D}{kT}\right); \quad E_L = E_D \pm \Delta E$$

the equation derived for the measure of the advantage factor will be

$$g = \frac{\exp(-E_L/kT) - \exp(-E_D/kT)}{\exp(-E_L/kT) + \exp(-E_D/kT)} \approx \frac{\Delta E}{kT} \tag{14}$$

Formally, the measure of the AF is defined similarly for both mechanisms, but it should be emphasized that the sense of ΔE for them is different.

In the analysis of the scenario of mirror symmetry breaking the question of the maximum attainable degree of assymetry this or that scenario can give is of great importance. This is associated with the requirement of the necessity of chiral purity as a prerequisite stage, preceding the formation of self-replicating systems.

In the scenario of a successive and continuous evolution usually the basic processes involved are such as destruction of isomers (IV)

$$L \xrightarrow{k_d^L} A; \quad D \xrightarrow{k_d^D} A \tag{15}$$

their simple synthesis (I)

$$A \xrightarrow{k_s^L} L; \quad A \xrightarrow{k_s^D} D \tag{16}$$

or autocatalytic synthesis (II)

$$A + L \xrightarrow{k_o^L} 2L; \quad A + D \xrightarrow{k_o^D} 2D \tag{17}$$

of isomers from an achiral substrate in the presence of the AF (k_i^L, k_i^D being the rate constants of the corresponding reactions). According to the classification given in Section 4, reactions (15) and (17) relate to the processes of neutral type, whereas reaction (16) relates to the process of racemizing type. Within the framework of an abiogenic scenario of mirror symmetry breaking, in addition to reactions (15) and (17) it is necessary to consider the processes of racemizing type, corresponding to these reactions, as well. For reaction (15) this the process of induced racemization (V)

$$L \underset{k_r}{\overset{k_r}{\rightleftarrows}} D \tag{18}$$

and for reaction (17) this is either the process of induced racemization (18) or an "erroneous" reaction of synthesis, occurring because non-absolute stereoselectivity of the chemical reactions (III)

$$A + L \xrightarrow{k_m} L + D; \quad A + D \xrightarrow{k_m} L + D \tag{19}$$

For the reaction of simple synthesis (16), even without taking into account additional racemizing processes, we obtain the following value for the maximum attainable asymmetry of the system (see Table 2 of Section 4):

$$\eta_{\mathbf{max}} = g$$

For reactions (15) and (17), with racemizing processes (18) and (19) taken into account, the maximum attainable asymmetry of the system may be represented as (see Table 3 and the solution of Equation (11)):

$$\eta_{\mathbf{max}} = \frac{g}{k_R}\left(1 + \sqrt{1 + \frac{g^2}{k_R^2}}\right)^{-1} \approx \frac{g}{k_R} \tag{20}$$

where k_R is the effective constant of the process of racemizing type.

Thus, $\eta_{\mathbf{max}}$ within the framework of the "evolutionary" scenario is defined either directly by the value g, or by the ratio g/k_R.

5.1. "Selection Rules" for AF

As stated above, a great diversity of asymmetry sources has been proposed. In reality, however, not all of them are suitable to play the role of the AF. A physical source of the AF must meet definite symmetry requirements.

It should be recalled that the property of chirality in the simplest case is described by the pseudoscalar quantity χ which changes its sign as one goes from the right-handed system of coordinates over to the left-handed one, i.e., upon relfection:

$$\hat{\sigma}\chi = -\chi$$

where $\hat{\sigma}$ is the operator of inversion. Such a quantity may be constructed as a scalar product of an axial vector $(\vec{A})$ and a polar vector $(\vec{P})$, related with the chiral object:

$$\chi = (\vec{A} \cdot \vec{P})$$

Indeed, it is well known that optical activity (natural) is observed only in chiral molecules, pertaining to point groups containing no mirror-turnable elements (for which both electric and magnetic dipole transitions are allowed):

$$\langle i|\vec{d}|j\rangle\langle j|\vec{m}|i\rangle \neq 0$$

Here $\vec{d}$ and $\vec{m}$ are operators of the electric and magnetic moments of the transition $i \rightarrow j$ respectively.

Consequently, any combination of the axial and polar fields, in principle, may be regarded as a possible AF source. However, the requirements to be met by an AF source are still more stringent. The fact is, that, as has been shown by an analysis[60-66], a combination of static fields, of which one is described by an axial vector and the other, by a polar vector (for instance, $\vec{E}$ and $\vec{B}$, or a gravitational field $\vec{G}$ and a centrifugal force), is not a "true" AF (at least for equilibrium processes) in thesense that the effect of such an AF on chemical transformations in the chiral system does not lead to asymmetry, i.e., to the inequality of the concentrations of the L and D isomers. To illustrate this, we shall consider a simple example along the same lines as in Ref. 64.

Let us consider the transformation of achiral molecules B into L and D isomers in the presence of an AF caused by a combination of static polar field P and axial field A: $D \rightleftharpoons B \rightleftharpoons L$. The state of the molecule in the presence of the AF will be symbolized as $(M, \vec{P}, \vec{A})$, where $M = L, D$. We shall introduce two operations of symmetry, mirror reflection ($\hat{\sigma}$) with respect to the plane containing vectors $\vec{A}$ and $\vec{P}$, and time inversion ($\hat{T}$). Since

$$\hat{\sigma}\vec{P} = \vec{P}, \quad \hat{\sigma}\vec{A} = -\vec{A}, \quad \hat{\sigma}L = D, \quad \hat{\sigma}D = L,$$
$$\hat{T}\vec{P} = \vec{P}, \quad \hat{T}\vec{A} = -\vec{A}, \quad \hat{T}M = M,$$

then

$$(L, \vec{P}, \vec{A}) \xrightarrow{\hat{\sigma}} (D, \vec{P}, -\vec{A}) \xrightarrow{\hat{T}} (D, \vec{P}, \vec{A}) \tag{21}$$

Equation (21) signifies that L-isomers have the same energy as D-isomers. Hence, the isomers have the same partition functions too, and therefore their equilibrium concentrations will be equal in the presence of such an AF as well.

"True" AFs (in accordance with the terminology proposed by L. Barron) are only such physical fields, radiations, etc., which possess the property of "helicity", i.e. which "can exist in two enantiomeric forms and transform in its antipode under the effect of spatial inversion, but do not change upon time inversion in combination with any spatial rotation"[61-63]. To such AFs there belong the following ones among these mentioned above: circularly polarized light, polarized products of β-decay, weak neutral currents (see Table 4).

Table 4
Physical Advantage Factor

Type of Advantage Factor	True (+) or False(−)	g
Circularly polarized light (CPL)	+	$10^{-2} - 10^{-4}$
Static magnetic field (SMF)	−	
Static electric field (SEF)	−	
Gravitational field (GF)	−	
SMF and SEF	−	$\chi_i(\vec{E} \cdot \vec{B})$
Rotation (Coriolis force) and gravitational field (GF)	−	$\chi_j([\vec{w}\,\vec{v}]\vec{G}) \sim \chi_j(\vec{\Omega}\vec{G})$
SMF and GF	−	$\chi_m(\vec{B}\vec{G})$
Rotation and SMF and SEF	+	$\chi_n(\vec{w}[\vec{E}\vec{B}])$
Rotation and SMF and GF	+	$\chi_p(\vec{B}[\vec{w}\vec{G}])$
SMF and linearly polarized light (LPL)	+	$\chi_r(\vec{B}[\vec{E}\vec{B})]) \sim \chi_r(\vec{B}\,\vec{k})$
Weak neutral currents (WNC)	+	$\chi_q Z^5 10^{-20} \frac{1}{kT} \sim 10^{-17}$
Longitudinally polarized β-particles	+	$h_e \frac{\sigma_L - \sigma_D}{\sigma_L + \sigma_D} \sim 10^{-11}$

Symbolism:
χ - factors determined by molecular structure
h_e - helicity of β-particles = $\langle \hat{s}\hat{p} \rangle$ ($\hat{s}, \hat{p}$ – operators of spin and momentum)
$\sigma_{L,D}$ - cross-sections of interactions of polarized β-particles with molecules.

5.2. Local AFs

(a) Circularly Polarized Light (CPL)

Asymmetry of the reaction products in photolysis of a racemate with the aid of CPL was found more than 50 years ago, and up till now this method of asymmetric synthesis has been used in laboratory practice (cf., e.g., Ref. 68). Sunlight, circularly polarized in the twilight near the surface in the IR region of the spectrum (800 nm), amounts to 0.1% of the total flux[69]. However, in the UV region which is most reactive for photochemical processes, no circular polarization was detected, sufficient for the asymmetric photolysis, i.e. its magnitude amounts to less than 0.01%. It should also be emphasized that circular polarization of sunlight is a consequence of parity-conserving electromagnetic interactions in the Earth's atmosphere and therefore CPL may be only a local AF and, on the average, over the entire surface of the Earth, the rotation of the latter being taken into account, the effect caused by CPL must disappear.

(b) Magnetic Field, Gravitational Field, etc.

Insofar as such sources as magnetic fields, gravitational fields, centrifugal forces arising in vortices, etc., fail to meet even the simplest necessary requirement to AF sources (a pseudo-scalar cannot be set in correspondence to them), they cannot be regarded as the AFs.

(c) Combinations of Axial and Polar Fields

In accordance with Section 5.1, in chemical reactions proceeding under equilibrium conditions asymmetry cannot be obtained through the agency of such AFs.

There are grounds to believe that the conclusions derived in the consideration of equilibrium processes under the effect of such AFs are also applicable to the processes occurring under the conditions of kinetic control[64], though an opposite opinion is also voted[63,62]. This question, however, requires further studies[63].

Consequently, great care should be exercised in dealing with the reported asymmetry of the reaction products obtained in experiments with such AFs, for example, reactions in collinear electric and magnetic fields[47], reactions in a rotating vessel[48,50,52]. We should like to note that in thoroughly conducted experiments[51] no asymmetry of the products of reaction in the rotating vessel was detected, this being in agreement with the results of theoretical analysis.

An AF proposed recently[70], caused by the interaction of the environment-induced "chiral" magnetic moment of molecules found on the surface of the ocean with the magnetic field of the Earth, belongs to this class of AFs.

Suppose that the AF of the type discussed for nonequilibrium processes is "true". C. Mead and A. Moscowitz[64], proceeding from the experimental conditions of simple synthesis in a rotating vessle[50], estimated the value of the effect and demonstrated that in the given case $g \sim 10^{-17}$. If we take into account tha the rotation speed of the reaction vessel was $\sim 1.4 \times 10^4$ r.p.m., this being, no doubt, much greater than the rotation speed in atmospheric vortices or than the speed of rotation of gas streams in volcanic eruptions, under natural conditions $g \ll 10^{-17}$. Consequently, an AF with such an amplitude could not play any role in breaking the mirror symmetry of the "primordial soup."

In conclusion of our brief discussion of the hypothesis of local advantage factors, we should like to draw the readers' attention to the fundamental difficulties inherent in it. In the first place, the maximum attainable asymmetry does not exceed (even in experiments) the values $g < 10^{-2}$ (experiments on photolysis in a bundle of CPL), which in the light of the results set forth in Section 2 is obviously too small a value from the standpoint of the problem of the origin of chiral purity. In the second place, so far not a single real (local) AF has been proposed, whose distribution on the Earth would possess an asymmetry in the sense specified above. It should also be noted that one of the "oldest" and most popular AF sources – enantiomorphic crystals of quartz, on which stereoselective sorption of molecules-isomers can occur – is represented on the Earth, as shown by detailed investigations on the and abundance of such crystals, by a racemic mixture[71], in spite of numerous assertions that the abudnance of *L*-quartz is 1% higher than of its *D*-antipode. This fact, as well as low stereoselectivity of the interaction of molecules with the surface of quartz and also of clays, is a serious argument against the hypothesis of Cairns-Smith[152] about the decisive role of these AFs in the origin of chiral purity.

5.3. Global AFs

The discovery of parity nonconservation in weak interactions of elementary particles[72,73] furnished a physical basis for the assumption that the asymmetry of the biosphere in a consequence of parity nonconservation in weak interactions.

Two possible mechanisms of such an influence of parity nonconservation are under discussion in the literature.

One of these is associated with so-called neutral currents. As far back as 1925 Le Bell[74] made a bold, though at that time purely speculative suggestion that atoms possess internal chirality. The physical support for such an idea was found only much later and associated with the role of weak interactions in the interactions of the electrons of atomic shells with nucleons – with the existence of neutral currents[75]. Parity nonconservation is associated with vectorial neutral boson Z_o[76]. Because of the neutral currents, the electronic structure of the atom may possess rather slight internal chirality. (At present, experimental investigation of the optical activity of atoms in pairs of heavy metals, conditioned by neutral currents, is under way[77]). Right-and left-handed molecules-mirror isomers, because of such chirality of the atoms, may, as has been mentioned earlier, differ slightly in energy (theoretical estimates of the possible energy non-equivalence of the antipodal molecules will be given below). Accordingly, the rate constants of chemical reactions (synthesis, destruction, polymerization, etc.) in which enantiomers participate may be different, thought to a very small extent. Experimental detection of such differences, apparently, is not yet feasible, but many authors have suggested that the physico-chemical processes supposedly underlying chemical evolution are capable of accumulating and enhancing such slight differences, and this, in principle, might lead to a macroscopic excess of one of the isomeric forms of the organic substnace[58,78–80].

Another plausible mechanism is associated with the effect of radioactive radiations on the processes of chemical transformations (synthesis or destruction) of mirror isomers. Parity nonconvservation in β-decay leads, as is known, to the predominant appearance of longitudinally polarized electrons (or positrons). Such electrons (positrons), either directly interacting with chiral molecules, or generating bremsstrahlung with a circularly polarized component in the substance, selectively interacting with the antipodes, may lead to the appearance of an excess of one of the antipodes in the course of physicochemical transformations in the chemical system.

(a) Weak Neutral Currents

Weak neutral currents (WNC) perturb the stationary states of atoms and molecules, "cancelling" the energy degeneration for the antipodal states of the chiral molecule. Indeed,

$$\langle L|\hat{H}_{WNC}|L\rangle = \langle D|\hat{\sigma}^{-1}\hat{H}_{WNC}\hat{\sigma}|D\rangle = -\langle D|\hat{H}_{WNC}|D\rangle$$

and the energy levels of two antipodes differ, because of WNC, by the value

$$\Delta E_{WNC} = 2\langle L|\hat{H}_{WNC}|L\rangle$$

where $\hat{H}_{WNC}$ is the potential describing the contribution of WNC, $\hat{\sigma}$ is the operator of spatial inversion.

Using a model of an atom in a chiral potential, it is possible to assess the order of magnitude of ΔE_{WNC}[81–83]:

$$\Delta E_{WNC} = qZ^5 \times 10^{-20} a.u.$$

where Z is the atomic number and q is the factor of structural asymmetry. The potential $\hat{H}_{WNC}$ may be represented as a sum of three potentials of pair interactions[82]:

$$\hat{H}_{WNC} = \hat{H}^{ne}_{WNC} + H^{pe}_{WNC} + H^{ee}_{WNC}$$

where $\hat{H}^{ee}_{WNC}$ is the electron-electron, $\hat{H}^{pe}_{WNC}$ is the proton-electron, $\hat{H}^{ne}_{WNC}$ is the neutron-electron potentials. The potential H^{ne}_{WNC} gives a predominant contribution to the splitting of the ΔE_{WNC} energy of chiral molecules-antipodes because of WNC. Two other potentials, $\hat{H}^{pe}_{WNC}$ and $\hat{H}^{ee}_{WNC}$, depend on $(1 - 4\sin^2\theta_W)$ i.e. on the Weinberg angle. Experimental values of $\sin^2\theta_W$ lie within the range of 0.215-0.23[84]; consequently, $\hat{H}^{ee}_{WNC}$ and $\hat{H}^{pe}_{WNC}$ actually make an essentially smaller contribution than $\hat{H}^{ne}_{WNC}$ does. Details concerning the calculations of ΔE_{WNC} can be found in Refs. 82 and 83.

S. Mason and G. Tranter[82,83] calculated ΔE_{WNC} caused by WNC for several organic compounds, including some amino acids[82] (alanine, valine, serine), peptides[83], and possible precursors of sugars: tetrahydrofurans[85]. The authors succeeded in demonstrating that WNC shift the energy levels of the antipodes of amino acids in such a way that L-isomers possess a smaller energy than D-isomers. A similar result was obtained also for the precursors of sugars: D-isomers have an energy advantage over L-isomers. This means that the sign of chirality of the biosphere and the sign of chirality of the WNC-induced AF coincide. The energy difference of the antipodes (both amino acids and tetrahydrofurans) amounts to about 10^{-20} a.u. (10^{-14} J/mole). Thus, the measure of such an AF amounts to

$$g \approx \frac{\Delta E_{WNC}}{kT} \tag{22}$$

at $T = 300$K.

It is easy to understand that the AF caused by WNC is extremely small. Nevertheless, until recently the hypothesis that in the course of synethsis of polymers even such a weak AF as WNC may lead to the dominance of one of the isomeric forms was very popular. The essence of this hypothesis, advanced by the Japanese chemist Yamagata[78], is as following. Suppose that in a chemical system a process of formation of left (L) and right (D) polymers of length n occurs in accordance with the scheme

$$\begin{gathered} L \xrightarrow{p_L} L + L \xrightarrow{p_L} \ldots \xrightarrow{p_L} nL; \\ D \xrightarrow{p_D} D + D \xrightarrow{p_D} \ldots \xrightarrow{p_D} nD \end{gathered} \tag{23}$$

In the present of AF the probabilities p_L and p_D somewhat differ:

$$\begin{aligned} p_L &= p_o(1+g); \\ p_D &= p_o(1-g) \end{aligned}$$

Solution of the kinetic equations corresponding to scheme (23) leads to the following ratio of the left and right-handed polymers of the length n:

$$\frac{N_L^{(n)}}{N_D^{(n)}} \approx \exp(gn)$$

It is easy to see, however, that even for very large molecules ($n = 10^6$) for the real value $g \sim 10^{-17}$, chiral polarization in such a system will not exceed the value

$$\eta_{\max} = \frac{1-\exp(gn)}{1+\exp(gn)} \approx gn \sim 10^{-11} \tag{24}$$

(earlier it was thought that parity nonconservation in weak interactions might lead to the values $g > 10^7$). Moreover, as demonstrated in Ref. 153, taking into account inverse reactions in Scheme (23), which actually always occur alongside of direct reactions, and taking into account the final stereoselectivity, lead to

$$\frac{N_L^{(n)}}{N_D^{(n)}} = 1$$

i.e., the system remains racemic.

Let us assess $\eta_{\max}$ in a different manner. Let the rate constant (probability p_o) of insertion of an isomer of foreign chirality into the chain being synthesized does not exceed 10^{-6} to 10^{-8} (this being an underrated estimate for the chemical processes, characteristic of biochemical reactions in contemporaneous organisms). Then, from Equation (24) we obtain

$$(g/p_o) = 10^{-11} \text{ to } 10^{-9}.$$

Thus, within the framework of "simple", evolutionary hypotheses the problem of the origin of chiral purity of the biosphere because of the AF caused by WNC has not been solved.

(b) β-Decay

The mechanism of the effect of products of β-decay on chiral molecules, most popular today, is their asymmetrical radiolysis under the effect of longitudinally polarized electrons and positrons (cf., e.g., Refs. 86-88). The radiolysis occuring in the decay of such nuclei as ^{14}C, ^{40}K, ^{235}U and ^{26}Al, abundant in the earth's crust, in the ocean, and in the biosphere, is of greatest interest.

The asymmetry of the radiolysis of antipodal molecules in their interaction with longitudinally polarized β-radiations may be represented in the following form[89,90]:

$$A_R = |h(e^{\pm})| H_R(E, Z) \tag{25}$$

Here $h(e^{\pm}) = \langle \hat{s}_i \cdot \hat{p}_i \rangle$ – the helicity of the i-th electron (positron) – the degree of correlation between the spin $\hat{s}$ of the particle and its momentum $\hat{p}$, Z is the atomic number of the nucleus, while

$$H_R = \frac{\sigma^+(L) - \sigma^-(L)}{\sigma^+(L) + \sigma^-(L)} = \frac{\sigma^{\pm}(L) - \sigma^{\pm}(D)}{\sigma^{\pm}(L) + \sigma^{\pm}(D)} \tag{26}$$

–assymetry of the interaction of longitudinally polarized β-particles with L and D isomers ($\sigma^{\pm}$ is the cross-section for interaction of molecules with particles of positive (negative) "helicity"). In the range of energies 100 KeV $\leq E \leq$ 500 KeV, H_R (in the non-relativistic Born approximation) may be represented as[91]:

$$H_R = \frac{q'_R(\alpha Z)^2}{2E \;\; ln(2E)} \tag{27}$$

where q'_R is the factor of molecular dissymetry in the radiolysis of molecules (depending on the structure of particular molecules). For $E = 100$ KeV (typical value of the energy of electrons in β-decay), Equation (27) reduces to

$$H_R \approx 10^{-5} (dZ)^2 q_R \tag{28}$$

If we take q_R to be 10^{-2} to 10^{-3}, (i.e. the amplitude of circular dichroism at the absorption bands of amino acids) for the asymmetry of H_R in the radiolysis of amino acids ($Z = 6$) we obtain

$$H_R \approx 10^{-10} \div 10^{-11} \tag{29}$$

Thus, even in the case of 100% polarization of the β-particles

$$10^{-11} < A_R < 10^{-10} \tag{30}$$

In reality $|h(e^{\pm})| = f(E) \ll 1$ and $A < 10^{-11}$. It should also be taken into account that $h(e)$ grows as E increases, whereas the reactivity of β-particles decines in this case, the asymmetry of the reaction products being thereby diminished still further.

Ya. B. Zel'dovich and D.B. Saakyan[92] obtained similar results for the asymmetry of radiolysis of chiral molecules by β-particles:

$$H_R = \frac{\alpha^2 v I_m(\vec{d} \cdot \vec{m})}{c|\vec{d}|^2} \tag{31}$$

where $\vec{d}$ and $\vec{m}$ are the electric dipole and magnetic dipole moments in the transition of a molecule from one state to another in the course of the radiolysis reaction, v is the velocity of the electron. Transitions at strong absorption bands, which dominate in the reaction cross-section, lead to the folloiwng estimate of the value

$$(\vec{d} \cdot \vec{m})/d^2 \sim 10^{-6}$$

In this case for the electrons with $E = 100$ KeV we obtain again

$$H_R \sim 10^{-11}$$

It is easy to see that A_R practically coincides with g, i.e., for radiolysis under the effect of longitudinally polarized particles of β-decay the estimate for the AF amplitude is

$$g = A_R \sim 10^{-10} \div 10^{-11}. \tag{32}$$

Asymmetrical radiolysis of antipodes is described by the following scheme of reactions

$$L \xrightarrow{k_d^L} A; \quad D \xrightarrow{k_d^D} A. \tag{33}$$

However, an effective radioracemization also proceeds in this case.

$$L \underset{K_r}{\overset{k_r}{\rightleftarrows}} D.$$

According to Table 3 and Equation (20), the maximum attainable value of chiral polarization in such a process depends on the ratio of the AF measure

g to the effective rate constant of the racemizing process $K_R = 2k_r/k_d$ in the following manner

$$\eta_{\max} = \frac{g}{K_R}\left(1+\sqrt{1+\frac{g^2}{K_R^2}}\right)^{-1}. \tag{34}$$

The value K_R for radiolysis can be estimated easily. The rate constant of the radiolysis varies for different amino acids from $5 \times 10^{-9} s^{-1}$ (in the case of β-decay of ^{40}K and ^{14}C) to $5 \times 10^{-10} s^{-1}$ for the "natural" nuclear reactor (of Oklo type)[93]. The rate constant of radioracemization amounts to $\sim 10^{-4} - 10^{-12} s^{-1}$ (see Ref. 94). Thus, $K_R \sim 10^{-5} - 10^{-2}$. Consequently, from Equation (34) we have the following estimate for the value $\eta_{\max}$:

$$\eta_{\max} \simeq \frac{g}{K_R} \sim 10^{-9} \div 10^{-11}. \tag{35}$$

More detailed calculations (on the basis of Equation (34), with taking into account the temperature dependence of the reaction rate constants, are given in Ref. 93 and presented in Fig. 15.

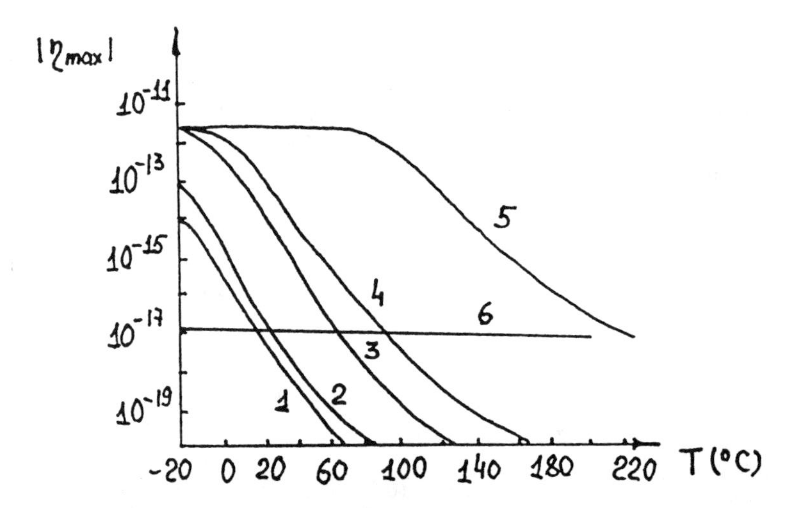

Fig. 15. $\eta_{\max}$ vs. T^0 plot for radiolysis of alanine in aqueous solution under the effect of natural sources of radiation[93]. 1 – ^{40}K, ^{14}C in oceans; 2 – average radioactivity of the earth's crust; 3 – uranium deposits; 4 – ^{26}Al; 5 – "natural" uranium reactor (^{235}U) (of Oklo type); 6 – WNC.

The estimate given by Equation (35) leads us to conclude that the AF caused by parity nonconservation in β-decay could not be the source of strong mirror symmetry breaking in the course of prebiological evolution within the framework of the evolutionary scenario.

Basing on the results of the present Section, one can make a conclusion that those sensational results on the radiolysis of racemates, which were reported now and then in the literature[95,96], were artifacts, as it was admitted later by the authors themselves (see, e.g., Ref. 97) or they were convincingly disporved (see, for exmaple, Ref. 98).

To sum up our discussion of the hypothesis about the effect of the AF on mirror symmetry breaking in the course of evolution of the primordial organic medium, it can be concluded with assurance that this hypothesis faces such fundamental difficulties, that very serious doubts arise as to its being able of serving as the basis for the scenario of the origin of chiral purity of the biosphere. It is quite obvious that the racemizing and neutral processes considered within the framework of the given hypothesis, even in the presence of the AF, are not capable of breaking the initially racemic state of the organic medium. The theoretical analysis and experimental data available today support this conclusion all the more.

On the other hand, by the middle of the 1970s, experimental data were, accumulated, indicative of the instability of racemic state in stereoselective, cooperative interaction of molecules-mirror isomers even in low-viscosity liquids. At the same time a new trend in studying nonequilibrium processes and self-organization phenomena began to grow up: synergetics. This was the basis for making a step towards new concepts in the problem of mirror symmetry breaking in the course of prebiological evolution on the Earth.

6. SPONTANEOUS MIRROR SYMMETRY BREAKING –SCENARIO OF "CHIRAL CATASTROPHE"

However attractive the idea of explaining the "left-handedness" of the bioorganic world by the breaking of symmetry between the left and right in the world of elementary particles may be, the results set forth in the preceding Section rule out such a "translation" of asymmetry, at least in the framework of the "evolutionary scenario", i.e. with a gradual and continuous accumulation of the AF in the course of biopoiesis.

An alternative to the scenario of gradual accumulation of the AF is spontaneous breaking of the mirror symmetry of the prebiological organic medium, self-organization of optical isomers in the process of chemical evolution.

Mirror symmetry breaking and the formation of chirally pure forms of the organic substance, from this point of view, was the consequence of the development and strengthening of such fluctuations in the course of prebiological evolution, which corresponded to an initially insignificant predominance of one of the enantiomers. This idea was first expressed by K. Pearson[99] in a discussion of the well-known work of F.R. Japp "Stereochemistry and Vitalism"[100]. Pearson suggested that molecular formations at the early stages of evolution might casually have a nonracemic composition. Subsequent selection of molecules of the same type of chirality from the racemic environment accounted for the appearance of a macroscopic excess of one of the antipodes.

The hypothesis of spontaneous mirror symmetry breaking, according to which under certain conditions the mirror-symmetrical state becomes unstable, and a spontaneous mirror symmetry breaking, a jump-like transition to the chirally ordered state, takes place, has been seriously substantiated in recent years[36,101–107]. According to this theory, the process leading to the self-organization of chirally pure forms of the organic substance is based on cooperative interactions of fluctuations of the antipodal composition of the medium in the course of the ensemble of physico-chemical transformations, which determined the origin of chiral purity of the prebiological world as a prerequisite stage of the transition of matter to the living form. Thus, the problem under consideration lies within the boundaries of the problem of spontaneous symmetry breaking, which has been extensively considered in modern natural science.

Concepts of the spontaneous symmetry breaking play an important role in the theory of such phenomena as phase transitions leading to ferromagnetism, superconductivity, hyperfluidity[108,109], to the origin of coherent radiations[37]. The same concepts are at the root of the present-day ideas concerning the properties of physical vacuum[110], the course of evolution of the Universe within the framework of the "hot" model[111], and also form the basis for the contemporary theory of interactions of elementary particles[112]. Spontaneous symmetry breaking is the basis for the theory of the origin of spatial and time ordering in chemical and biological processes, developed by the school of I. Prigogine[38], and the basis for the new trend in science: synergetics[37]. Such crucial problems in biology as morphogenesis and differentiation, as well as the problems of population dynamics, actually come down spontaneous symmetry breaking[113,114]. Even the origin of life was considered from this standpoint[115].

In addition to the unified concept, these problems are united by a common mathematical tool: the theory of bifurcation[37,38,116].

On the basis of the concepts and methods of the modern theory of nonequilibrium processes (dissipative structures) it has become possible to apply general physical principles of mirror symmetry breaking in the evolution of the organic substance[36,102–107]. From this standpoint the causeof deracemization of the organic primordial medium should not be sought in any *a priori* differences in the dynamics of the evolution of left- and right-handed mirror-isomeric molecules. Symmetry breaking stems from the instability of the racemic state. Because of this instability, any fluctuational deviations in the ratio of the concentration of the isomers from the racemate, no matter how small such deviations may be, after being autocatalytically enhanced, are capable of leading to practically complete predominance of one of the isomeric forms in the final state of the evolutionary process at the stage of prebiological evolution.

The hypothesis of spontaneous mirror symmetry breaking may be represented by the following scheme:

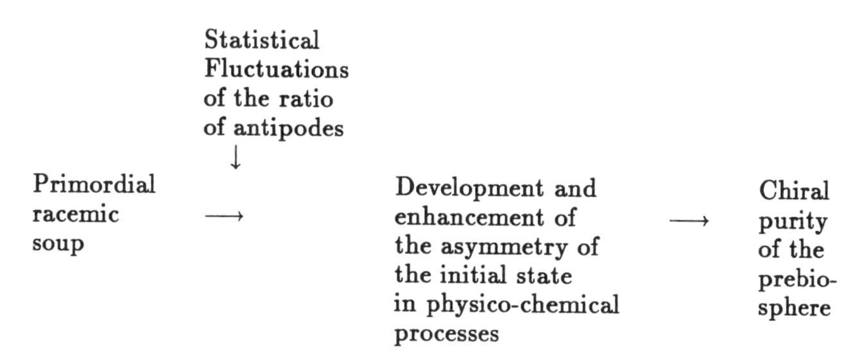

It should be emphasized that according to the hypothesis of spontaneous deracemization, a repetition of the whole ensemble of events which have led to the appearance of life on the Earth (including chiral purity), with equal success is capable of leading to such a biosphere, which employs *D*-amino acids and *L*-sugars.

It should be noted that in recent years this scenario has been attracting the most notice, even among the proponents of the idea of the decisive effect of the AF of fundamental nature (WNC, polarized radiations), who invoke the idea of spontaneous deracemization for the "enhancement" of the AF in the course of the nonequilibrium phase transition from the racemic state to the chirally ordered medium[117,118].

Spontaneous mirror symmetry breaking in crystallization of solutions of optically active compounds was observed already by L. Pasteur[119]. He was also the first to discover that definite conditions had to be fulfilled for this to take place: if cyrstallization occurs at a temperature below the critical one (for a given compound), chirally pure crystals precipitate; if the temperature is above the critical one, the crystals precipitated from solution prove to be racemic.

A theoretical explanation of this phenomenon was given in Ref. 9, where on the basis of the theory of regular solutions spontaneous resolution of a racemic mixture of antipodes was demonstrated to be caused, in the presence of critical conditions, by the stereoselectivity (nonsimilarity) of the interactions of the molecules-isomers of different chirality (the energy of the interaction of the isomers are not equal: $E_{LL} = E_{DD} \neq E_{LD}$). The transition from combined crystallization to the separate one is associated with a transition of "disorder-order" type.

The idea of the dominant role of stereoselective, cooperative interactions of the antipodes, leading to the origin of a positive feedback in relation to the chiral ordering, underlies the scenario of spontaneous breaking of the mirror symmetry of the organic substance in the course of chemical evolution. Already at the earliest stages of prebiological evolution a specific "phase transition" took place, which deracemized the "primordial soup."

6.1. Scheme of Spontaneous Mirror Symmetry Breaking

In the last few decades possible schemes of transformations of molecular isomers, simulating spontaneous deracemization, have been studied very intensively (see, e.g., Refs. 120-125). However, the general requirements to be met by the schemes of such type, irrespective of their concrete realizations and details, were specified in Refs. 36, and 102-107. Referring to Table 2 of Section 4, one can easily see that the scheme of spontaneous symmetry breaking must contain a deracemizing "block"

$$L + D \rightarrow A \tag{36}$$

or

$$\begin{aligned} A + 2L &\rightarrow 3L \\ A + 2D &\rightarrow 3D \end{aligned} \tag{37}$$

(see blocks VIII and X in Table 2). It should be pointed out that reactions (36) and (37) may imply complicated sequences of multistage transformations, effectively "reducible" to them.

The scheme which has found maximum favor was first suggested by Frank[126] In the comtemporary literature this scheme is usually written down as (see Table 3):

$$A + B \xrightarrow{k_1^L} L; \quad A + B \xrightarrow{k_1^D} D \tag{38a}$$

$$L + A + B \underset{k_{-2}^L}{\overset{k_2^L}{\rightleftharpoons}} 2L; \quad D + A + B \underset{k_{-2}^D}{\overset{k_2^D}{\rightleftharpoons}} 2D \tag{38b}$$

$$L + D \xrightarrow{k_3} C \tag{38c}$$

where $k_i^{L,D}$ are the rate constants of the corresponding reactions. (In the scheme initially suggested by Frank stage (38a) and the inverse reactions at stage (38b) were missing). Using this scheme, we shall demonstrate the major specific features of the scenario of spontaneous deracemization (bifurcation scenario).

In variables η and $\theta = \frac{k_2}{2k_1}(x_L + x_D)$, the dynamic equations corresponding to the model (38) will have the form ($k_i^L = d_i^D$)

$$\begin{aligned} \frac{d\eta}{d\tau} &= -\frac{\rho}{\theta}\eta + a\theta(\eta - \eta^3) \\ \frac{d\theta}{d\tau} &= \rho + \rho\theta - b\theta^2 - a\theta^2(1 - \eta^2) \end{aligned} \tag{39}$$

Here $\tau = k_2 Qt; \rho = \frac{x_A x_B}{Q}; Q = \frac{4k_1 k_{-2}}{k_2^2(k_3 - k_2)}; a = \left(\frac{k_3}{k_{-2}} - 1\right); b = \left(\frac{k_3}{k_{-2}} + 1\right)$. Attention should be paid to the fact that the equation for η has a characteristic structure of the form $\alpha(\theta)\eta - \beta(\theta)\eta^3$. As a governing parameter we shall choose $\rho = \frac{x_A x_B}{Q}$ (x_A and x_B being the concentrations of achiral precursors). It is easy to see that there exists a value $\rho = \rho_c$ such that in the region $\rho < \rho_c$ (subcritical region) the only stable state of the system (38) is racemic ($\eta = 0$). Upon reaching the bifurcation point $\rho = \rho_c$ this state loses stability and in the supercritical region $\rho > \rho_c$ there appear two stable mirror-conjugated states $\eta_{\pm}^{(s)} = \pm\sqrt{1 - \frac{\rho_c}{\rho}}$ (the bifurcation diagram is presented in Fig. 16).

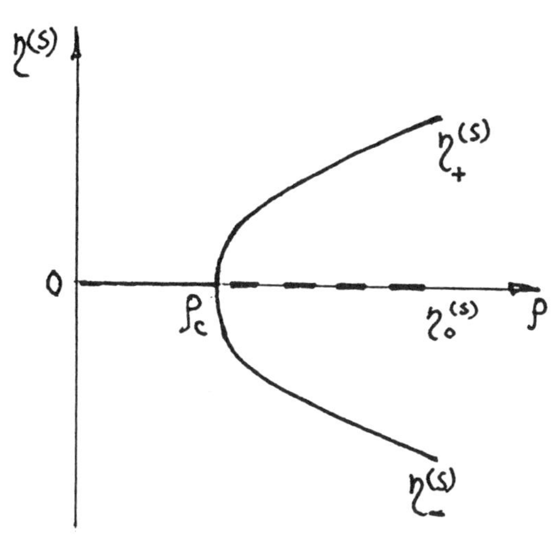

Fig. 16. Bifurcation diagram of Equation (39).

It should be emphasized that the formation of the asymmetrical states $\eta_{+}^{(s)}$ and $\eta_{-}^{(s)}$ (deracemization of the system) is not associated with the effect of the AF (as, for example, in the system of neutral type), but it is caused by the dynamic properties of the system itself, by the processes of transformations or mirror isomers. If the system starts from the racemic (on the average) state, then the probability of detecting the system in the state $\eta_{+}^{(s)}$ is equal to the probability of detecting it in the state $\eta_{-}^{(s)}$ as well: $\eta_{+}^{(s)} = -\eta_{-}^{(s)}$. Thus, in the scenario of spontaneous mirror symmetry breaking the "sign" of the chiral purity of the prebiosphere is the result of memorizing the random choice of the sign of fluctuations sign $\eta^{(s)}$ = sign η_0 of the initial state[36,106,107].

6.2. "Ampliciation" of the AF in the Course of Spontaneous Deracemization

In recent years the subject of extensive discussions has been the question of whether the AF caused by WNC could deterine the "sign" of the chiral purity of the biosphere (see Section 5, and, for example, Refs. 36, 90 and 127-132, as well as the references cited in those publications). The point of view expressed in such discussions is that systems capable of spontaneous deracemization are capable of "amplifying" even such weak AFs. Moreover, the authors of Refs. 130-131 put forward a hypothesis of an abnormally high sensitivity of cooperative chiral systems of type (38) is "slow passage of the critical point." It is assumed that as a result of such dynamics of the system, the AF, though small ($g \sim 10^{-17}$), even in the presence of fluctuations, will be able to determine the sign of chirality of the final state of the system. Thus, physically the question about "amplification" of the AF in systems of deracemizing type comes down to the question of competition between the AF and statistical fluctuations.

The dynamic equations corresponding to the system (38) in the presence of the AF take the form

$$\frac{d\eta}{d\tau} = \frac{\rho}{\theta}(g - \eta) + a\theta(\eta - \eta^3)$$
$$\frac{d\theta}{d\tau} = \rho + \rho\theta - b\theta^2 - a\theta^2(1 - \eta^2) \tag{40}$$

where $g = (k_1^L - k_1^D)/(k_1^L + k_1^D)$, other symbols being the same as in Equation (39). The bifurcation diagram corresponding to Equation (40) (Fig. 17) differs from the bifurcation diagram for the case of AF $= 0$. An analysis shows that the effect of the AF brings about essential changes $\eta^{(s)}(\rho)$ only in the "strong-field" domain $|\rho/\rho_c - 1| \sim g^{2/3}$. Beyond this domain ("weak field") the changes of $\eta^{(s)}(\rho)$ are small $\sim g$. It is easy to see that the AF singles out one branch of steady stationary states, namely, $\eta_+^{(s)}$ at $g > 0$. Consequently, if a chemical system that starts from the racemic (on the average) state reaches $\eta_+^{(s)}$ in spite of the presence of fluctuations, it is natural to consider the role of the AF to be decisive in the formation of the "sign" of the chirally ordered state. In Refs. 132 and 133, it is shown that the mechanisms of AF "amplification" in the "strong-field" and "weak-field" domains are essentially different.

In the "weak-field" domain ($\rho \gg \rho_c$), though the racemic state gets into the region of attraction of the stationary state $\eta_+^{(s)}$, nevertheless at $\rho \gg \rho_c$ the boundary of the region of attraction of an alternative state $\eta_-^{(s)}$ is at a distance $\sim g$ from the value $\eta = 0$. Therefore, because of the smallness of g, the problem of AF amplification becomes first of all a problem of "preparation" of the initial state. It is easy to understand that an effective amplification can be expected only if the amplitude of fluctuations in the initial state is much smaller than g [127]. Since the amplitude of fluctuations of chiral polarization depends on the stock of chiral material $\sqrt{\sigma_\eta} = N_\chi^{-1/2}$, at $g \leq 10^{-17}$ amplification is possible only in "global"-scale systems[36,127]:

$$N_\chi \gg g^{-2} \sim 10^{34}. \tag{41}$$

If otherwise, i.e. $N_\chi < g^{-2}$, the probabilities of attaining stationary states $\eta_+^{(s)}$ and $\eta_-^{(s)}$ differ by the value $(g/\sqrt{\sigma_\eta}) \ll 1$. The meaning of the above

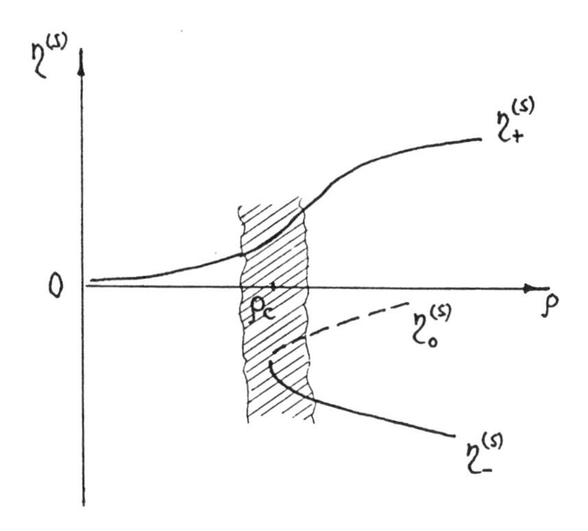

Figure 17. Bifurcation diagram of system (38) in the presence of AF. Strong-field domain is hatched.

condition (41) is simple: the amplitude of fluctuations of chiral polarization in the racemic medium suppresses the AF in smaller-scale systems, and the sign of chiral polarization of the final state depends on the fluctuations rather than on the AF.

The above estimate (41) has been obtained in the approximation of a spatially homogeneous system. In the weak-field domain, however, the characteristic time of the system evolution is small, and possible heterogeneity of such large systems should be taken into consideration. Ya. B. Zel'dovich and A.S. Mikhailov[135] studied the effect of WNC on the cooperative chiral system of type (38) with allowance for spatial diffusion. They showed that if at the earliest stage of evolution the medium happened to be casually subdivided into "domains" with different "signs" of chirality, the WNC-conditioned AF cannot withstand diffusion and it is incapable of determining the chirality of the whole prebiosphere.

In the "strong-field" domain, i.e., at $|\rho/\rho_c - 1| \lesssim g^{2/3}, |\eta| \lesssim g^{1/3}$, the time scale of variations of the varaibles η and θ differs strongly: the variation rate of the overall concentration of the antipodes $(X_L + X_D)$ is essentially higher than the variation rate of chiral polarization. Therefore, θ reaches its stationary state faster than η. This allows, by using the procedure of adiabatic elimination of the "rapid" variable, to reduce the system of Equations (40) to one equation that describes the dynamics of the parameter of order η (for details cf. Ref. 133):

$$\frac{d\eta}{d\tau} = -A\eta^3 + A\left(\frac{\rho}{\rho_c} - 1\right)\eta + Cg \tag{42}$$

where $A = \frac{k_3 - k_2}{2k_3}$; $C = \frac{k_3 + k_2}{2k_{-2}}$. The stochastic equation, corresponding to Equation (42), within the framework of the Langevin approach, has the form

$$\frac{d\eta}{d\tau} = -A\eta^3 + A\left(\frac{\rho}{\rho_c} - 1\right)\eta + Cg + \sqrt{\varepsilon_\eta}\xi_\eta(\tau) \tag{43}$$

where $\sqrt{\varepsilon_\eta}\xi_\eta(\tau)$ is "white noise" with zero average and amplitude $\sqrt{\varepsilon_\eta}$. We should like to stress, however, that the description of the dynamics of the system (40), undergoing evolution from the racemic state, on the basis of Equation (43) is possible only within the "strong-field" domain $|\rho/\rho_c - 1| \lesssim g^{2/3}, |\eta| \lesssim g^{1/3}$ and at times $\tau < g^{-2/3}$. Beyond this domain the system must be analyzed on the basis of the system of Equations (40) with fluctuations taken into account.

Before passing over to studying the amplification mechanism in the strong-field domain, we should like to note that the racemic state $\eta = 0$ gets into the region of attraction of the stationary state $\eta_+^{(s)}$, but it is at the distance $\sim g^{1/3}$ from the boundary $\eta_0^{(s)}$ of the attraction region $\eta_-^{(s)}$, and not at the distance $\sim g$ from it, as in the case of the weak field. Consequently, if the amplitude of fluctuations of the initial racemic state $\sqrt{\sigma_0} < g^{1/3}$, the system as a whole proves to be within the region of attraction of $\eta_+^{(s)}$, singled out by the AF. In principle, the condition $\sqrt{\sigma_0} < g^{1/3}$ can be easily satisfied. The amplitude of the fluctuations of chiral polarization $\sqrt{\sigma_\eta}$ is determined by the initial number of chiral particles: $\sigma_0 = N_\chi^{-1}(N_\chi = N_A\theta V$, where N_A is Avogadro's number and V is the volume of the system). For laboratory-scale systems ($N_\chi \sim 10^{24}$ and $g \sim 10^{-17}$) we have $\sqrt{\sigma_o} \sim 10^{-12}$, this being a fortiori smaller than the value $g^{1/3} \sim 10^{-6}$. Thus, the problem of the AF amplification in the strong-field domain is reduced to analysis of the dynamics of the average value of the chiral polarization $\overline{\eta}(\tau)$ and its variance $\sigma(\tau)$.

The evolution of chiral polarization in the domain embracing almost the entire range $|\eta| \lesssim g^{1/3}$ (with the exception of the vicinity $\sim g^{2/3}$ of stationary states $|\eta^{(s)}| \sim g^{1/3}$) is described by an equation appreciably more simple than Equation (43):

$$\frac{d\eta}{d\tau} = Cg + \sqrt{\varepsilon_\eta}\xi_\eta(\tau). \tag{44}$$

The Fokker-Planck equation corresponding to Equation (44) has the form

$$\frac{\partial}{\partial\tau}[P(\eta,\tau)] = -\frac{\partial}{\partial\eta}[CgP(\eta,\tau)] + \frac{1}{2}\varepsilon_\eta\frac{\partial^2}{\partial\eta^2}[P(\eta,\tau)]. \tag{45}$$

It should be noted that for $g = 10^{-17}$ this equation is inapplicable only in a very narrow neighborhood of $\eta^{(s)}$ values, which comes to merely 10^{-6} of the entire range of η variations in the strong-field domain.

From Equation (45), we derive equations for the average value of chiral polarization $\overline{\eta}(\tau)$ and its variance $\sigma(\tau)$

$$\frac{d\overline{\eta}}{d\tau} = Cg; \quad \frac{d\sigma}{d\tau} = \varepsilon_\eta \tag{46}$$

whence

$$\overline{\eta}(\tau) = Cg\tau; \quad \sigma(\tau) = \varepsilon_\eta \tau; \quad (\overline{\eta}_0 = \sigma_0 = 0) \tag{47}$$

Consequently, the "signal-to-noise" ratio $N(\tau) = \overline{\eta}(\tau)/\sqrt{\sigma(\tau)}$, which characterizes AF "amplification" ("signal") against the background of fluctuations ("noise"), is related to time as follows (cf. also Ref. 131):

$$N(\tau) = \frac{Cg}{\sqrt{\varepsilon_\eta}}\sqrt{\tau} \tag{48}$$

Thus, in the strong-field domain the mechanism of AF "amplification" consists in the AF accumulation and is equivalent in the physical sense to the scenario of "gradual evolution." Indeed, as demonstrated in Refs. 132 and 133, AF "amplification" against the background of fluctuations in systems of neutral type proceeds in exactly the same manner. This conclusion appears to be natural rather than unexpected, since almost in the entire strong-field domain the AF is dominant in the system dynamics. The contribution of nonlinear η^3 (deracemizing) terms (as, by the way also of linear terms) far from the stationary states is negligibly small. Consequently, cooperative systems do not have an anomalously high sensitivity to the AF in the vicinity of the critical point.

In the strong-field domain the AF amplification, as shown by Equation (48), depends not only on the values g and ε_η, but also on the residence time T of the system in this domain. For a given AF with the measured g, the time T should not exceed the value $\tau \sim g^{-2/3}$. The maximum attainable value of the "signal-to-noise" ratio $N_{\mathbf{max}}$ is determined as

$$N_{\mathbf{max}} = \frac{Cg}{\sqrt{\varepsilon_\eta}}\sqrt{T} = \frac{Cg^{2/3}}{\sqrt{\varepsilon_\eta}}. \tag{49}$$

In contrast to the neutral systems where the "accumulation" effect is realized only when the racemizing processes are sufficiently weak[132,133], the possibilities of the AF amplification in the strong-field domain are limited by the duration of the system residence in this domain, which has a scale $|\rho/\rho_c - 1| \sim g^{2/3}$ and for very small AFs is extremely narrow. This results in the necessity of strict fixing of the system parameters at large times $\sim T$ for an effective amplification. The arising problem could seemingly be obviated by means of passing the vicinity of the critical point (of the strong-field domain) by varying the control parameter with a certain rate γ. Let us estimate the rate of system travel through this domain, at which the system residence time in it, T, proves to be sufficient for the AF "accumulation" ($N > 1$). Assuming $\tau = \Delta\rho/\gamma$, from Equation (48), we obtain ($\Delta\rho \sim g^{2/3}, \rho_c \sim 1$)

$$N = \frac{Cg^{4/3}}{\sqrt{\varepsilon_\eta}}\gamma^{-1/2}. \tag{50}$$

It should be emphasized that since the accumulation effect is determined only by the time T (at preset g and ε_η) the value γ cannot be selected arbitrarily: this value is rigidly preset by the time $\tau \leq T$.

When analyzing the question of the AF amplification in the scenario of the origin of the chiral purity of the prebiosphere, necessary for the subsequent stage of biopoiesis-formation of self-replicating systems, it should be borne in mind that in this case we are concerned with such an amplification which must result in the chirally pure final state of the system ($\eta^{(s)} = 1$). This implies that at least the final stage of evolution of the chemical system must occur in the weak-field domain. It would seem that with slow passage of the critical point one could expect an AF accumulation sufficient for fixing the system in the vicinity $\eta_+^{(s)}$. However, the strong-field domain for the WNC-conditioned AF is so narrow that with the rate of passage ensuring AF accumulation the chirally pure state cannt be attained. An analysis of Ref. 131 shows that even with the chemical system parameters favorable to the hypothesis of slow passage, during the time of the existence of the Earth ($t = 4.5 \times 10^9$ years), with the rate of passage ensuring sufficient AF accumulation, the system will prove to be in the region where the attainable value of the chiral polarization will not exceed the small value 10^{-2}, this being, certainly, unacceptable from the evolutionary standpoint. The hypothesis of "slow" passage requires additional assumptions concerning quite specific conditions that could have ensured a sharp transition of the system from the strong-field domain far beyond the critical point. As long as the grounds for such assumptions are lacking, the decisive role of the WNCs in the origin of the biomolecular chirality not only by way of evolution, but also by way of biofurcation should be regarded unfounded.

Thus, it should be admitted that "amplification" does not "save" the AF hypothesis of the decisive role of the global AF in establishing the chiral purity, and the fact that exactly *L*-amino acids and *D*-sugars being used in the Earth's biosphere, although they maybe singled out by the WNC-caused AF, is casual. In the course of chemical evolution on the Earth, apparently, a biosphere based on *D*-amino acids and *L*-sugars could originate with equal success.

7. PHYSICAL CONDITIONS OF DERACEMIZATION OF PREBIOSPHERES

The critical parameters responsible for the transition to the chirally pure prebiological medium are conditioned by its state, i.e., by its physico-chemical characteristics. Indeed, the governing parameter introduced in Section 6 has a sufficiently complicated structure:

$$\rho = \rho(x_A, x_B, \ldots, \{k_i^{L,D}\})$$

It depends on the concentrations of the initial products (substrate) A and B, while through the constants of elementary interactions it depends on the stereoselectivity of the processes with the partiicpation of the antipodes, on the temperature of the medium, pH, etc. Therefore, the possibility of deracemization is determined by the state and physical parameters of the initially racemic "primordial soup."

In the scenario of nonequilibrium phase transition (bifurcation scenario) one can formulate a criterion of the capability of the system to irreversible breaking of mirror symmetry, i.e., to the origin of prerequisites for the nascence of life in the system. This criterion is the expectation time of the onset of irreversible deracemization of the prebiological medium[106,107,136].

The existence of an expectation time other than zero follows from the following qualitative considerations.

Suppose that the system (medium) has reached critical conditions. Then, as a result of some individual fluctuation at the moment of time $t_{cr} + \tau_0$ (t_{cr} being the time of attaining critical conditions and τ_0 being the characteristic expectation time of an individual fluctuation in the medium) the development of chiral polarization begins, which, in principle, may end in complete deracemization of the medium. If during the time necessary for this to happen no other fluctuations will arise, or if they arise rarely, and the distribution corresponding to the fluctuations of chiral polarization is narrow (the probability of the first fluctuation being soon followed by another large fluctuation with an opposite sign is small), the state with broken symmetry, generated by the first fluctuation will be extended over the entire available medium; in this case the expectation time $\tau_{ex} = \tau_0$. If the fluctuations are frequent and their distribution is broad (the probability of any fluctuation being soon followed by another fluctuation, capable of "turning" the symmetry breaking in an opposite direction is high), chiral polarization will oscillate in a random manner about the zero value (racemic state) till a large "critical" fluctuation arises at a certain moment of time $t_{cr} + \tau_{ex}$ and initiates steady evolution to the chirally pure state of the medium.

Since in the vicinity of the phase transition point (both for the phase transition of the first and of the second kind) the phases differ little from each other, large-size nuclei of a less symmetrical phase may be formed in a more symmetrical phase. The estimate τ_{ex} will be obtained under the following assumptions. Let in the initially symmetrical medium "nuclei" of a new phase with broken symmtry originate as a result of fluctuations. Let the characteristic time of the origin of such a nucleus be denoted by τ_0 and the variance of the fluctuations of the number of chiral particles in the nucleus, by σ_θ^2. The dynamics of chiral polarization after t_{cr} is assumed to obey the simplest equation for the racemizing processes in the vicinity of the critical point (with the AF vanishingly small):

$$\frac{d\eta}{d\tau} = \eta - \eta^3.$$

For the estimate, the taking into account of the set of such fluctuations may be reduced to the replacement of η by its average value and introducing into the right-hand side of the equation of a "racemizing" term, stemming from the stochastic behavior of the processes in the system. It can be shown that in the given case this term is equal to $-(\sigma_\theta/\kappa_0\tau_0)$[137]. From the bifurcation equation for the thus modified dynamic equation $d\eta/d\tau$ we find that the deracemization will start, if (on the average over the system)

$$\Lambda = \frac{\kappa_0\tau_0}{\sigma_\theta} > 1 = \Lambda_{cr}$$

On the other hand, if on the average $\Lambda \ll 1$, for each individual nucleus this parameter being a random value, namely $\Lambda_i = \lambda_i\Lambda$ (λ_i bieng a certain random function with unit average and variance), the onset of symmetry breaking is the formation of the nucleus with $\lambda_{cr} \leq 1/\Lambda$. Since λ_{cr} for the distribution λ_i is a large value, it may be assumed to obey the statistics of extreme values, giving the following relation between the amplitude of the fluctuation and its expectation time[138]:

$$x - \overline{x} = \sigma_x ln(t/\tau_0).$$

Thus, for the expectation time of the onset of deracemization we have:

$$\tau_{ex} = \tau_0 \exp\left(\frac{\sigma_\theta}{\kappa_0\tau_0}\right). \tag{51}$$

In the time interval between t_{cr} and $t_{cr} + \tau_{ex}$ the prebiological medium exists as a "metastable" phase with the preserved symmetry and growing mass of chiral material, which is characterized by the value θ.

It is apparent that the values determini τ_{ex}, i.e. σ_θ, τ_0 and κ_0 depend on the state of the medium. The relation of these values with the state of the system has been analyzed in Ref. 136. In the model developed there:

$$\begin{aligned}\sigma_\theta &= [(4/3)\pi R_{cr}^3\rho_0]^{1/2}\\ \tau_0 &= (4/3)\pi R_{cr}^3 \mathcal{T}\\ \kappa_0 &\simeq 4\pi\omega\varepsilon R_{cr}^2 I\end{aligned} \tag{52}$$

where R_{cr} is the critical size of the nucleus, $\mathcal{T}$ is the rate of nucleation, I is the flux of θ through the surface of the nucleus. Assuming that the laws of nucleation of a new phase are applicable in the given case[139], for τ_{ex} we derive the equation (for calculation details cf. Ref. 136):

$$\tau_{ex} = \tau_0 \exp\left[\frac{\delta}{\omega\varepsilon}\left(\frac{\rho}{\rho_0}\right)^{\mu}\left(\frac{\Delta H}{k_B T}\right)^{\nu}\right] \tag{53}$$

where ρ and ρ_0 are the densities of the chiral substance in the medium and nucleus respectively, ΔH is the molecular bond energy in the nucleus, δ and μ are nondimensional values, depending on the state of the substance in the medium: for gaseous phase $\delta = 10^3, \mu = 1$; for solution $\delta = 10^{-1}$, $\mu = 2/3; \nu = 9/4, \omega = \exp(\Delta H/k_B T)$ is so-called "adhesion" coefficient in the theory of nucleation; $0 < \varepsilon < 1$ is the coefficient characterizing the selectivity of interaction of the molecules of the nucleus with the molecules of the medium.

We shall now comment upon some of the estimates which follow from Equation (53). Firstly, for the early stages of the evolution of the Universe – for the hadron, lepton, and baryon eras – we obtain $\tau_{ex} > 10^{10^{14}} s$, which is not commensurable with the time of existence of the Universe ($t \sim 10^{17} s$). Naturally, such a value just should be expected.

Secondly, in the scenario of the origin of life from the "primordial soup" on a planet of terrestrial type (at $\rho \sim 10^{21}$ part./cm^3, $T^0 = 300$ K) $\tau_{ex} \sim 10^{13} - 10^{14} s$, which accounts for only a few percent of the time of prebiolofical evolution. In the same scenario there exists an upper limit on the molecular bond energy in the nucleus, within which τ_{ex} does not exceed the time of the existence of the Earth, namely at $(\Delta H/k_B T) \lesssim 30, \tau_{ex} < 4 \times 10^9$ years. It should be noted that this estimate of H corresponds to the energy of hydrogen bonds. Thus, even simplest estimates point with sufficient accuracy (taking into account the roughness of the assumptions made) that the physical conditions on the early Earth are most adequate for the formation of living systems during those times, on the basis of those interactions, and from those particles which are observed in reality (the estimates of particle sizes, minimizing the expectation time give the value $\sim 10^{-7}$ cm).

Expectation time is a criterion of the possibility of the orgin of life under these or those conditions, on these or those cosmic objects. Indeed, the parameters of the medium, characteristic of a given cosmic object (for instance, a planet, a gas-and-dust cloud, etc.) being known, one can estimate the expectation time of the onset of irreversible racemization of the medium. If this time exceeds the existence time of the cosmic object in question, it should be admitted that the origin of life there is impossible, since the deracemization of the medium, which is a prerequisite for the appearance of living structures, will not occur.

The concept of the expectation time of the onset of mirror symmetry breaking allows one to approach a solution of one more extensively debated problem: whether the appearance of life on the Earth was the consequence of a single event or it was the result of competition of several, independently originated prebiospheres. This question may be attacked from the standpoint of deracemization of the prebiosphere. The appearance and coexistence of areas with different signs of the chirality of the organic substance in each of the areas is equivalent to the origin of a plurality of competing prebiospheres, i.e. to the "multiplicity" of the acts of incipience. On the other hand, if the process of deracemization, generated by the "critical fluctuation" has embraced the whole planet, the "act of incipience" was unique.

Let us consider the following mental experiment. Suppose the "primordial soup" to have reached the critical state necessary for a transition to chiral ordering and suppose it occupies two intercommunicating areas (say, the northern and southern hemispheres). Suppose also that the AF is acting in each of the areas, the nature and measure of the AF being identical for the both areas, whereas the AF signs are opposite. The effect of the AF will lead to an excess of one of the antipodes – "one's own" for each area. Nevertheless, if stirring of the substance is sufficiently intensive, the medium on the whole remains racemic. Since such a medium is in the state which is unstable with respect to the fluctuations of chiral polarization, the very first critical fluctuation that appears after the time τ_{ex}, will initiate the process of deracemization. Let even the critical fluctuation be such that the stock of chiral material θ meets the condition of the AF domination in the formation of the sign of symmetry breaking $N_\chi > g^{-2}$. However, insofar as the appearance of the critical fluctuation in this or that area is equiprobable, it should be admitted that, with equal probability, the sign of the chirality of the prebiosphere will be determined by the sign of the AF of any of the areas, i.e., it will be casual for the prebiosphere as a whole. It is easy to understand that the result of this mental experiment is dependent neither on the number of the areas considered nor on the nature of the local AF. Since "colonization" of the medium by the critical fluctuation will take $\sim 10^2 - 10^4$ years (stirring due to currents, etc.), this being substantially less than the expectation time of the critical fluctuation following the first one ($\tau_{ex} \sim 10^7$ years), it can be said with certainty that the deracemization of the prebiosphere is the result of a single event, rather than the consequence of a plurality of local acts of deracemization (irrespective of the presence or absence of the AF).

8. "COLD PREHISTORY OF LIFE"

In the preceding Sections of our review we discussed different physical aspects of the origin of chiral purity of the prebiological medium within the framework of socalled "hot" (terrestrial) scenario of the origin of life. In the present Section we shall consider a totally different approach to the problem of the origin of life: so-called "cold" (cosmic) scenario which presupposes the possibility of at least some stages of biopoeisis to proceed in the cosmic space.

In this connection we shall refer to some recent publications which have revived interest to the, one would think, completely forgotten hypothesis of panspermia, advanced by S. Arrhenius in the beginning of our century. As has been stated in the Introduction, recent geological discoveries strongly "compressed" the time allocated by nature for the prebiological evolution, actually, by the factor of 10: from 2 billion down to 200 million years. This gave rise to doubts as to so small a period of time being sufficient for the prebiological stage of biopoeisis: from small molecules to the formation of the structures capable of self-replication. On the other hand, diverse and very complicated organic compounds have been found in space, to amino acids and oligomers inclusive[13,14]. All this led to the revival of cosmic scenarios of the origin of life – from the hypothesis of "directed panspermia" of F. Crick[140] to the hypothesis of "lifeclouds" of F. Hoyl and N. Wickramasinghe[141]. The first hypothesis of actually a version of the "hot" scenario, though extraterrestrial rather than terrestrial. The second hypothesis is principally different in character: it transfers the major stages of the biopoiesis, the formation of self-replicating structures among them, into open space, to be more exact, into dark gas-and-dust clouds. The possibility of "cold prehistory of life" was first pointed to by V.I. Gol'danskii (see Refs 142, 143 and the references cited there), who proceeded from the low-temperature quantum limit of the rate of chemical reactions, discovered by him. The abovementioned astrophysical discoveries and the experimentally supported existence of the other than zero rate of chemical transformations at low temperatures urge one to treat the hypothesis of the "cold prehistory of life" with greater attention.

An extremely schematic outline of the cosmic scenario based on the contemporary ideas of the chemistry of deep cold and on the data concerning the composition of the interstellar medium may be presented as follows.

In dark gas-and-dust clouds (lifetime, $\sim 10^5 - 10^6$ years; characteristic dimensions, $\sim 10^{17} - 5 \times 10^{19}$ cm; $T^0, \sim 20$ K; overall weight of dust, $\sim 10^{-2}$ of the weight of gas; dust particles of 10^{-5} cm in size consist of a silica or graphite nucleus covered with a "dirty" ice "jacket" containing CO, NH_3, H_2O, HCN, etc.) due to the chemical processes stimulated by the action of long-range protons, hard UV and γ-radiation, mainly in the "dirty ice" on the surface of dust particles all the more complicated organic compounds may be synthesized – up to precursors of biologically important molecules.

As a gravitational instability develops, differentiation of the substance and the formation of a protostar occur in the dark gas-and-dust cloud. From the gas-and-dust disk forming around the star planetosimals accrete, growing up to planets. Consequently, the orgnaic compounds formed in the gas-and-dust cloud may get onto the planet through the agency of two processes: first, in the course of accretion (though upon the birth of the star and upon warming-up of the planet the, apparently, undergo sufficiently strong destruction) and, secondly, already after the formation of the planet due to the adsorption on its surface from the surrounding space. The organic compounds which have thus got onto the planet could serve as the "initial raw material" for the formation of the "primordial soup."

However, this hypothesis as such and is conceivable versions "work" only in the case if under the conditions of cosmic space (low temperatures, hard radiations, small densities of the substance, etc.) at least in principle a situation is realized, in which mirror symmetry breaking of the organic substance becomes possible. We should like to note that for the hypothesis of Hoyle and Wickramasinghe about "living interstellar clouds" this requirement becomes more severe: for the origin of living systems not merely mirror symmetry breaking, but complete deracemization of the medium is necessary (otherwise the key property of the living, viz., self-replication, cannot be formed).

It is necessary to point to an essential difference of the problem of mirror symmetry breaking in the "hot" and "cold" scenarios. In the first of them chemical processes proceed sufficiently fast on the evolutionary time scale and in the course of chemical transformations molecules-isomers preserve the 'sign" of chirality (the very idea of the chirality of an individual molecule has a completely definite sense). In the "cold" scenario the rates of chemical transformations are extremely low and in view of effective tunnelling racemization or racemization induced by hard radiation, whose characteristic times are comparable with the rates of the chemical processes themselves, the molecules can change the "sign" of their chirality more than once. Thus, the process $L \rightleftharpoons D$ in the "cold" scenario may lead to the "loss" of the idea of the "sign" of the chirality of the molecule-isomer at the characteristic times (see Section 3).

Consequently, the first problem associated with mirror symmetry breaking in the "cold prehistory of life" is the problem of stabilization of chirality of the molecules of mirror isomers.

8.1. Stabilization of Chirality of Antipodes in "Cold" Scenario

Let us now consider the possibilities existing for the stabilization of the chirality of molecules within the framework of the "cold" scenario. It will be convenient to analyze this problem in terms of two-level systems[144].

We shall consider first the behavior of an isolated chiral molecule, on condition that only tunnel transitions between the L and D states of the molecule are possible. Let the axis x correspond to the position of some atom (atomic group) or to a certain other collective coordinate of the molecule. In the symmetric potential $(+x)$ corresponds to the L state of the molecule, $(-x)$ corresponds to the D state. By following the same procedure as in Ref. 145, we shall define the operator $\hat{x}$ of the position of the atom in the chiral potential:

$$\hat{x} = |L\rangle\langle L|\hat{x}|L\rangle\langle L| + |D\rangle\langle D|\hat{x}|D\rangle\langle D| \tag{54}$$

but

$$\langle D|\hat{x}|D\rangle = -\langle L|\hat{x}|L\rangle = -x_0$$

where $-x_0$ corresponds to the minimum of the D-well in the chiral potential. In Ref. 145 it is demonstrated that for the isolated molecule the middle position of the $\overline{x}$ molecule in the chiral potential will satisfy the equation of an oscillator

$$\frac{d^2\overline{x}}{dt^2} + (2\Omega)^2\overline{x} = 0 \tag{55}$$

where $\Omega = (1/\hbar)\langle L|\hat{H}|D\rangle$ is the characteristic frequency of tunnel transitions between the L and D states and $\hat{H}$ is the Hamiltonian of the molecule. (The dependence of Ω on the parameters of the molecule and on the potential is given in Section 3). After the period of time $\tau \sim 1/2\Omega$ the particle, which was at the moment of time $t = 0$ in the state L, will be found, due to the tunnel transition, in the state D. Thus, at the times $t > \tau$ the probability of detecting the molecule either in the state L or in the state D will be equal to 1/2.

However, in some papers[144–146] it was shown that if chiral molecules interact with an optically inactive medium, which is a deeply chilled low-density gas, where only two-particle collisions are dominating, under these conditions "stabilization" of the chirality of the molecules-isomers is possible at the times appreciably exceeding τ. The interaction of the isomer with the molecules of the medium results in that the chiral particle behaves not as a free oscillator, but as an oscillator with damping. In terms of the position of the atom in the chiral potential the equation for $\overline{x}$ takes up the form[144,146]

$$\frac{d^2\overline{x}}{dt^2} + \lambda\frac{d\overline{x}}{dt} + (2\Omega)^2\overline{x} = 0 \tag{56}$$

where λ is the damping parameter, dependent on the characteristics of the medium:

$$\lambda = N_0\frac{\pi}{K^2}v\sum_{J,i}\frac{(2J+1)w_i^J}{(2j_1+1)(2j_2+1)}\langle Ji|S_L^{J\dagger}S_D^J - 1|Ji\rangle$$

N_0 is the numerical density of the atoms of the medium; K is the relative momentum in the center-of-mass system; j_1 and j_2 are the internal angular momenta of the partners in collision; w_i^J are statistical weights; $S_{L,D}$ are scattering matrices; v is the velocity of the atoms of the medium.

As a result of interaction with the medium, the particle which was initially in the state L, for instance, will be "stabilized" in that state:

$$\overline{x}(t) = \overline{x}(0)\cos(2\Omega t)\exp\left(-\frac{\Omega^2}{\lambda}t\right). \tag{57}$$

Thus, though racemization does take place, the characteristic time of this process $\tau_r = \lambda/\Omega^2$ increases sharply, compared with the racemization time $\tau = 1/2\Omega$ of the isolated particle. The racemization "rate constant" $R = \tau_r^{-1}$ is temperature-dependent in a very non-trivial manner (Fig. 18). As a result, even small molecules, interacting with the medium at low temperatures, will be for a long period of time be in the state with a definite chirality, whereas transitions between the states with the opposite chirality (tunnelling racemization) will be suppressed.

8.2. Problem of Deracemization of the Medium in "Cold" Scenario

It is understandable that the very fact of the stabilization of the chirality of molecules at low temperatures due to the interaction with the medium is very attractive for the "cold" scenario. At the same time, however, it should be borne in mind that this involves the problem of deracemization of the medium as a whole. Indeed, at the initial moment of time each of the molecules-isomers is in the state with a definite chirality, i.e. it is either in the L state or in the D state. Nevertheless, an ensemble of such molecules will be, in all probability,

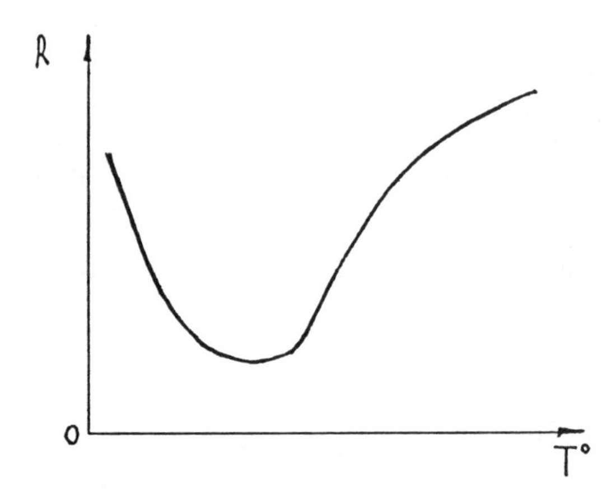

Fig. 18. Racemization rate constant vs. temperature for scenario of "cold prehistory" of life[145].

in the racemic state, and stabilization of the chirality of each molecule will in this case work to the benefit of racemization.

An excess of molecules of one particular chirality (for instance, of L-isomers) may arise, in principle, due to the AF (WNC, β-decay), since the potential loses symmetry because of parity nonconservation, and the L-state for amino acids and D-isomers of the precursors of sugars becomes more advantageous energetically (see Refs. 82 and 83). We should like to note that the amplitude of the AF increases by 1.5-2 orders of magnitude at the low temperatures characteristic of gas-and-dust clouds, and the entropy factor diminishes under these conditions.

It should be remembered, however, that "amplification" of the AF in the scenario of "gradual evolution" is possible only in the processes of neutral type, i.e. either in the process of stereoselective destruction ($L \xrightarrow{k^L} A, D \xrightarrow{k^D} A$) under the effect of polarized radiations or in the stereoselective autocatalytic synthesis ($A + L \xrightarrow{k^L} 2L, A + D \xrightarrow{k^D} 2D$), where the difference in the constants is caused by WNC (see Sections 4 and 5). The first of these processes, i.e. destruction has a very essential disadvantage: in the course of destruction the chiral material is destroyed, the decrease in the number of the molecules being exponentially time-dependent

$$\theta(t) = \theta(0)\exp(-kt); \quad k = \frac{1}{2}(k^L + k^D)$$

Consequently, destruction leads to a rapid decrease of the number of chiral molecules; this, with the rate of the synthesis processes under cosmic conditions being very low, makes such a process unsuitable for the cosmic scenario.

The process of autocatalytic synthesis in real conditions (in the presence of unavoidable racemizing processes), as shown in Section 4, may lead only to a small excess of one of the isomers: $\eta_{\max} \sim g/K_R$ (g is the measure of the AF and K_R is the effective rate constant of the racemizing process). But even if we leave out of account the question of possible η_{max}, it should be emphasized that the process of autocatalysis under cosmic conditions can hardly take place. The accumulation of molecules on the surface of dust particles due to long-term sorption does not save the situation, since the molecules are fozen into the ice "jacket"; their mobility is thereby sharply limited (the probability of bimolecular reactions and complicated chains of reactions, necessary for the autocatalysis to be realized, is small).

Therefore, one should hardly expect that in the scenario of the "cold pre-history of life" the problem of breaking of the mirror symmetry of the organic substance can be solved within the framework of the scenario of "gradual evolution" under the effect of the AF. The problem which arises here is the same as in the "hot" scenario: searching for the processes of deracemizing type, which lead to spontaneous mirror symmetry breaking. This problem within the framework of the "cold" scenario is still more non-trivial than in the "hot" scenario. Nevertheless, one should not refute a possibility, that the chemistry of deep cold, cosmo-chemistry-fields which have just started developing–in combination with the ideas of self-organization of nonequilibrium processes may offer possbile (and unexpected) solutions of this problem. This will, no doubt, extend our concepts concerning the pathways of biopoiesis.

9. CRITICAL LEVELS OF DERACEMIZATION OF THE BIOSPHERE

The analysis of the effects of global impacts on the environment is increasingly urgent, since the development of civilization has reached a level at which the anthropogenic impact can very decisively affect the function of the biosphere (see, for example, Refs. 148 and 149). However, the influence of long-term, large-scale impacts on one of the fundmanetal properties of life – its chiral purity – has yet to be analyzed. In this Section we would like to discuss the existence of critical levels of disturbance of the chiral purity of the biosphere during long-term racemizing global impacts. One catastrophic aftereffect of these impacts may be the collapse of living nature to the state preceding its appearance.

As shown earlier (Section 2), the chiral purity of the organic medium is a necessary condition for the birth of life on the Earth – the appearance of self-replicating systems. The chiral purity of systems whose replication is based on the property of complementarity must be maintained during their evolution. Otherwise the system loses the ability for self-reproduction. This means that at the stage of formation of the early biosphere the disturbed mirror symmetry of the organic environment was inherited by self-replicating systems, was maintained by them, and was fixed as a fundamental property of the bioorganic world[1,103]. Thus, the contemporary biosphere as an open system located in a mirror-symmetrical environmental must maintain a characteristic chiral purity. This is why we deem it necessary to investigate the effects of additional racemizing impacts upon this property of the biosphere.

Such an investigation requires the construction of an adequate model. This is so because the fact of the chiral purity of the contemporary biosphere by itself does not contain sufficient information for modelling the dynamic laws responsible for such a property. The lack of information can, in our opinion, be remedied by analyzing the reasons for the appearance of chiral purity and the dynamic laws responsible for the appearance and maintenance of this feature throughout the evolution, starting with the stage of the appearance of life.

The required model can be constructed in the following manner. The biosphere is represented in the form of two interacting subsystems P and A, one of which (P) is responsible for the production of chiral substances at the expense of the achiral substance S and energy arriving from outside. At the present stage of evolution subsystem P may include, in particular, the plant world, producing chiral material by means of photosynthetic processes. The second subsystem (A) consists of the objects α using the chiral material of subsystem P for their own reproduction. In this case the consumption of one of the antipodal forms of the chiral substrate (for specificity, L) results in the self-reproduction of α, while the consumption of another form (D) results in the loss of this ability by object α, its death $\alpha \to \alpha'$, and the transition of α to the system P in the form of a racemate. (The processes of the racemization of bioorganic matter and their efficiency in relation to the external conditions were considered, for example, in Refs. 94 and 150.

An analysis of the question of the maintenance of chiral purity in the biosphere is conveniently done in terms of the total amount of chiral substrate $\theta = L + D$ and the parameter of the chiral organization of the subsystem P – its chiral polarization $\eta = (L - D)/(L + D)$. A racemic state corresponds to the value $\eta = 0$, while a chirally pure state of subsystem P, to $\eta = 1$.

Dynamic laws controlling the behavior of subsystem P can be formulated if they are considered, as indicated above, as inherited from the stage of evolution at which there occurred the spontaneous breaking of the mirror symmetry of the organic environment and the formation of its chiral purity. This enables us to use the results of Refs. 36 and 103 in which it was shown that such an event can occur only in a bifurcation system.

In the simplest model we examined[28,154] in which subsystem P is represented by the typical scheme analyzed in detail in Ref. 36, the dynamic equations have the following (dimensionless) form:

$$\frac{d\alpha}{d\tau} = \theta(\eta - \varepsilon)\alpha - \left(\frac{1}{T}\right)\alpha \tag{58}$$

$$\frac{d\alpha'}{d\tau} = (1 + \varepsilon)\theta(1 - \eta)\alpha + \left(\frac{1}{T}\right)\alpha - k_0 k'\alpha' \tag{59}$$

$$\frac{1}{k'}\frac{d\eta}{d\tau} = \left[\delta\theta - k_0\frac{\alpha'}{\theta} - k_r\right]\eta - \delta\theta\eta^3 + \frac{1}{k'}\varepsilon(1 - \eta^2)\alpha \tag{60}$$

$$\frac{1}{k'}\frac{d\theta}{d\tau} = k\theta - \theta^2 + \delta\theta^2\eta^2 - \frac{1}{k'}\theta(1 - \varepsilon\eta)\alpha + k_0\alpha' \tag{61}$$

where $\tau = (k_1 + k_2)t/2$ (k_1 and k_2 are the constants of the rates of replication and death of objects α during the consumption of the L and D isomers, respectively), T is the dimensionless lifetime of α, k_r is the adjusted constant of the rate of racemization of the material of subsystem P, k is the adjusted stock of the achiral raw material S (and of energy) while δ, ε, k_0 and k' are ancillary parameters of the model. To describe the chiral organization of the biosphere, the quantity H is used, related to the variables introduced above in the following manner: $H = (\alpha + \theta\eta)/(\alpha + \theta)$.

An answer to the question concerning the existence of critical levels of the chiral purity of the biosphere is found from an analysis of the steady-state solution of the system of Equations (58)-(61). There are three possible stages in the examined system. First, the one we observe today, state I: $\alpha > 0, \theta > 0, H = 1$. Second, the state in which the entire biosphere is represented by subsystem P in a chirally polarized state, state II: $\alpha = 0, \theta \neq 0$, and $H = \eta \neq 0$. Finally, state III, characterized not only by the absence of subsystem A but also by complete racemism of subsystem P: $\alpha = 0, \theta > 0$, and $H = \eta = 0$. Such a state of the organic environment is characteristic of the prebiological stage of evolution on the Earth before the mirror symmetry breaking.

Negative anthropogenic impacts may influence both subsystem A and subsystem P. The direct impact on subsystem A is trivial. A reduction of the characteristic lifetime τ of objects α to some critical value τ_c results in the disappearance of subsystem A with the retention of all the properties of subsystem P (transition from state I to state II). However, an impact on subsystem P, even without a direct influence on subsystem A, may result in effects equally dramatic for subsystem A. As shown by an analysis of the model, there exists a critical level of disturbance of the chiral purity of the biosphere, i.e., a critical level of "pollution" of the biosphere with an unnatural isomeric form of molecules, upon the attainment of which the existence of subsystem A is impossible. This occurs when the impact of subsystem P exceeds some critical level determined by the racemization of the chiral substance, by the stock of achiral material, etc., i.e. it is aimed specifically at those parameters of subsystem P that are responsible for the formation and maintenance of the chiral purity of the biosphere. (This situation is illustrated by Figs. 19 and 20, and corresponds to a transition from branch (1) to branch (2).) We term such a transition the *catastrophic loss of chiral purity*. It should be noted, however, that even if such a catastrophe occurs, it is still in some sense "reversible" in relation to the possibility of an "evolutionary restoration" of the situation. After the cessation of impacts on the biosphere its chiral purity will be restored and, consequently, the appearance of subsystem A is possible in principle (although it is not entirely apparent that it will consist of the same objects).

Impacts on subsystem P may have even more catastrophic effects for the biosphere as a whole. In addition to the first critical level of impacts, leading to the diappearance of subsystem A, there exists a second critical level, upon the attainment of which not only subsystem A is destroyed, but complete racemization of subsystem P occurs as well (see Fig. 19, k_1 — k_r^2, Fig. 20, transition from branch (2) to branch (3)). The biosphere disappears and the entire system is in the state preceding the birth of life – the racemic state of the environment. In the sense indicated above, such a transition is completely irreversible: for the evolutionary process to begin, Nature "must wait" for a spontaneous breaking of the mirror symmetry in the prebiological racemic

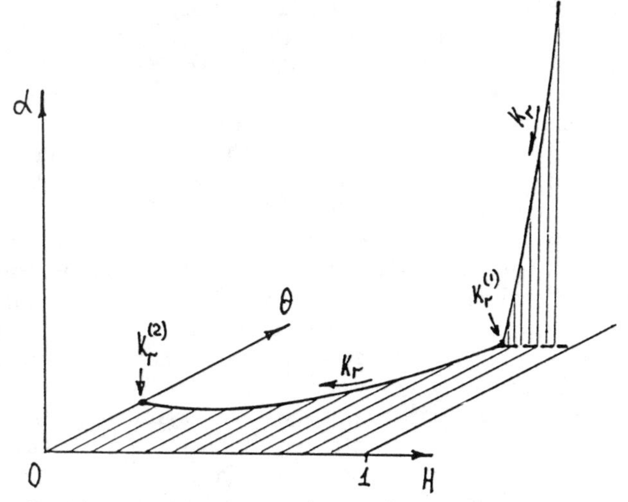

Fig. 19. Steady-state conditions (α, H, θ) vs. racemization of chiral substrate: $k_r^{(1)}$ is the first critical value of racemization rate constant k_r: $k_r^{(2)}$ is the second critical value of k_r; the direction of k_r increase is indicated by arrows.

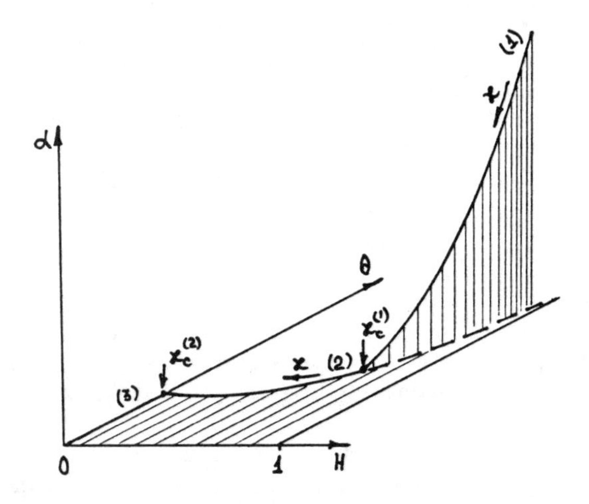

Fig. 20. Stead-state conditions (α, H, θ) vs. achiral stock κ: $\kappa_c^{(1)}$ is the first critical value of κ_c; $\kappa_c^{(2)}$ is the second critical value of κ_c; the direction of κ decline is indicated by arrows.

environment. (An estimate of the expectation time for such an event under terrestrial conditions was obtained in Section 7.) The attainment of the second critical level of impacts and the transition of the biosphere to state III may be termed the *collapse of living nature.*

It is now clearly apparent that long-term global impacts can result in the breaking of the biogeocenosis all over the Earth. We should like to point out, however, that additional impacts racemizing the biosphere may affect the fundamental properties of living nature and upon the attainment of the critical levels result in real disappearance of life on the Earth.

10. CONCLUSION

It was not our objective to cover all the aspects of the problem of biomolecular chirality. For instance, a very interesting problem of the biological significance of the chiral purity of most important biomolecules for the functioning of organisms under normal and pathological conditions, the problem of maintaining chiral purity in contemporary organisms and of the consequences of the organism contamination with "unnatural" isomers, proved to be beyond the scope of our survey. At present, the medico-biological aspects of the "asymmetry of protoplasm" attract all the greater attention of experts in most diverse fields of knowledge.

The objective pursued in this review was to demonstrate the fruitfulness of physical approaches to the problem of the origin of chiral purity of the biosphere and to show that such approaches allow one not only to formulate different hypotheses about the mechanisms of mirror symmetry breaking but also to analyze them both qualitatively and quaititatively.

It is worth pointing out that this problem which only recently seemed to be purely biological requires recruiting most diverse approaches for its solution: from the ideas of the microworld physics (weak interactions and parity nonconvservation, quantum tunnelling) to the ideas of the mathematical and statistical physics, of the theory of nonlinear processes, chemical physics, etc. Resorting to these ideas and methods has made it possible, in our opinion, to outline the ways for settling some crucial questions in the problem of deracemization of the biosphere and of the origin of life.

For instance, the application of the fundamental concepts of chemical physics – the science dealing with the kinetics of chemical reactions and the chemical structure of the substance – made it possible to make a most important conclusion that the origin of the chiral purity of the major "structural constituents" of the living – of sugars and amino acids – took place at the stage of prebiological evolution and is a necessary condition, without which self-replication cannot arise. Thus, chiral purity is the relic property the biosphere had inherited from the stages of the chemical evolution of the organic medium on the primitive, prebiological Earth.

The analysis of the processes in nonequilibrium chiral systems provided the possibility for a critical estimation of numerous hypotheses which associated the origin of the chiral purity with the effect of the factor of advantage and for deriving a conclusion that the processes of "gradual evolution" are unable to ensure, under the effect of the AF, strong breaking of the mirror symmetry at the stage of prebiological evolution.

Today the most acceptable scenario of the origin of the chiral purity of the prebiosphere, of its deracemization, is the scenario of spontaneous mirror symmetry breaking, which relates the origin of this most important property of the living with the processes of self-organization of chirality in the "primordial soup." The very fact of the existence of the chiral purity of the biosphere should be regarded as an evidence of the "experiment" successfully carried out by Nature on the absolute asymmetrical synthesis of organic compounds with the 100% asymmetrical yield in the course of the process of self-organization! Thus, the formation of the chiral purity of the prebiosphere may be called *"Biological Big Bang"* which paved the way for the further stages of evolution.

The results obtained in the analysis of the role of the AFs allow one to discuss the possibilities of detecting small AFs in cooperative chemical systems. In principle, there exist two possibilities: detection in the weak-field domain and detection in the strong-field domain (see Section 6). The first of these (with which chirally pure states $\eta \sim 1$ are attained) is principally limited by the minimal necessary size of the chemical chiral system ($N_\chi > g^{-2}$). Under laboratory conditions ($N_\chi \sim 10^{24}$) it can be realized for not too small AFs only ($g > 10^{-12}$). The second possibility is associated with the mechanism of AF "amplification" in the vicinity of the critical point. Its realization depends, first of all, on whether such conditions as rigid fixing of the system parameters in the vicinity of the critical point (slow variations of the control parameter in the domain with large fluctuations) and the time required for the "accumulation" of asymmetry to the experimentally observed value can be satisfied in the experiment. Indeed, if the value g is such that the time required for the experiment (the time of the system residence in the strong-field domain) $T \sim g^{-2/3}$ is reasonable, then, by choosing an appropriate low rate of passage, one can attain the effect of AF "accumulation." In principle, this may prove also possible for systems in which the number of particles is appreciably smaller than g^{-2}. But for this to be the case, for instance at $g = 10^{-17}$, it would be necessary to be in a very "narrow" domain near the critical point ($10^{-12}\rho_c$) during the time $\sim 10^{11}$ s.

It should be noted that experiments on modelling the processes of the origin of asymmetry (symmetry breaking) in chiral systems up till now have been pracitcally limited to experiments with different AFs (and to some studies on the relation of the matrix synthesis of oligonucleotides with the chiral composition of the medium). The studies were concerned, mainly, with the processes of destruction or catalytic synthesis. We believe, that experimental search should also be carried out towards realization of systems, in which chains of physico-chemical transformations establish positive feedbacks in respect of chiral polarization of products – systems with spontaneous mirror symmetry breaking.

It would be also of interest to carry out experiments aimed at investigating the processes of stabilization of the chirality of molecules-isomers in solids and in the gaseous phase at low temperatures, which would simulate the conditions of the "cold" scenario.

It would be possible to extend the life of investigations of this problem, which are of both practical and theoretical interest, but we believe that extensive joint efforts of physicists and chemists engaged in various fields are required here. We should like to stress that in this problem formulated by L. Pasteur, in the problem of the origin of the chiral purity of the biosphere, the number of questions still remaining to be solved exceeds the number of the solved ones in spite of one hundred years of investigations.

ACKNOWLEDGEMENTS

The authors express their gratitude to Drs. V.A. Avetisov and Yu. A. Berlin for valuable discussions of and assistance in preparing the paper.

REFERENCES

1. L. Pasteur, Recherches sur la Dissymetrie Moleculaire (1860), Oeuvres de Pastuer, Vol. 1, ed., Pasteur Valery-Radot (Masson, Paris, 1922), p. 314.
2. P. Curie, J. Physique Theor. Appl. Ser. 3, 393 (1894).
3. Ld. Kelvin, Baltimore Lectures (Clay, London, 1904), p. 618.
4. P. Jordan, Die Physik und das Geheimnis des Organischen Lebens (Braunschweig, F. Vieweg u. Sohn, 1947).
5. E. Fischer, Untersuchungen über Kohlen Hydrate und Ferment (Verlag von J. Springer, 1909).
6. J.D. Bernal, The Origins of Life (G. Weidenfield and Nicholson, London, 1967).
7. W. Bonner, in: Exobiology, ed., Ponnamperuma (North-Holl., Amsterdam, 1972), p. 170.
8. E. Broda, Origins of Life 14, 391 (1984).
9. L.L. Morozov, E.I. Fedin and M.I. Kabachnik, Russ. J. Phys. Chem. 47, 2193 91973); 47, 2200 (1973); 47, 2210 (1973).
10. L.L. Morozov, Ph.D. Thesis, Moscow, (1975), (in Russian).
11. L.L. Morozov, A.A. Vetrov, M.S. Vaisberg and V.V. Kuz'min, Doklady Phys. Chem. 247, 875 (1979).
12. A.A. Vetrov, L.L. Morozov, Russ. J. Phys. Chem. 58, 614 (1984).
13. J.M. Greenberg, Origins of Life 14, 25 (1984).
14. W.H. Irwine and Å Hjalmarson, Origins of Life 14, 15 (1984).
15. Earth's Earliest Biosphere: Its Origin and Evolution, ed., J.W. Schopf (Princeton University Press, 1983).
16. Search of Universal Ancestor (NASA Sci. & Tech. Inform. Branch, 1985).
17. Ch.B. Thaxtot, W.L. Bradby and R.L. Olsen, The Mystery of Life's Origin (Phil. Library, NY, 1984).
18. R. Shapiro, Origins: A Sceptic Guide to the Creation of Life in Earth (Univ. Press, NY 1986).
19. V.I. Goldanskii, V.A. Avetisov and V.V. Kuz'min, Dikl. Biophys. 290, 734 (1986).
20. V.I. Goldanskii, V.A. Avetisov and V.V. Kuz'min, FEBS Lett. 207, 181 (1986).

21. B.L. Bass and T.R. Cech, Nature 308, 820 (1984).
22. C. Gurrier-Takada, et al., Cell, 849 (1983).
23. A.J. Zang and T.R. Cech., Science 231, 470 (1986).
24. G.F. Joyce, et.al., Nature 310, 602 (1984).
25. G.M. Visser, Thesis D.Sci., Leiden Univ., (1986).
26. G.M. Visser, et.al., Recl. Tav. Chim. Pays-Bas 103, 141 (1984).
27. V.A. Avetisov, S.A. Anikin, V.I. Goldanskii and V.V. Kuz'min, Dokl. Biophys. 282, 115 (1985).
28. V.I. Goldanskii, E.A. Anikin, V.A. Avetisov and V.V. Kuz'min, Comments Mol. Cell. Biophys. 4, 79 (1987).
29. L.L. Morozov, E.I. Fedin, IX Ampere Congr. Abstr. (Heidelberg, 1976).
30. D. Kemp, in: Peptides, Ed., E. Ross and J. Meienhofer, Vol. I (Academic Press, NY-London, 1979).
31. J. Bada, R. Schroeder, Naturwiss 62, 71 (1975).
32. S.J. Jacobson, C.G. Wilson, H. Rapoport, J. Org. Chem. 39, 1074 (1974).
33. J.L. Bada, Adv. Chem. Ser. 106, 309 (1971).
34. J.L. Bada, S.L. Miller, Proc. Conf. on Chiral Symmetry Breaking, Rouen, 1986 (C.U. Newsletter No. 12, Strassbourg, 1986).
35. F. Hund, Z. Phys. 43, 805 (1927).
36. L.L. Morozov, V.V. Kuz'min and V.I. Goldanskii, in: V.P. Skulachev (ed.) Sov. Sci. Rev. Ser. D. Physicichem. Biol., Vol. 5 (Harwood Acad. Publ. NY-London, 1984) p. 357.
37. H. Haken, Synergetics (Springer, Berlin-Heidelberg-NY, 1978).
38. G. Nicolis and I. Prigogine, Self-Organization in Nonequilibrium Systems (J. Wiley & Sons, NY, 1977).
39. H. Kuhn, Angew. Chem. 84, 838 (1972).
40. H. Kuhn, in: Synergetics Proc. International Workshop of Synergetics, Bavaria (Springer, Berlin, 1977), p. 325.
41. L. Pasteur, Bull. Soc. Chim. Fr. 41, 215 (1884).
42. V.A. Kizel, Physical Causes of Dissymetry of Living Systems, Moscow, Mir., Nauka Publishers, 1985 (in Russian).
43. S.F. Mason, Molecular Optical Activity and the Chiral Discrimination (Cambridge, Cambridge, 1982).
44. H. Kagan, G. Valavoine and R. Moradpur, J. Mol. Evol. 4, 41 (1974).
45. B. Norden, Nature 266, 567 (1977); J. Mol. Evol. 11, 313 (1978).
46. C.D. Tran, J.H. Fendler, in: Origin of Optical Activity in Nature, D.C. Walder (ed.), (Elsevier, Amst.-NY-Oxf., 1979), p. 53.
47. P. Gerike, Naturwiss. 62, 38 (1975).
48. K. Piotrowska, et.al., Naturwiss. 67, 442 (1980).
49. L. Mörtberg, in: Origins of Optical Activity in Nature, D.C. Walker (ed.), (Elsevier, Amst.-Oxf.-NY, 1979), p. 101.
50. W. Rhodes and R.C. Dougherty, J. Amer. Chem. Soc. 100, 6247 (1978).
51. K.L. Kovacs, L. Keszthelyi and V.I. Goldanskii, Origins of Life 11, 93 (1981).
52. R.C. Dougherty, Origins of Life 11, 71 (1981).
53. N.B. Baranova and B.Ya. Zel'dovich, Molec. Phys. 38, 1085 (1979).
54. G. Wagniere and A. Meier, Chem. Phys. Lett. 93, 78 (1982).
55. N. Hokkyo, Origins of Life 14, 447 (1984).

56. F. Vester, T. Ulbricht and H. Krauch, Naturwiss 46, 59 (1959).
57. T. Ulbricht, Origins of Life 6, 303 (1975).
58. V. Letokhov, Phys. Lett. 53*A*, 275 (1975).
59. D.W. Rein, J. Mol. Evol. 4, 15 (1974).
60. P.G. de Gennes and C.R. Hebd. Seances, Acad. Sci. Ser. B. 270, 891 (1970).
61. C.A. Mead, A. Moscowitz, H. Wynberg and F. Meuwese, Tetrahedron Lett., No. 12, 1063 (1977).
62. A. Peres, J. Amer. Chem. Soc. 102, 7389 (1980).
63. L.D. Barron, Mol. Phys. 43, 1395 (1981).
64. C.A. Mead and A. Moskowitz, J. Amer. Chem. Soc. 102, 7301 (1980).
65. L.D. Barron, J. Amer. Chem. Soc. 108, 5539 (1986).
66. L.D. Barron, Chem. Soc. Rev. 15, 189 (1986); Chem. Phys. Lett. 123, 423 (1986); Biosystems 20, 7 (1987).
67. W. Kuhn and F. Braun, Naturwiss. 17, 227 (1929).
68. J.D. Morrison and H.S. Mosher, Asymmetric Organic Reactions (American Chemical Society, Washington, DC, 1976).
69. J.R.P. Angel and R. Liting, Nature 238, 389 (1972).
70. G. Gilat, Chem. Phys. Lett. 121, 9 (1985); Chem. Phys. Lett. 121, 13 (1985); Chem. Phys. Lett. 125, 129 (1986).
71. A.B. Vistelius, Zap. Mineral. Ob-va. 79, 191 (1950) (in Russian); G.G. Lemlein, Morphology and Genesis of Crystals, Moscow (1973) (in Russian).
72. T-D. Lee and C-N. Yang, Phys. Rev. 104, 254 (1956).
73. C-S. Wu, et.al., Phys. Rev. 105, 1413 (1957).
74. J. Le Bell, Bull. Soc. Chim. (4) 37, 353 (1925).
75. Ya. B. Zel'dovich, Sov. Phys. JETF 36, 964 (1959).
76. R. Walgate, Nature 303, 473 (1983).
77. L.M. Barkov and M.S. Zolotarev, Sov. Phys. JETF 79, 713 (1980); P. Buksbaum and E. Commins, L. Hunter, Appl. Phys. *B*28, 280 (1982); T. Emmons, J. Reeves and E. Torston, Phys. Rev. Lett. 51, 2089 (1983); G.N. Borich, et.al., Sov. Phys. JETF 87, 776 (1984).
78. Y. Tamagata, J. Theoret. Biol. 11, 495 (1966).
79. Ya.B. Zel'dovich, Sov. Phys. JETF 67, 2357 (1974).
80. D.W. Rein, R.A. Hegstrom and P.G.H. Sandars, in: Origins of Optical Activity in Nature, D.C. Walker (ed.) (Elsevier-NY-Oxf-Amst., 1979), p. 21; K. Wagener, J. Mol. Evol. 4, 77 (1974).
81. R.A. Hegstrom, D.W. Rein and P.G.H. Sandars, J. Chem. Phys. 73, 2329 (1980).
82. S.F. Mason and G.E. Tranter, Mol. Phys. 53, 1091 (1984; Proc. R. Soc. Lond. *A*397, 45 (1985); J. Chem. Soc. Chem. Comm. 117, (1983).
83. G.E. Tranter, Chem. Phys. Lett. 115, 286 (1985); Chem. Phys. Lett. 120, 93 (1985); J. Theoret. Biol. 119, 467 (1986).
84. T.P. Emmons, J.M. Reeves and E.N. Fortson, Phys. Ref. Lett. 52, 86 (1984); L.F. Abbot and R.M. Barnett, Phys. Rev. *D*19, 3230 (1979).
85. G.E. Tranter and A.J. MacDermott, Chem. Phys. Lett. 130, 120 (1986).
86. Y-C. Jean and H.J. Ache, in: Origins of Optical Activity in Nature, D.C. Walker (ed.), (Elsevier, Amst.-Oxf.-NY, 1979), p. 67.

87. W. Bonner, Origins of Life 14, 383 (1984).
88. D.W. Gidley, et.al., Nature 297, 639 (1982).
89. R.A. Hegstrom, Nature 297, 643 (1982).
90. L. Keszthelyi, Origins of Life 14, 375 (1984).
91. M. Inokuti, Rev. Mod. Phys. 43, 297 (1971).
92. Ya.B. Zel'dovich and D.B. Saakyan, Sov. Phys. JETF 78, 2232 (1980).
93. R.A. Hegstrom, A. Rich and J. VanHous, Nature 313, 391 (1985).
94. L. Keszthelyi, J. Czege, Cs. Fajszi, J. Posfai and V.I. Goldanskii, in: Origins of Optical Activity in Nature, D.C. Walker (ed.), (Elsevier, Amst.-Oxf.-NY, 1979), p. 229.
95. W. Bonner, et.al., Origins of Life 15, 103 (1985).
96. A.S. Garay, Nature 219, 338 (1969); W. Bonner, M.A. Van Dort and M.R. Yearian, Nature 258, 419 (1975); W. Darge, A. Laczko and W. Thiemann, Nature 261, 522 (1976).
97. W. Bonner, M.A. VanDort and M.R. Yearian, Nature 280, 252 (1979).
98. V.I. Goldanskii and V.V. Khrapov, Sov. Phys. JETP 16, 582 (1963); L.A. Hodge, F.B. Dunning and G.K. Walters, Nature 280, 251 (1979); W. Bonner, in: Origins of Optical Activity in Nature, D.C. Walker (ed.), (Elsevier, NY-Oxf.-Amst., 1979), p. 5.
99. K. Pearson, Nature 58, 452 (1898).
100. F.R. Japp, Nature 58, 452 (1898).
101. M.I. Kabachnik, L.L. Morozov and E.I. Fedin, Dokl. Phys Chem. 230, 1135 (1976).
102. L.L. Morozov, Dokl. Biochem. 241, 481 (1978).
103. L.L. Morozov, Origins of Life 9, 187 (1979).
104. L.L. Morozov, D. Sci. Thesis, Moscow, (1979), (in Russian).
105. L.L. Morozov and V.E. Kulesh, Dokl. Biochem. 248, 1263 (1979).
106. L.L. Morozov and V.I. Goldanskii, in: Self-Organization, V.I. Krinsky (ed.), (Springer, NY, 1984), p. 224.
107. L.L. Morozov and V.I. Goldanskii, Vestnik AN SSR No. 6, 54 (1984) (in Russian).
108. L.D. Landau and E.M. Lifshitz, Statistical Physics, Pt. 1, (Pergamon Press, Oxford, 1986).
109. H.E. Stanley, Introduction to Phase Transitions and Critical Phenomena (Clarendon Press, Oxford, 1971).
110. A.D. Dolgov and Ya. B. Zel'dovich, Sov. Phys. Uspekhi 130, 559 (1980).
111. L.B. Okun', Leptons and Quarks (Moscow, Nauka, 1983), (in Russian).
112. Ya.B. Zel'dovich and I.D. Novikov, Structure and Evolution of the Universe (Moscow, Nauka, 1975), (in Russian).
113. M.V. Vol'kenshtein, General Biophysics Vol. II (Academic Press, London, 1983); Yu.M. Romanovskii, N.V. Stepanova and D.S. Chernavskii, Mathematical Biophysics (Moscow, Nauka, 1984), (in Russian).
114. B.N. Belintsev, Sov. Phys. Uspekhi 141, 55 (1983).
115. F. Dyson, J. Mol. Evol. 18, 344 (1982).
116. V.I. Arnold, Additional Chapters of the Theory of Ordinary Differential Equations, Moscow, Nauka, 1978 (in Russian).
117. R.A. Hegstrom, Nature 316, 749 (1985).
118. G.E. Tranter, Nature 318, 172 (1985).

119. L. Pasteur, C.R. Hebd. Sean. Acad. Sci. Paris 26, 535 (1848).
120. A. Hochstim, Origins of Life 6, 317 (1975).
121. L.C. Harrison, J. Mol. Evol. 4, 99 (1974); L.C. Harrison, in: Origins of Optical Activity in Nature, D.C. Walker (ed.), (Elsevier, NY-Oxf.-Amst., 1979), p. 125.
122. P. Decker, in: Origins of Optical Activity in Nature, D.C. Walker (ed.), (Elsevier, NY-Oxf.-Amst., 1979), p. 109.
123. J. Szamosi, Origins of Life 16, 65 (1985).
124. A. Klemm, Z. Naturforsch. 40a, 1231 (1985).
125. W-M. Lio, Origins of Life 12, 205 (1982).
126. F. Frank, Biochim, Biophys. Acta 11, 459 (1953).
127. L.L. Morozov, V.V. Kuz'min and V.I. Goldanskii, Origins of Life 13, 119 (1983).
128. L. Keszthelyi, Biosystems 20, 15 (1987).
129. D.K. Kondepudi and G.W. Nelson, Phys. Rev. Lett. 50, 1023 (1983).
130. D.K. Kondepudi and G.W. Nelson, Physica 125A, 465 (1984).
131. D.K. Kondepudi and G.W. Nelson, Nature 314, 438 (1985).
132. V.A. Avetisov, V.V. Kuz'min and S.A. Anikin, Chem. Phys. 112, 179 (1987).
133. V.A. Avetisov, V.V. Kuz'min and S.A. Anikin, Sov. Chem. Phys. 6, (1987).
134. L.L. Morozov, V.V. Kuz'min and V.I. Goldanskii, Sov. Phys. JETF Lett. 39, 344 (1984).
135. Ya.B. Zel'dovich and A.S. Mikhailov, Sov. Chem. Phys. 5, 1587 (1986).
136. L.L. Morozov, V.V. Kuz'min and V.I. Goldanskii, Dokl. Biophys. 274, 55 (1984); Dokl. Biophys. 275, 71 (1984).
137. M. Mangel. J. Chem. Phys. 69, 3697 (1978); Physica 97A, 616 (1979).
138. E.G. Gumbel, Statistics of Extremes (Columbia Univ. Press, NY, 1958).
139. R. Strickland-Constable, Kinetics and Mechanisms of Crystallization (Academic Press, NY, 1968).
140. F. Crick, Life Itself (Simon & Schuster, NY, 1981).
141. F. Hoyle and N.C. Wickramasinghe, Lifecloud (Harper & Row, NY-London, 1978); Nature 306, 420 (1983).
142. V.I. Goldanskii, M.D. Frank-Kamenetskii and I.M. Barkalov, Science 182, 1344 (1973).
143. V.I. Goldanskii, Nature 269, 583 (1977); Nature 279, 109 (1979).
144. R.A. Harris and L. Stodolsky, Phys. Lett. 78B, 313 (1978).
145. R.A. Harris and R. Silbey, J. Chem. Phys. 78, 7330 (1983).
146. M. Simonius, Phys. Rev. Lett. 40, 980 (1978); R.A. Harris and L. Stodolsky, Phys. Lett. 116B, 464 (1982).
147. L.L. Morozov, V.V. Kuz'min and V.I. Goldanskii, Sov. Phys. JETP Lett. 39, 414 (1984).
148. V.F. Krapivin, Yu.M. Svirezhev and A.M. Tarko, Mathematical Modelling of Global Biospheric Processes (Moscow, Nauka, 1982), (in Russian).
149. "Nuclear War: The Aftermath", Ambio, Special Issue 11, No. 2/3 (1982).
150. M.P. Boehm and J.L. Bada, Proc. Natl. Acad. Sci. USA 81, 5263 (1984); G. Blaschke, Angew. Chem. Int. 19, 13 (1980).
151. M. Eigen, Naturwiss. 58, 465 (1971).

152. A.G. Cairns-Smith. Genetic Takeover and the Mineral Origins of Life (Cambridge Univ. Press, Cambridge, 1982).
153. Cs. Fajsvi and J. Czege, J. Theoret. Biol. 88, 523 (1981).
154. V.A. Avetisov, S.A. Anikin, V.I. Goldanskii and V.V. Kuz'min, Dokl. Biophys. 283, (1985).

QUANTUM MECHANICS AND MACROSCOPIC REALISM

A.J. Leggett
Department of Physics
University of Illinois at Urbana-Champaign

It is a great pleasure to dedicate this paper to Hans Frauenfelder on the occasion of his sixty-fifth birthday. Although its topic is at first sight rather remote from the biomolecular problems which have been the main subject of his recent research, I believe that from a long-term perspective there may be some deep and surprising connections. In particular, workers in both areas are forced to face up squarely to the problem of describing consistently the interface between the microscopic system which is the principal subject of the investigation – which one knows must be described, at least in principle, by the quantum formalism – and its environment, which, one hopes, can be given a classical, and hence "realistic," description.

The quantum measurement paradox essentially consists in the fact that the linear formalism of quantum mechanics in some sense keeps all possibilities open, whereas in macroscopic everyday experience a single definite outcome is selected. Consider, for example, a (microscopic) system which can make transitions from some initial state A to one of two states B or C, then from B to either D or E and from C to either E or F; thus, state E can be reached from A via either B or C. (A typical physical realization of such a situation would be experiments using a neutron interferometer.) By blocking the path through the state C we can determine the probability $P_{A\to B\to E}$ that the system passes from A to E via B, and similarly by blocking the path through B we can determine $P_{A\to C\to E}$. However, if we measure the probability $P_{A\to E}$ when neither path is blocked, we find that in general it is not equal to the sum of the above two quantities:

$$P_{A\to E} \neq P_{A\to B\to E} + P_{A\to C\to E} \tag{1}$$

a typically quantum-mechanical interference effect, which is of course mirrored in the formalism through the idea of a probability amplitude. Thus, from an intuitive point of view it is difficult to avoid the conclusion that at the intermediate stage both possibilities, B and C, were in some sense left open, and have some kind of "reality". On the other hand, if at this stage we choose to "measure" which of the states B and C corresponds to the "real state" of the system (e.g., in the case of the neutron interferometer, by putting some kind of counter behind each of the intermediate-stage crystals) then we always find one result or the other.

Now I want to emphasize that at the *microscopic* level this state of affairs, while perhaps surprising, is not necessarily a major difficulty for the interpretation of the formalism. The reason is that, as Niels Bohr repeatedly emphasized, there is no need to attribute any particular ontological states to microscopie entities such as neutrons in the first place: it is perfectly possible, and internally consistent, to regard the formalism of quantum mechanics at this level as simply a recipe for predicting the results obtained when a macroscopic measurement apparatus is coupled to the microsystem in various ways. Since, once an apparatus to detect whether the system passed through state B or C is employed, the physical conditions are no longer the same as in its absence, it is not particularly surprising that when this is done the interference effect is no longer visible (so that the conclusion that *under these conditions* the system "preserved both possibilities" can no longer be drawn); indeed, as is well known, a consistent application of the formalism of quantum mechanics to the coupled system formed by the microsystem and the apparatus, *plus* the use of the standard measurement axioms on the states of the latter, yields predictions in complete agreement with the experimental results.

This, however, raises a much deeper and more troubling question — one which, to be sure, is dismissed by many physicists as "merely philosophical," whatever that is supposed to mean, but is to my mind no less worrying for that. In applying the standard quantum measurement axioms at the level of the macroscopic apparatus, we in effect have to assume either that the mere fact of observation by a conscious human being creates a "reality" which was not there in its absence, or that by the time the microscopic superposition has been amplified to the level of macroscopic states of an apparatus, the universe "really is" in one or other of the states correponding to different outcomes of the measurement, irrespective of the act of human observation. The first alternative is not obviously logically absurd, but leads to an anthropocentric, indeed possibly solipsistic, view of nature which I suspect is alien to most physicists. The second, on the other hand, implies a radical discontinuity between the interpretations of the quantum formalism adopted at the microscopic and macroscopic levels. Let us suppose (of course oversimplifying grossly!) that microscopic state ψ_1, if it occurs, is amplified by some suitable device (counter, electronics etc.) to macroscopic state Ψ_1, and ψ_2 similarly to Ψ_2. Then, by virtue of the linearity of the quantum formalism, the linear superposition $a\psi_1 + b\psi_2(|a|^2 + |b|^2 = 1)$ is amplified to the corresponding superposition of macrostates:

$$a\psi_1 + b\psi_2 \rightarrow a\Psi_1 + b\Psi_2 \qquad (2)$$

and, if we look at the dynamics of the amplification process in detail, there seems no point at which the application of the formalism takes us across any kind of qualitative discontinuity. Nevertheless, we wish to interpret the left-hand side as implying that both the branches of the superposition retain some degree of "reality", whereas on the right-hand side, in each particular event, only one of them is realized. These arguments are developed in much more detail in numerous papers in the literature: see, for example, Reference 1.

While this difficulty has been appreciated for more than fifty years, the general opinion has been that it has no experimental implications (which is, perhaps, what is meant by calling it "merely philosophical"). In Niels Bohr's heyday, and indeed for decades thereafter, this was a very reasonable point of view. For to demonstrate any kind of experimental consequence of the apparent paradox one needs a situation where one can see the characteristically quantum-mechanical interference phenomena at the level of macroscopic measuring instruments: and with perhaps 8 orders of magnitude in geometrical size, and 23 in number of particles involved, separating the latter from the atomic systems for which quantum mechanics was originally developed, it was not surprising that this seemed completely out of the question. However, one of the more exciting devlopments in condensed-matter physics over the last decade or so has been the realization that, under certian special and carefully engineered conditions, it is not necessarily hopeless to see quantum effects in systems which by some reasonable criterion are indeed as "macroscopic" as the measurement apparatus from which we are accustomed to "read off" results in a classical way. I have discussed this topic in a number of recent review-type articles[2,3,4,5] and will here only summarize briefly the current situation.

To observe interference between macroscopically distinct states we need, as a minimum, a variable which by some reasonable criterion is itself macroscopic but which is controlled by an energy which is of a *macroscopic* order of magnitude. While there are a number of candidates, the most clear-cut example of such a variable is the Cooper-pair phase, or the trapped magnetic flux, in various kinds of superconducting device based on the Josephson effect, and I shall confine myself here to those systems, on which experiments have been done in this context, in many laboratories throughout the world over the last decade or so. The first thing one can try to do is to verify that, when applied to the dynamics of such variables, the quantum-mechanical formalism gives predictions in agreement with experiment, even for phenomenon which have no classical analog, such as tunnelling through classically forbidden regions or the quantization of energy levels. In trying to make the correct quantum predictions, one is faced with a difficulty which is in some sense peculiar, or nearly so, to macroscopic systems: namely, they interact dissipatively with their environments, and while the effects of this dissipation are readily observed in the classical motion, its microscopic basis is often unknown in detail. Until it is understood how to handle this kind of situation within the quantum framework, no firm prediction can be made about the "characteristically quantum" behavior, if any, of realistic macroscopic systems, and indeed a major fraction of theoretical work in this area has been devoted to this question. I think most workers in this area would probably agree that we have by now reached some understanding of this problem, so that the major qualitative aspects of the theoretical predictions are not generally regarded as controversial. Given this, one can compare them with the experimental results; the outcome of this comparison is that both the phenomena mentioned above (tunnelling through a barrier, and quantization of energy levels) do indeed occur, as predicted, and the general features are indeed precisely as expected from the theory. (Some experiments have in fact given excellent quantitative as well as qualitative agreement with theory, but in others there remain quantitative discrepancies in some respects.) Overall, it seems that the experimental results obtained to date are consistent with the hypothesis that the quantum formalism continues to apply unmodified at this level.

This, however, does not in itself answer the even more intriguing question: Are "common-sense" ideas about the macroscopic world thereby excluded? It was, after all, the conviction that such ideas, even if they fail at the microlevel, should continue to apply at the level of macroscopic objects such as cats or counters which generated the quantum measurement paradox in the first place. So it is interesting to ask the question: At the *macro*-level, does Nature believe in realism (as we tend to in going abut our everyday affairs)? Rather surprisingly, it turns out that an *experimental* investigation of this question, at least if suitably interpreted, is not necessarily out of the question. This is discussed in detail in Refs. 3 and 5, and I will just briefly summarize the principal qualitative points here.

One seeks a system which has available to it two (and only two) macroscopically distinct states, and can move between them by tunnelling (so that the probability of finding it anywhere between them is exponentially small); such a system is formed, under appropriate conditions, by the superconducting device known as a rf SQUID (a close superconducting ring interrupted by a single Josephson junction). The two states are defined to correspond to values ± 1 of some dichotomic variable Q. The relevant experiment consists of measuring, on a "time ensemble" of such systems, various two-time correlations of the form $\langle Q(t_i)Q(t_j)\rangle$, where it is understood that no measurement is made between times t_i and t_j. (Thus, two different series of experiments are necessary to measure (e.g.) $\langle Q(t_1)Q(t_2)\rangle$ and $\langle Q(t_2)Q(t_3)\rangle$.) We then compare the experimentally obtained results with the predictions of (a) quantum mechanics and (b) a class of alternative, "common-sense" theories which embody the property of "macro-realism", defined by the following two axioms: (1) At (nearly) all time, irrespective of whether or not it is observed, the system is in a definite macroscopic state (i.e. $Q(t) = \pm 1, \textit{at any } t$). (2) Measurements of the value of $Q(t)$ can, in principle at least, be made in such a way that they do not affect the subsequent dynamics of the system (i.e. so that the sub-ensemble of systems on which a determination of Q has been made at time t_2 has the same statistical behavior for times $t > t_2$ as the subensemble on which it has not). Needless to say, quantum mechanics fails to satisfy the conjunction of (1) and (2), though whether it can be regarded as satisfying (1) alone is a moot point[6,7]. The crucial point is that it can be shown by simple algebra[5] that any theory satisying axioms (1) and (2) must predict certain inequalities for combinations of the correlation functions $\langle Q(t_i)Q(t_j)\rangle$ which are manifestly violated by the predictions of quantum mechanics for a totally isolated system, and also violated, though somewhat less strongly, by its predictions for a non-isolated system provided that the degree of damping is below a critical value which, though small, is probably attainable. Thus, if the results of a suitably performed experiment of this type are in fact consistent with the predictions of quantum mechanics, they must ipso facto be *inconsistent* with those of any theory embodying macro-realism. At least one such experiment of this type is currently being set up[8].

To conclude, let me return briefly to the possible long-term connection, mentioned at the beginning, with Hans' work in biophysics. We are very used, in the traditional areas of physics, to a mode of description in which the basic object under study, be it an atom, an elementary particle, a piece of copper wire or whatever, can at least be *described* in its own right, even if its *interactions* with its environment play an essential role in its behavior. But we should not expect to be able to carry on with this comfortable mode of description forever; after all, Bell's theorem, and the experiments related to it, have shown dramatically that, if we take the quantum-mechanical world-view seriously, the whole idea of assigning individual descriptions to "component" (?) parts of a complex system must at some stage break down, and it should be no surprise to find that this may eventually affect the description of even a macroscopic system. Indeed, if the experiment described above should come out in favor of quantum mechanics, we may well have to conclude that in some sense the SQUID ring interacting with a measuring apparatus is simply *not the same object* as it is in isolation, even though it is by most people's reasonable criteria a fully macroscopic system. In other words, we should have to question the very *meaning* of separating the world into the "system" under study and its "environments" – the very basis of the reductionist ethic. While the reasons are perhaps rather different – arising from the necessity of emphasizing the concept of function rather than from the intrinsic nature of the formalism employed – I believe that this point of view comes much more naturally to workers in biology and biophysics than to the practitioners of the more traditional areas of physics, and one may hope and expect that this general area, and the one I have discussed in this paper, will indeed flow towards one another in the future. Needless to say, if they do I am sure that Hans will be one of the leaders of the trend, and it is a pleasure to conclude by wishing him many more years of happy and productive research.

REFERENCES

1. A.J. Leggett, in Quantum Implications, ed. B.J. Hiley and F.D. Peat, London, Routledge and Kegan Paul, 1987.
2. A.J. Leggett, In Proc. NATO ASI on Percolation, Localization and Superconductivity, ed. A.M. Goldman and S. Wolf, Pergamon, New York, 1984.
3. A.J. Leggett, in Directions in Condensed Matter Physics, ed. G. Grinstein and G. Mazenko, World Scientific, Singapore, 1986.
4. A.J. Leggett, in The Lesson of Quantum Theory, ed. J. De Boer, E. Dal and O. Ulfbeck, North-Holland, Amsterdam, 1985.
5. A.J. Leggett, in Chance and Matter (Proc. 1986 Les Houches Summer School), ed. J. Souletie, J. Vannimenus and R. Stora, North-Holland, Amsterdam, 1987.
6. L.E. Ballentine, Phys. Rev. Lett. 59, 1493 (1987).
7. A.J. Leggett and Anupam Garg, Phys. Rev. Lett. 59, 1621 (1987).
8. C.D. Tesche, in New Techniques and Ideas in Quantum Measurement Theory, ed. D.M. Greenberger, Ann. N.Y. Acad. of Sciences, Vol. 480, New York, 1986.

PROTEINS, DYNAMIC SOLUTES WITH SOLVENT PROPERTIES

Anders Ehrenberg
Depratment of Biophysics, University of Stockholm
Arrhenius Laboratory, S-106 91 Stockholm, Sweden

INTRODUCTION

At the beginning of this century, proteins were considered to be some kind of colloidal substances with variable particle size. Since pure solid metals, e.g. gold, could also be brought into a colloidal state, it is not surprising that protien colloids were considered to consistof rigid particles. Enzymes were known from their activity but they were supposed to be small molecules, for which proteins could function as carriers or support, or just were a contaminating nuisance.

Not until the 1920s did it become clear that proteins are substances with well defined molecular sizes, specific for each protein. The experiments that led to this conviction were made largely by Svedberg who developed the analytical ultrcentrifuge and together with Pedersen and other collaborators examined a number of proteins which at that time could be obtained in a reasonably pure form[1]. From the hydrodynamical behavior of the molecules it was even possible to estimate their shape, usually pictured as a rotational ellipsoid.

PROTEIN CRYSTALS

At about the same time Sumner succeeded in obtaining a protein in crystalline form[2]. This was a beautiful confirmation of the view that a specfiic protein species consists of molecules of a precise size, shape and structure. This finding of course did not contradict but rather underscored the view that the protein molecules were rigid with just the shape that fitted into a crystal lattice. Furthermore, it was not merely a pure protein that Sumner had crystallized, it was an enzyme, urease, and his claim was that this enzyme was a pure protein and nothing else. However, apparently it took several years until the idea of enzymes as proteins had penetrated through the scientific community.

MYOGLOBIN - "THE HYDROGEN ATOM" OF BIOCHEMISTRY

Shortly thereafter Theorell crystallized myoglobin, prepared from horse muscle[3]. After the war two crystallographers from Bragg's laboratory undertook to try to determine the three-dimensional structures of two proteins by means of x-ray diffraction from the protein crystals. Kendrew worked on myoglobin, prepared from sperm whale which could grow into nice, large crystals, and Perutz worked on the four times larger hemoglobin. It took some time to solve the problems in this enterprise but the emerging picture of myoglobin[4,5] showed a number of marvellous details (Figure 1). There are eight stretches of α-helices forming an apparently irregular grid or cage around the heme. The

α-helices themselves have a rather compact appearance and are held together by hydrogen bonds, and the helix stretches are kept in place by the turns of loops in the peptide chain and interactions between the amino acid side chains, which fill up the space between and around the helices. The heme group that binds oxygen reversibly was found to be imbedded within the α-helices network without direct access to the binding site from the surrounding medium. This was the first time that such intimate details of a biological macromolecule could be visualized and discussed. Because of this, and because of the salient spectroscopic properties of the active heme group of myoglobin and the apparent simplicity of the overall reaction of reversible oxygen binding, no other protein has been so intensively investigated, and research results on no other protein have helped to formulate so many new ideas and concepts about the nature of proteins. For such reasons myoglobin is sometimes called the hydrogen atom and hemoglobin, with its four myoglobin-like subunits, the hydrogen molecule of biochemistry.

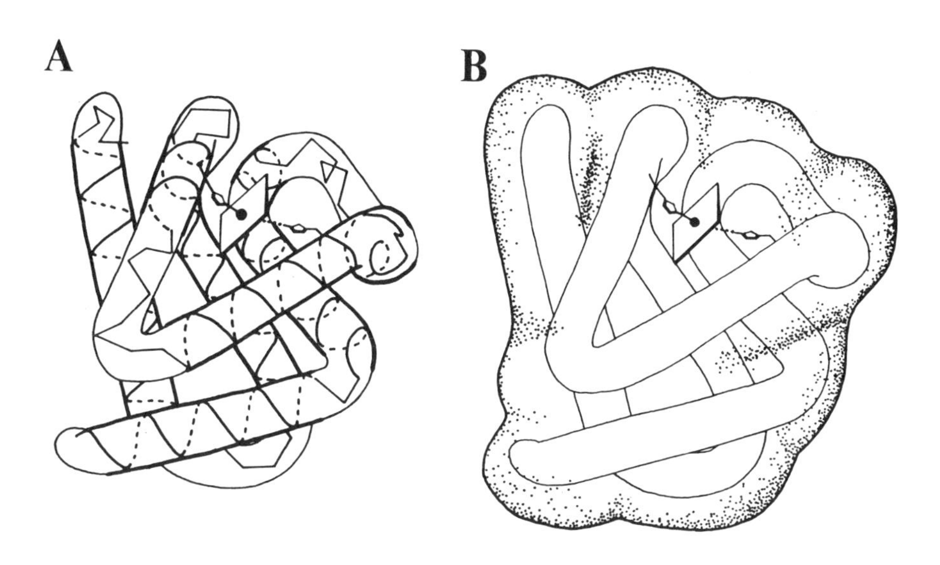

Figure 1. Structure of myoglobin as determined by x-ray crystallography[4,5]: *A* shows the peptide backbone with boundary lines drawn to guide the eye. The boundary lines are made thick around the α-helix stretches. In *B* the volume occupied by the amino acid side chains has been indicated. Ligands of ferromyoglobin (O_2, CO) and of ferrimyoglobin (e.g. H_2O, N_3^-) bind to the iron below the heme plane and are positioned between the iron and the distal histidine (dotted line).

As time went on, more proteins were crystallized and their three-dimensional structures were solved by means of x-ray crystallography. Three-dimensional protein models were built in the laboratories and the rigid appearance of these models helped to conserve the view of protein molecules as solid and rigid entities.

LEAVING THE PROTEIN "STONE AGE"

However, already at about the same time as Kendrew and Perutz published their crystallographic results[4,6], there were signs indicating that the protein "Stone Age" was coming to an end.

In 1958 Koshland suggested that a conformational adjustment takes place when a substrate binds to an enzyme[7]. The year thereafter Linderström-Lang and Schellman[8] in a review presented a picture of a protein in solution as a dynamic equilibrium between several possible conformations with concentration ratios in accordance with their relative free energies. This dynamic model of a protein molecule was based on observations on tritium exchange in the hydrogen bonds of the molecule. It took, however, many years until more direct and detailed evidence could be obtained. The dynamic nature of protein molecules has now been verified with great emphasis and explored in great detail, particularly by Hans Frauenfelder and his associates[9–11].

In their article Linderström-Lang and Schellman[8] furthermore proposed that protein dynamics may play a direct and important role in the enzyme catalytic reaction at the active site. This latter hypothesis still awaits direct experimental proof.

EPR RESULTS INDICATE CONFORMATIONAL MULTIPLICITY OF MYOGLOBIN

The mid 1950's saw the first applications of EPR, electron paramagnetic resonance, to biochemical problems. Two of the first objects to be investigated in detail were myoglobin and hemoglobin, viz. their ferric forms with derivatives. Here David Ingram's work on single crystals marked a breakthrough (for a review, see Ref. 12). These studies gave much detailed information about the electronic properties of the heme group and its ligands in the two heme proteins. One important result was also that the orientation of the heme group in the crystal could be quite accurately determined long before such details were available from the crystallographic structure[13,14]. This was based on the orientation of the anisotropic g-tensor. It was soon found that ferric myoglobin and other ferric heme proteins also could be studied by EPR in frozen aqueous solutions[15]. From a recorded powder spectrum the g-values could be read of from the turning points and the peak of the absorption.

As EPR results were collected on myoglobin and other EPR responsive metallo-proteins, some peculiarities became evident. In the single crystals of ferric myoglobin and its derivatives, the observed EPR linewidths were unusually broad and varied strongly with the orientation[16,17]. A random misorientation of the heme plane and g-axes could only explain part of the observed effects[16,17]. A more general explanation seems to be a random variation of the g-values caused by variations in the bonds and contacts between the heme and the protein[18-20]. This effect is sometimes referred to as g-strain[20,21].

Similar but more vague results were obtained from frozen solutions. Our experience with several heme proteins collected through the years[22] and the work later published by others on heme proteins[19] and other systems[23] showed that linewidths could vary with solvent and freezing conditions. solvent mixtures with glycerol, for instance, had a tendency to induce sharper lines. In some cases more than one g-value could be resolved in a certain "direction" for a sample that according to normal biochemical criteria, such as visible light absorption and electrophoresis at room temperature, should be homogeneous. Such observations suggested that when a protein solution was frozen the protein molecules could be fixed in different conformations and that the conformations that were fixed could depend on the composition of the solution and on the freezing conditons, e.g. the speed of freezing.

When the dynamic protein picture based on kinetic observations on ferrous heme proteins was developed by Frauenfelder and collaborators[9], it became apparent that the concept of many coexisting conformational substates being present in a frozen protein solution had the potential to explain the EPR observations. Any factor that influences the distribution of substates also can give changes in EPR linewidths and *vice versa.*

The important message for the present discussion is that there is evidence that both ferric and ferrous heme proteins in frozen solutions are fixed in several substates. Only at a higher temperature the substates interconvert between each other. Depending on the barriers for these interconversions, a whole hierarchy of substates exists[10,11].

O_2 CAN PASS THROUGH AND BE DISSOLVED IN THE PROTEIN MOLECULE

A crucial point in the work by Frauenfelder and collaborators is the possibility for CO or O_2 to penetrate from the solvent matrix through the protein to the imbedded heme group and there bind to the iron[9]. This pass through the protein is generally supposed to follow a specific route with up to four successive barriers, which may be resolved at suitable temperatures. However, there is also other evidence showing that O_2 may diffuse through a protein molecule in a non-specific more general sense. Lakowicz and Weber have shown[24,25] that oxygen may quench the flourescence of tryptophan residues, which are imbedded inside the protein without direct contact with the surrounding solvent. This quenching is remarkably effective. The conclusion was that the atoms and amino acid side chains of a protein in solution may execute rapid structural fluctuations on a time scale on nanoseconds which makes it fairly easy for the oxygen molecule to diffuse through the protein matrix.

It is of great interest for our discussion that similar quenching studies have been made by several groups[26-28] on myoglobin with the heme replaced by iron-free protoporphyrin. The transport of O_2 from the solvent to the interior of the protein, where it could collide with and quench the excited porphyrin, should be quite analogous to the transport of O_2 (or CO) from the solvent into the heme pocket, where it could rebind to the iron of the previously photodissociated heme group. However, the O_2 molecule does not need to start from the solvent. In several instances it might already be inside the protein molecule when the porphyrin is excited, or it will not leave the myoglobin molecule completely upon photodissociation. The possibility should also be considered that there could be two, or even more, oxygen molecules simultaneously inside a particular protein molecule. This means that the protein matrix has a certain capacity to dissolve molecules like O_2 and CO.

A PROTEIN ANALOGY WITH LIPID MEMBRANES

The secondary structures of the proteins are the α-helices and the β-sheets. The peptide backbones of these structures are rather rigid and the hydrogen bonds holding them together are very stable, i.e. they are only very slowly exchangeable with the solvent protons. It seems unlikely that diffusion of O_2 or CO could go rapidly through those structures. They function rather as barriers for the diffusion.

The peptide side chains that fill the space between the secondary structural elements, for instance between and around the α-helices of myoglobin (Figure 1), have been shown from x-ray Debye-Waller factors (temperature factors)[29–31] and by molecular dynamics simulations[31,32] to be flexible and mobile. The majority of the side chains facing the interior of the protein are hydrophobic. These parts of the protein will hence have properties somewhat similar to the lipid core of a biological lipid membrane.

The interior of a biological lipid bilayer is known to have liquid-like properties, where small non-polar molecules may be dissolved and diffuse rapidly. This interior membrane liquid is highly anisotropic.

There are important differences, of course, between the protein interior and the lipid bilayer: The peptide side chains are shorter than the hydrocarbon chains of the membrane lipids, and the α-carbon atoms, the first atom of each amino acid side chain, have relatively fixed positions in the protein, whereas the lipids in a membrane are free to exchange position with neighboring lipid molecules, i.e. there is lateral diffusion in the membrane. Nevertheless, there is enough evidence to conclude that small nonpolar molecules dissolved in a protein are most likely to be found in the regions of hydrophobic side chains.

In a protein molecule like myoglobin the liquid-like volume of the amino acid side chains forms a network through the molecule, a network which is complementary to that of the backbones of the α-helix stretches. This network has a continuous three-dimensional structure but is highly anisotropic in each point. When one discusses the diffusion of O_2 or CO through this matrix it seems most proper to do it in terms of two- or even one-dimensional diffusion. Such diffusion processes are known to be more effective than the three-dimensional one[33,34]. This might explain the rapid apparent diffusion of O_2 through the protein calculated from the fluorescence quenching experiments[25].

SMALL POLAR MOLECULES AND IONS MAY DIFFUSE THROUGH THE PROTEIN MOLECULE ALSO

However, not only apolar small molecules like O_2 and CO may pass through the protein matrix. Also charged or polar molecules may penetrate from the surrounding to the interior of the protein. As an example we may consider the azide ion, N_3^-, which is known to bind to the ferric heme iron of oxidized myoglobin in competition with H_2O:

$$\text{myoglobin } Fe^{3+}H_2O + N_3^- \rightleftarrows \text{myoglobin } Fe^{3+}N_3^- + H_2O$$

The x-ray structure of ferrimyoglobin azide[35] shows that N_3^- is trapped in the heme pocket like O_2 in oxyferromyoglobin. The protein must open itself in order for the azide to dissociate off. We can assume that the same is valid in aquoferrimyoglobin for the dissociation of H_2O. Hence the binding of N_3^- to ferrimyoglobin may be considered as a gated reaction. The protein must open itself up to let H_2O dissociate from the iron and diffuse out to the surrounding. Only then can N_3^- find its way to the heme pocket and bind to the iron, in competition with rebinding H_2O[36].

VISCOSITY DEPENDENCE OF THE THROUGH-PROTEIN DIFFUSION

It is a matter of great interest to compare the details of the diffusion in the two cases: the diffusion of a neutral apolar molecule, and the diffusion of an ion or a polar molecule. It has not yet been possible to study the binding of N_3^- in the same microscopic detail as in the case of CO and O_2 in the photodissociation rebinding experiments. However, we can compare the two types of transport processes in a global sense by comparing how the viscosity of the solvent medium affects the reactions.

From a modified Kramer's equation that should be applicable to these reactions[36–38] we can write down the following relationship between the reaction rate r and the viscosity η:

$$r \propto \eta^{-\kappa} \qquad (0 < \kappa < 1)$$

The reaction rate r is proportional to the viscosity η with an exponent $-\kappa$. Here κ can take values between zero and one. The κ-value of the rate of a reaction shows how strongly the rate determining motional modes of the system are coupled to the medium viscosity. If two reactions, for instance the two diffusion processes of CO and N_3^-, have widely different dependences on the solvent viscosity, i.e. different κ-values, the mechanisms must be different. On the other hand, if their viscosity dependences are equal, the mechanisms of the two processes could be the same, or the processes could have different mechanisms which accidentally happen to have similar overall viscosity dependences[36].

Frauenfelder and collaborators have determined the viscosity dependences over a wide range of viscosities and temperatures for the barrier passages on the route of CO from the solvent to the heme pocket of myoglobin[38]. The innermost barrier passage was assumed to be independent of viscosity, i.e.

$\kappa = 0$. The other passages have κ-values in the range 0.5 – 0.8. The slowest porcess at 260 K was found to have $\kappa = 0.6$. This process may be assumed to be rate determining at room temperature. In a recent work the binding of azide ions to ferrimyoglobin was studied by stopped flow technique[36]. The viscosity was varied in the range 0.8 to 10 cp and temperature in the range 285 to 310 K. Under these conditions the κ-value was determined to be 0.59 $\pm$ 0.02. These combined results suggest the possibility that the rate limiting steps of the two processes are the same or at least they are of the same nature.

AN EXTENDED ANALOGY WITH LIPID BILAYERS

The question we now have to address concerns how the viscosity effect is attenuated inside the protein so that κ-values in the range 0 to 1 are obtained. Again comparison with some results from experiments on lipid bilayers are helpful. We have studied by means of EPR the motion of the cholestane spin label in lipid bilayers and shown how to determine the rotational rate around the long axis of the molecule from spectral simulations[39,40]. This was recently made for lipid bilayers in aqueous suspensions where the viscosity of the surrounding aqueous phase was varied by glycerol additions[41].

The results show that the rotation of the cholestane spin label in a lipid bilayer in the gel state is very little affected by the medium viscosity. In the liquid crystalline state the viscosity influence is larger and temperature dependent. At and around the transition temperature between the gel and lipid crystalline phases the viscosity dependence has a peak. In these different situations the motional modes of the spin label are different and the coupling to the surrounding lipid molecules and to the medium are different. In the transition region the coupling to the medium is expected to be the strongest. The results show that different modes of motion in a given system may have different couplings to the medium viscosity and that the coupling of a particular mode might depend on the globular behavior of the system[41].

It has also been shown that lipids that are in immediate contact with an integral membrane protein can have a restricted mobility and that this effect dies out very rapidly so that two or three molecules away there is no notion of the disturbance[42,43].

In a protein molecule like myoglobin the diffusion of small molecules through the protein matrix must be directly dependent on the mobility of the amino acid side chains. Microscopically any diffusion is a series of transient trapping events, fluctuating barrier events and passing over barrier events. In a normal solvent these events are truly stochastic. In a constrained liquid of the type inside a protein, these event are localized, built into the structure of the protein, and in many cases strongly correlated at different positions. The amino acid side chains at the surface of the protein are in direct contact with the solvent and their motion is directly and fully affected by the solvent viscosity, i.e. $\kappa = 1$. A few steps away from the surface the motion of the side chains is not influenced by the solvent viscosity in the same way and κ should tend towards zero. However, the mobilities of the side chains in a region inside the protein are also dependent on the motion of the surrounding α-helices relative to each other. This motion of the α-helices may sometimes lock the positions of the side chains and immobilize them and occasionally relax the constraints for their motion more or less completely. This regional modulation of the side chain mobility may be coupled to a "breathing" motion of the surface of the molecule which should in turn couple to the solvent viscosity and tend to increase the κ-value inside the protein. We note here that different types of motion in a protein are coupled differently to the medium viscosity and the interior local motion of side chains may depend simultaneously on different modes of motion in the rest of the protein. In addition the local interior friction of viscosity should be frequency dependent.

SOME CONCLUDING COMMENTS

In this communication it was first outlined how our knowledge of the dynamic nature of the proteins has developed. It was then examined how this knowledge could help us to understand how small molecules and ions can be dissolved in and transported diffusively through the protein matrix. The experimental studies of the viscosity dependence of the interior dynamics in the proteins and the transport processes through the proteins should be helpful means to extend our knowledge about the nature of these phenomena. However, the viscosity influence itself is a complicated phenomenon about which we are far from an understanding of all the details.

ACKNOWLEDGEMENTS

Parts of this work was inspired and composed during a stay at the Institute for Theoretical Physics, University of California, Santa Barbara, California, Marhc 1987, and a stay at the Department of Physics, University of Illinois at Urbana-Champaign, Urbana, Illinois, April-May, 1987. The work was supported by a grant from the Swedish Natural Science Research Council.

REFERENCES

1. T. Svedberg, and K.O. Pedersen, The Ultracentrifuge, Clarendon Press, Oxford (1940).
2. J.B. Sumner, J. Biol. Chem. 69, 435 (1926).
3. H. Theorell, Biochem. Z. 252, 1 (1932).
4. J.C. Kendrew, R.E. Dickerson, B.E. Strandberg, R.G. Hart, D.R. Davies, D.C. Phillips and V.C. Shore, Nature 185, 422 (1960).
5. R.E. Dickerson, in *The Proteins*, H. Neurath, ed., Vol. II, Academic Press, New York, p. 603 (1964).
6. M.F. Perutz, M.G. Rossman, A.F. Cullis, H. Muirhead, G. Will and A.C.T. North, Nature 185, 416 (1960).
7. D.E. Koshland, Jr., Proc. Natl. Acad. Sci. USA 44, 98 (1958).
8. K.K. Linderström-Lang and J.A. Schellman, in *The Enzymes*, P.D. Boyer, H. Lardy and K. Myrbäck, eds., Vol. I, Academic press, New York, p. 443 (1959).
9. R.H. Austin, K.W. Beeson, L. Eisenstein, H. Frauenfelder and I.C. Gunsalus, Biochem. 14, 5355 (1975).

10. H. Frauenfelder, in *Structure, Dynamics and Function of Biomolecules*, A. Ehrenberg, R. Rigler, A. Gräslund and L. Nilsson, eds., Springer, Heidelberg, p. 10 (1987).
11. A. Ansari, J. Berendzen, S.F. Bowne, H. Frauenfelder, I.E.T. Iben, T.B. Sauke, E. Shyamsunder and R.D. Young, Proc. Natl. Acad. Sci. USA 82, 5000 (1985).
12. D.J.E. Ingram, Biological and Biochemical Applications of Electron Spin Resonance, Adam Hilger Ltd., London (1969).
13. J.E. Bennet and D.J.E. Ingram, Nature 177, 275 (1956).
14. D.J.E. Ingram and J.C. Kendrew, Nature 178, 905 (1956).
15. A. Ehrenberg, Arkiv f. Kemi 19, 119 (1962).
16. G.A. Kelcké, D.J.E. Ingram and E.F. Slade, Proc. Roy. Soc. *B*169, 275 (1968).
17. P. Eisenberger, and P.S. Pershan, J. Chem. Phys. 47, 3327 (1967).
18. R. Calvo and G. Bemski, J. Chem. Phys. 64, 2264 (1976).
19. W.R. Hagen, J. Magn. Reson. 44, 447 (1981).
20. D.O. Hearshen, W.R. Hagen, R.H. Sands, H.J. Grande, H.L. Crespi, I.C. Gunsalus and W.R. Dunham, J. Magn. Reson. 69, 440 (1986).
21. J. Fritz, R. Anderson, J. Fee, G. Palmer, R.H. Sands, J.C.M. Tsibris, I.C. Gunsalus, W.H. Orme-Johnson and H. Beinert, Biochim. Biophys. Acta 253, 110 (1971).
22. A. Ehrenberg, unpublished observation.
23. W.R. Hagen and P.J. Albracht, Biochim. Biophys. Acta 702, 61 (1982).
24. J.R. Lakowicz and G. Weber, Biochem. 12, 4161 (1973).
25. J.R. Lakowicz and G. Weber, Biochem. 12, 4171 (1973).
26. B. Alpert and L. Lindqvist, Science 187, 836 (1975).
27. R.H. Austin and S.S. Chan, Biophys. J. 24, 175 (1978).
28. D.M. Jameson, E. Gratton, G. Weber and B. Alpert, Biophys. J. 45, 795 (1984).
29. H. Frauenfelder, G.A. Petsko and D. Tsernoglou, Nature 280, 558 (1979).
30. P.J. Artymiuk, C.C.F. Blake, D.E.P. Grace, S.J. Oatley, D.C. Phillips and M.J.E. Sternberg, Nature 280, 563 (1979).
31. J. Åqvist, W.F. van Gunsteren, M. Leijonmarck and O. Tapia, J. Mol. Biol. 183, 461 (1985).
32. M. Karplus and J.A. McCammon, Annu. Rev. Biochem. 52, 263 (1983).
33. G. Adam and M. Delbrück, in *Structural Chemistry and Molecular Biology*, A. Rich and N. Davidson, eds., Freeman, San Francisco, p. 198 (1968).
34. O.G. Berg and P.H. von Hippel, Annu. Rev. Biophys. Biophys. Chem. 14, 131 (1985).
35. L. Stryer, J.C. Kendrew and H.C. Watson, J. Mol. Biol. 8, 96 (1964).
36. A. Ehrenberg, in preparation.
37. H.A. Kramers, Physica 7, 284 (1940).
38. D. Beece, L. Eisenstein, H. Frauenfelder, D. Good, M.C. Marden, L. Reinisch, A.H. Reynolds, L.B. Sorensen and K.T. Yue, Biochem. 19, 5147 (1980).
39. J. Israelachvili, J. Sjösten, L.E.G. Eriksson, M. Ehrström, A. Gräslund and A. Ehrenberg, Biochim. Biophys. Acta 382, 125 (1975).

40. Y. Shimoyama, L.E.G. Eriksson and A. Ehrenberg, Biochim. Biophys. Acta 508, 213 (1978).
41. A. Ehrenberg, in preparation.
42. P.F. Knowles, A. Watts and D. Marsh, Biochem. 18, 4480 (1979).
43. J.C. Owicki, M.W. Springgate and H. McConnell, Proc. Natl. Acad. Sci. USA 75, 1616 (1978).

CYTOCHROME AND MYOGLOBIN

B. Chance[a,b], P.L. Dutton[b], M.R. Gunner[b]
K.S. Reddy[a,b], L.S. Powers[c] and K. Zhang[b]

[a]Department of Biochemistry and Biophysics
University of Pennsylvania

[b]Institute for Structural and Functional Studies
University City Science Center
Philadelphia, PA 19104

[c]AT&T Bell Laboratories
Murray Hill, NJ 07974

Firstly, my apologies for my appearance on arrival direct from beam line II-2 at the Stanford Synchrotron after a week of MbCO study. A second thing, I'd like to present my warmest wishes to Hans, who everybody knows is a special figure in my scientific life because ofhis remarkable contributions to protein dynamics, which are a continuous intellectrual challenge to all of us.

I'm going to follow a theme of oxygen and instrumentation; the former is essential to life, and the latter sparks the interest of physicists who can perceive the technological choices which can be made in the biophysical world. I choose to begin with the visible region plotting an energy route (Figure 1) first to the near infrared, then to X-rays, and finishing up at the radio frequencies where Paul Lauterbur has labored so successfully.

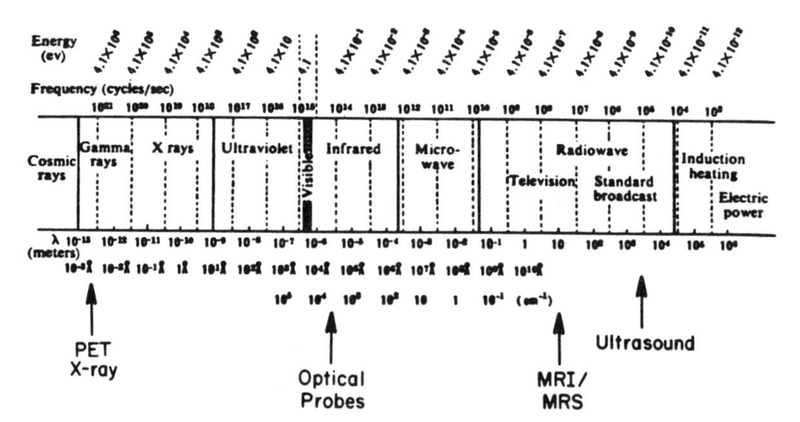

Figure 1. The energy spectrum for "non-invasive probes of body structure and function."

OPTICAL STUDIES OF THE CYTOCHROME CHAIN

The instrument used in early studies of respiratory enzymes is illustrated in Figure 2. David Keilin knew how to use the microspectroscope effectively; it is a techniques that is learned only with difficulty. He delineated the respiratory pigments: cytochrome *c* oxidase (aa_3), cytochromes *c* and *b* (Figure 3)[1]. Cytochrome *c*, already discussed here at this symposium, and cytochrome aa_3, were not accepted by Otto Warburg[2] and Keilin labored for about ten years to clarify the reality of cytochrome oxidase concept as a heme pigment and as a membrane protein. Neither Warburg nor Keilin did the key experiment of reconciliation, the action spectrum for the photolysis of the CO compound of cytochrome aa_3. If Keilin had used Warburg's photochemical technique or if Warburg has employed Keilin's heart muscle preparation in his potochemical studies, the CO-cytochrome a_3 band could have been visualized and ten years of unproductive polemics would have been avoided. It is important to communicate and to be flexible.

Current views of the respiratory chain from the complex of cytochrome aa_3, cytochrome bc_1 complex NADH-ubiquinone oxidoreductase and the associated dehydrogenases (Figure 4)[3]. This remarkable concatenation of metalloenzymes affects electron-transfer-driven charge separation that is harnessed to ATP synthesis. An X-ray diffraction crystal structure of none of the respiratory membrane proteins of the respiratory chain has been obtained, but great progress has been made in the electron transfer chain of bacterial photosynthesis. Dutton[4] has recently discussed light-induced electron transfer that occurs in the analogous intracytoplasmic membranes of photosynthetic bacteria. In this system, the photochemical reaction center protein of two species have been crystallized and an useful structure obtained. Michel[5] (Figure 5) has elucidated the X-ray structure of the *Rp. virids* reaction center-cytochrome *c* complex (it is now down to a 2.2 Å resolution), while the Argonne[6] and California[7] groups are working on the reaction center from *Rb. sphaeroides*. From these determinations has emerged the important structural parameters of cofactor distance and geometry; the most remarkable aspect of the structure is the long distance between the electron donor/acceptor cofactors in the reaction center and cytochrome *c* complex (Figure 5). DeVault and Chance[8] measured the tunneling time of the cytochrome *c* to the bacteriochlorophyll dimer $(BChl)_2$ of *C. vinosium*, a species with a reaction center cytochrome *c* composition as in *Rp. viridis*, and found it to be 2 msec. In 1966 they calculated that cytochrome *c* could be as far as 30 Å away from the rection center of $(BChl)_2$[8]; the measured center to center distance for *R. viridis* is close to 20 Å. This distance appears to be characteristic for "through the barrier" electron transfer. Tunneling also occurs in the reaction center itself not only from the light generated excited singlet state of $(BChl)_2$ to the bacteriopheophytin (BPh), but also from BPh^- to the primary ubiquinone -10 (Q_A), and from Q_A^- as an electron returns back to $(BChl)_2^+$. the work of Gunner and Dutton[9–11] is especially noteworthy since they have measured the rate of electron transfer from BPh^- to Q_A and from Q_A^- to $(BChl)_2^+$ for a wide variety of quinones of different redox potentials. An example of the dependence of rate form BPh^-

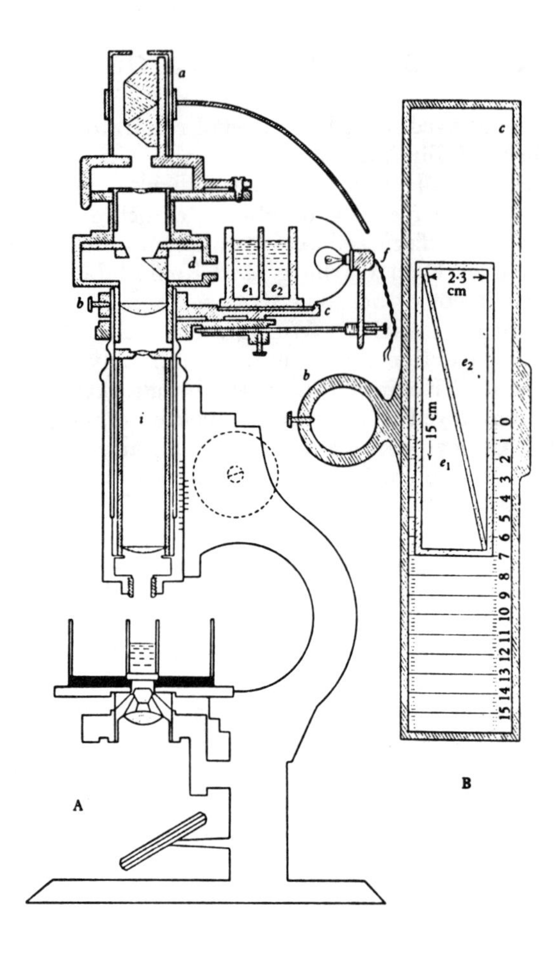

Figure 2. Microscope and double-wedge trough. (A): Microscope with the spectroscope ocular (*a*) and Ramaden ocular (*i*) inserted into microscope tube; (*b*) the ring which attaches the horizontal platform (*c*) upon which slides the double-wedge trough, showing its two compartments (e_1 and e_2) in front of the side aperture (*d*) of the spectroscope. The small lamp (*f*) illuminates the scale and the solutions in compartments (e_1 and e_2) of the trough. (*B*): Plan view of platform (*c*) and two compartments (e_1 and e_2) of the double-wedge trough[1].

to Q_A and from Q_A^- to $(BChl)_2^+$ on the free energy G^0 for the system at a variety of temperatures is shown in Figures 6a and 6b; this surprisingly, as far as we know, is the first report of an investigation of an electron transfer rate

Figure 3. David Keilin's representation of the absorption bands of cytochrome oxidase and its various ligated states[1].

as a function of *both* ΔG^0 and T. The dependence of the velocity constant upon ΔG^0 is calculated from the overall equation for electron tunneling for a non-adiabatic multiphonon decay process[9]. DeVault's summary [12] gives a complete description of the equation. The key factors are described in Figures 6a and 6b. Examinations of both experimental parameters in the same system has the effect of putting some strictures on the large number of possible interpretations of results from rate measurements found when only one parameter is varied. Principal points to emerge from this work are:

a) The reaction rates are essentially temperature independent over the entire range of ΔG^0 values studied so far; this shows the long standing explanation that the temperature independence of electron transfer in reaction centers due to the ΔG^0 being matched exactly to the reorganizational energy (λ), has no foundation. A corollary is that a reaction that displays a rate dependency on ΔG^0 needs not have a temperature dependence.

b) The reactions cannot be treated as coupled to simple solvent reorganization vibrational modes in a semiclassical formulation. The reactions are coupled to quantized modes of vibrations in a range of frequencies; these include high frequency vibrations characteristic of molecular skeletal modes such as those in the 1600 cm^{-1} range and middle protein vibrational modes around 150 cm^{-1}; however, there is as yet no evidence of coupling to very low frequency solvent vibrations in the 10 cm^{-1} range although they cannot be totally discounted (see Refs. 9-11).

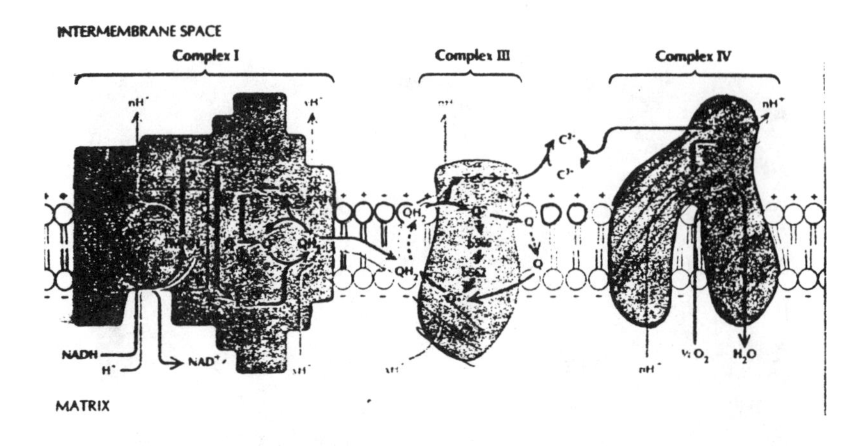

Figure 4. Wallace's representation of the components of the respiratory chain[3].

CYTOCHROME OXIDASE STRUCTURE/FUNCTION

Low angle X-ray diffraction, and high resolution optically processed EMG (Electron Microscopy) does given an image of many membrane proteins and they characteristically project far beyond the lipid bilayer as does the dimer of cytochrome oxidase (see Figure 4)[3]. It is characteristic of such membrane proteins to be more outside the membrane than within the membrane. The "anchors" of the protein to the membrane may be transmembrane "channels" for shuttling protons across the membrane to achieve charge separation. However, such hypothetic channels are not proved to be functional in cytochrome oxidase; in fact, some doubt exists on this point in bacterial rhodopsin as well, and "membrane Bohr effects" are under active consideration as an alternative to the channels.

An iron/copper binuclear center seems to be characteristic of those oxidases that bind oxygen, especially in cytochrome aa_3[13,14] and possibly in cytochrome o[15] (Figure 7, Ref. 13). Moreover, in collaboration with Dr. L.S. Powers, EXAFS, studies of the oxidized and reduced forms of iron and copper of cytochrome a_3 suggest that a conformation or state change that may be a key to energy conservation occurs on reduction of the copper and the iron atoms. This may be seen in the Fourier transform of the copper EXAFS data at 3.75 Å, and is not seen when the iron and copper are reduced[13,16]. Apparently cytochrome oxidase structure relaxes or "opens up" in order to accept its

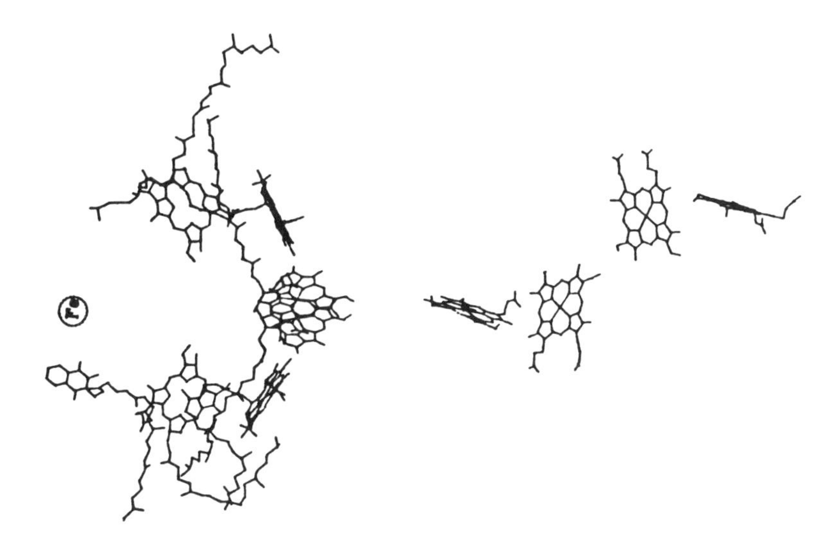

Figure 5. The X-ray crystal structure of the reaction center of *R-viridis* from Michel, et al.[5].

oxygen ligand – about which Anders Ehrenberg was talking about in some detail during this symposium[17]. Furthermore, there is a heavy atom, concluded by us to be sulfur, and thought by Scott to be chloride[18], which interfers with ligation in the ferric state and makes the site cryptic to ligands but accessible to electrons. Recent studies shows that mercury binds the bridging ligand and can be detected in the Cu EXAFS (L. Powers, unpublished observations). This is a characteristic reaction of thiol (S) groups in proteins. Such studies often involve dilute solutions (100 μM Fe) and the utmost of sensitivity and stability of the X-ray beam. In addition to large solid-angle low noise detectors, we tested a tracking system that follows the vertical position fluctuations of the X-ray beam when the tracking loop is closed, and very large position errors were compensated for in beam line II-2 at SSRL[18].

In order to achieve its meanifold functions, cytochrome oxidase is a veritable chameleon of structures and functions (Figure 8). There is a bridging atom which blocks ligation of the iron when cytochrome aa_3 functions as an electron acceptor. This structure resembles that of cytochrome c because of the six coordinate state with N and S ligand. The above mentioned structural change occurs on reduction of Fe + Cu, no backscattering from the copper or from the iron is found in the EXAFS data[13,16] (Extended X-ray Absorption Fine Structure), corresponding to an open floppy structure, appropriate to ligation with oxygen. On reoxidation to the ferric form, the sulfur bridge may not reform and this third structure may bind H_2O_2. In this form, cytochrome

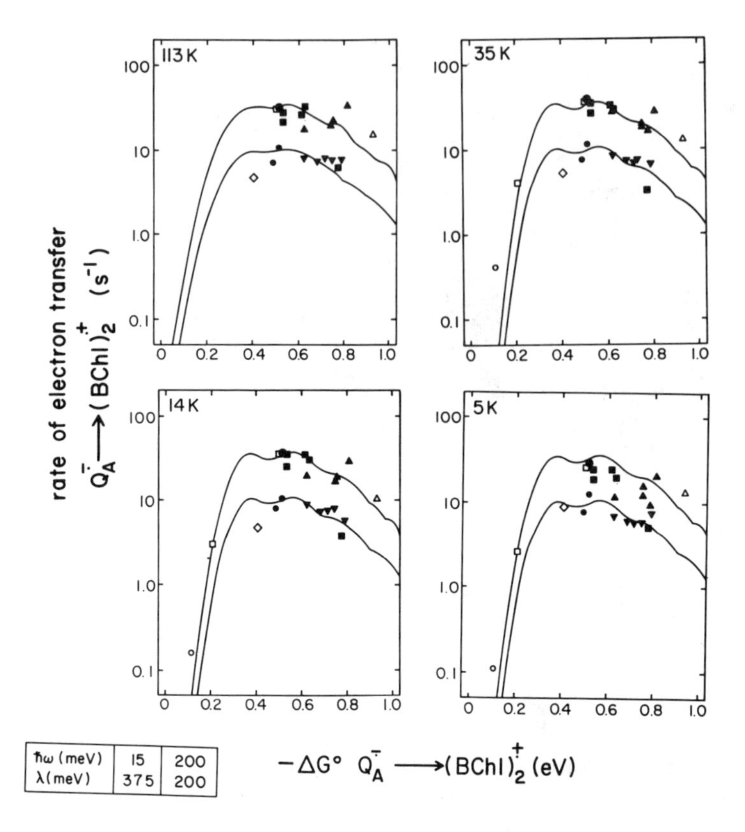

Figure 6A. Dependence of the rate of $(BChl)_2^+$ reduction of *Rb.sphaeroides* on the $-G^0$ between $(BChl)_2^+ Q_A^-$ and the ground state. The reaction $-\Delta G^0$ was measured *in situ* (filled symbols) or estimated from in vitro $E_{1/2}$ values of the quinones (date from Ref. 5). The various symbols are (o) BQs; (□) 1,4-NQs; (◇) 1,2-NQ; (∇) AQs substituted in the 1-position; (Δ) AQs substituted in the 2-position. Theoretical lines are calculated using the equation

$$k = \frac{2\pi}{\hbar^2 \omega_m} |V(r)|^2 \left[\sum_{q=0}^{\infty} \frac{e^{-s'} S'^q}{q!} \right] e^{-s(2\hbar+1)} \left(\frac{\bar{n}+1}{\bar{n}} \right)^{p/2} I \left[2S\sqrt{\bar{n}(\bar{n}+1)} \right]$$

with $\hbar\omega_m = 15$ meV, $\lambda_m = 375$ meV, and $\hbar\omega_l = 200$ meV, $\lambda_l = 200$ meV. Values of $V(r)$ are 4.8×10^{-8} eV (upper curve) and 1.4×10^{-8} eV (lower curve)[9].

oxidase acquires the properties of horseradish or yeast peroxidase, binding H_2O_2 to make a compound ES type of peroxide intermediates to cytochrome

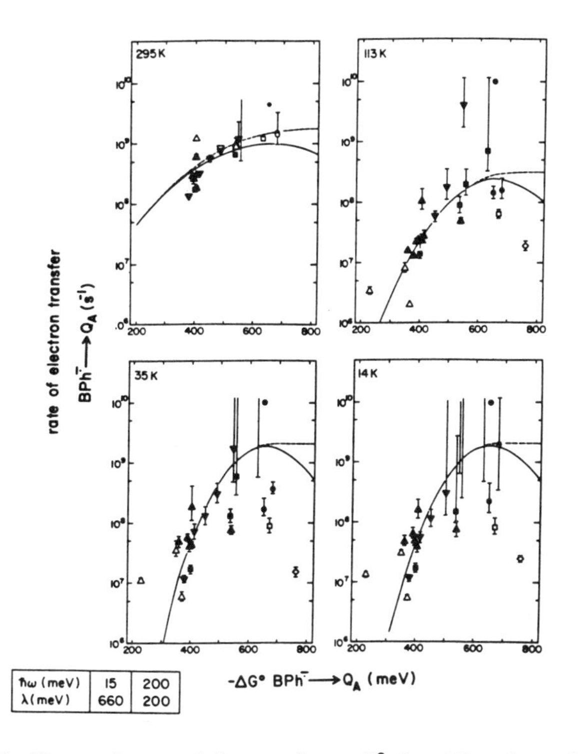

Figure 6B. Dependence of k_1 on the $-G^0$ for *Rb.sphaeroides.* The rates are calculated from Q using the equation:

$$\Phi = \frac{k_1}{k_1 + k_2}$$

At 295 K k_2 was assumed to be $7.7 \times 10^7 s^{-1}$, at lower temperature a value for k_2 of $3.3 \times 10^3 s^{-1}$ was used (date from Ref. 5). The quinones functioning as Q_A where (o) BQs; (□) NQs; (∇) AQ and 1-substituted AQs; (Δ) 2-substituted AQs. The error bars represent the variation in k_1 given the standard deviation of Φ_Q. (a) Data at 295 K. Open symbols represent values derived from picosecond measurements of the decay of $(BChl)_2^+$ BPh^-. Filled symbols are for k_1 calculated from Φ_Q. All $-G^0$s were calculated with $Q_A E_{1/2}$s determined *in situ.* (b) Data at 113 K; (c) 35 K; (d) 14 K. For (b-d) filled symbols imply the $-G^0$ was calculated using $Q_A E_{1/2}$ values determined *in situ* while for open symbols *in vitro* values were used. The theoretical lines were calculated using the equation from Figure 6A with $\hbar\omega_m = 15$ meV, $\lambda_m = 660$ meV, $\hbar\omega_l = 200$ meV, $\lambda_l = 200$ meV, $V(r) = 4 \times 10^{-1}$ meV at 295 K, 35 K and 14 K, and 1×10^{-1} meV at 113 K[9].

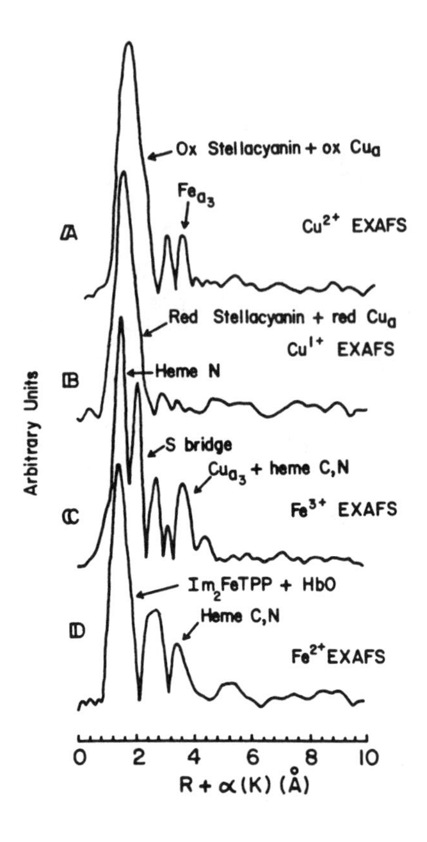

Figure 7. The radial distribution functions for the various redox states for the oxidized and reduced cytochrome oxidase as viewed from the EXAFS study of the copper (A&B) and the iron (C&D). The peaks represent back-scattering from shells of atoms surrounding the copper and iron atoms. The various identifications are the comparison model with which the iron and copper have been compared[13].

aa_3; which is ferric and six coordinate, with the S bridge being absent. This leads us to propose a series of reactions (Figure 9) in which there are oxy and peroxo intermediates, resembling the horseradish peroxidase compound II and an oxygen intermediate resembling that of oxyhemoglobin. At the end of the reaction cycle, cytochrome aa_3 can reform a sulfur bridge to reinstitute the cytochrome *c* configuration. So one of the important lessons taught by this membrane protein is that it is very versatile, and can assume a variety of configurations, and thus carry out manifold functions necessary to accept electrons, to reduce oxygen to water and to separate charge. Even with all this knowledge we do not know which specific step in the reaction separates charge or whether the conformation change observed is the key to a "Bohr" effect and thereby separates charges. In summary, cytochrome aa_3 is obviously a highly evolved multi-function enzyme which differs sharply from myoglobin and hemoglobin.

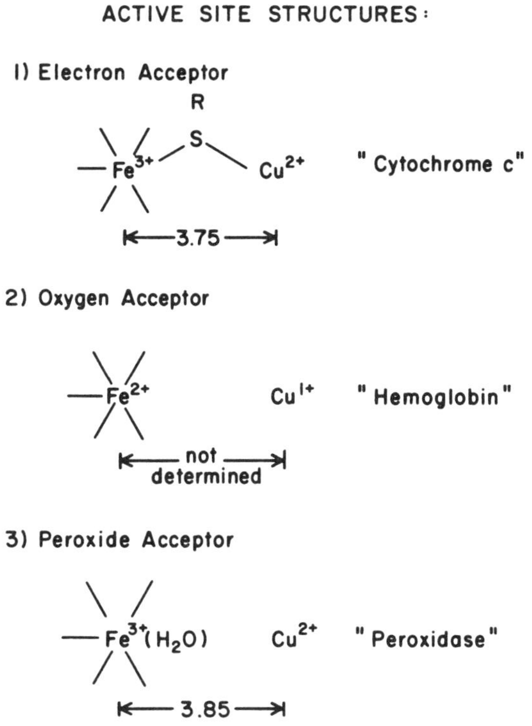

Figure 8. The three structures of cytochrome oxidase and their analogous functional modalities: cytochrome *c*, hemoglobin and peroxidase.

Equally important to animal life is the sequence of reactions that transport oxygen to cytochrome oxidase (Figure 10). As would be expected, phyiscal argument suggests that hemoglobin would bind oxygen most loosely, myoglobin with intermediate strength, and cytochrome oxidase most tightly. There is indeed a sequence of oxygen affinities. In fact, this sequence also applies to the electron transfer chain at the level of cytochrome *c*, and finally NADH. For example, cytochrome *c* responds to oxidation of cytochrome aa_3 at 0.2 μM O_2. However, these values for cytochromes are flux dependent such that the O_2 affinity is the ratio of the turnover of cytochrome (sec^{-1}) to the on-velocity constant for the cytochrome oxidase/oxygen reaction (M^{-1} sec^{-1}). In the case of myoglobin, i.e. oxygen carriers that equilibrate and do not turn over, the affinity is the ratio of the off-velocity constant to the on-velocity constant, no adjustment needs to be made for variable flux. Thus the entire

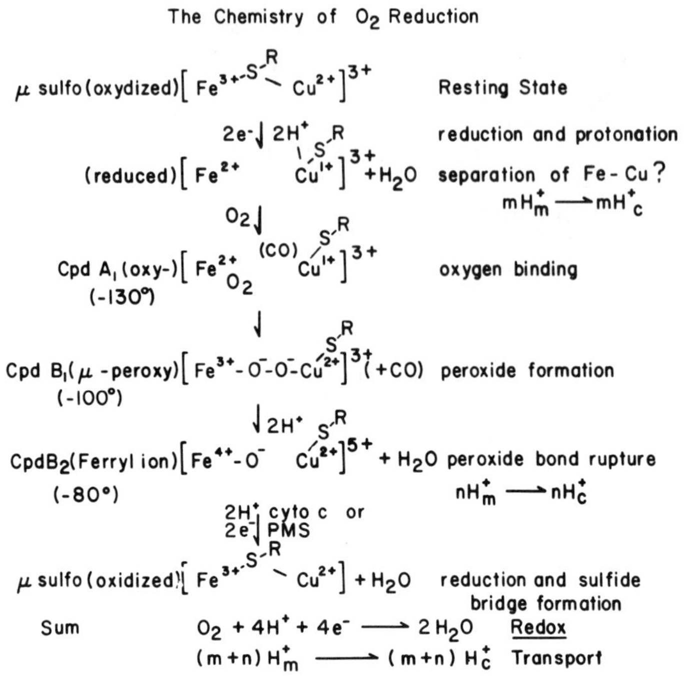

The Chemistry of O_2 Reduction

μ sulfo (oxydized)	$[Fe^{3+}\text{-}S(R)\text{-}Cu^{2+}]^{3+}$	Resting State
	$\downarrow 2e^-$ $2H^+$	reduction and protonation
(reduced)	$[Fe^{2+}$ $S(R)\text{-}Cu^{1+}]^{3+}$ $+H_2O$	separation of Fe - Cu? $mH^+_m \longrightarrow mH^+_c$
	$\downarrow O_2$	
Cpd A_1 (oxy-) (-130°)	$[Fe^{2+}O_2$ (CO) $S(R)\text{-}Cu^{1+}]^{3+}$	oxygen binding
	$\downarrow$	
Cpd B_1 (μ -peroxy) (-100°)	$[Fe^{3+}\text{-}O^-\text{-}O^-\text{-}Cu^{2+}(S\text{-}R)]^{3+}$ (+CO)	peroxide formation
	$\downarrow 2H^+$	
CpdB_2(Ferryl ion) (-80°)	$[Fe^{4+}\text{-}O^-$ $S(R)\text{-}Cu^{2+}]^{5+}$ $+ H_2O$	peroxide bond rupture $nH^+_m \longrightarrow nH^+_c$
	$\downarrow$ $2H^+$ cyto c or $2e^-$ PMS	
μ sulfo (oxidized)	$[Fe^{3+}\text{-}S(R)\text{-}Cu^{2+}]$ $+ H_2O$	reduction and sulfide bridge formation
Sum	$O_2 + 4H^+ + 4e^- \longrightarrow 2H_2O$	Redox
	$(m+n)H^+_m \longrightarrow (m+n)H^+_c$	Transport

Figure 9. A sequence epitomizing reaction mechanisms based upon the experimental results of cytochrome oxidase structure as obtained by EXAFS and other techniques. It should be noted that two of the three forms in Figure 8 are functional in this reaction sequence, namely – the cytochrome *c* and hemoglobin.

set of heme enzymes acts in a transport sequence from high O_2 levels of the blood to the low oxygen levels of the tissues. At higher fluxes the tissue oxygen gradients become high – several torr/micron and at the same time the oxygen affinity of cytochrome oxidase decreases from nanomolar to micromolar levels. Under these conditions myoglobin function can be of importance in the heart or working skeletal muscle.

While many questions remain to be answered, a veritable cornucopia of altered forms of cytochrome oxidase (Figure 11)[20–21] has emerged in the molecular biology of cytochrome oxidase. The further availability of DNA sequences[22–24] may increase this number in the near future. There are available more forms of cyrochrome aa_3 with altered activity, and hence structural properties than were previously available in the wild types in microbial forms such as *Saccharomyces* and *Rhodobacter*. As this meeting has pointed out, it is

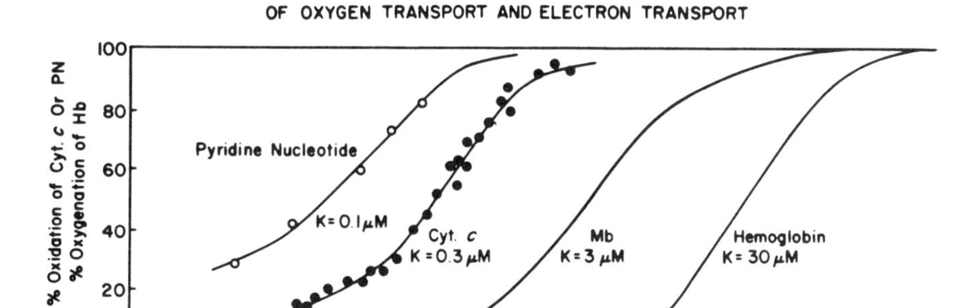

Figure 10. A schematic illustration of the different oxygen requirements of the oxygen/electron transport pigments.

a challenge and an opportunity to study the structure/function relations. Some forms have different activities; and some do not have any detectable activity due to key structural alterations. These may well be the most interesting!

StrainA	Genotype	Cytochrome Content (uM/g wet weight)				Cytochrome-c Oxidase	λ Max (nm)		O_2 reaction (initial phase)		Mid-edge feature
		c	c_1	b	aa_3	T.N.(sec^{-1})	$a_3^{2+}CO$	$(aa_3)^{2+}$	$t_{1/2}$(min)	Temp(^oC)	(eV)
JM43	COx5a+,COx5b+	65.0	65.9	56.0	15.3	65	592	610	5	-95	8889
GD5b	COx5a+,COx5b-	71.0	75.6	53.2	12.2	69	593	611	15	-95	8985
GD5a	COx5a-,COx5b+	75.4	70.0	53.5	4.6	108	-	-	-	-	-
GD5ab	COx5a-,COx5b-	70.1	73.7	59.9	0	0	-	-	-	-	-
RP3	COx5a-,COx5b+R	72.0	70.1	63.6	10.9	207	592	610	1.5	-105	-

(1) Patterson, E. et al in Cytochrome Systems Molecular Biology & Bionergetics (Papa, S. ed), Plenum Publ.(1987). (2) Andersson, B. & Akerband, H-E. in Electron Transfer Mechanism and Oxygen Evolution (Barber, J.ed) 1987. Supported in part by NIH Grants HL 31909 and GM 31992.

Figure 11. A portion of a library of cytochrome oxidase mutants having different subunit composition[19].

Specific features of four altered oxidases of type aa_3 have been studied by R.O. Poyton and collaborators[10–21] who have alterated their subunit content, particularly subunits Va and Vb as shown in (see Figure 11)[20]. First, the table underlines the now well established theorem, that nuclear coded subunits can control the cytochrome oxidase activity. Second, the amount of cytochrome aa_3, its overall activity, its reaction with CO and features of the Cu edge of cytochrome aa_3 are altered. Just how these changes will impact upon the ligands and the distances between Fe and Cu remains to be studied by EXAFS. A whole new chapter in cytochrome oxidase studies seems to be emerging.

STUDIES *IN VIVO*—CYTOCHROMES

In cytochrome studies *in vivo* (Figure 12) we encounter an additional problem, namely light scattering is very prominent due to the membranes of cells and organelles[25]. In order to compensate for light scattering changes, we have employed a more sophisticated development of Tyndall's differential spectrophotometer (Figure 12, Ref. 26) which time shares two different wavelengths by means of a vibrating mirror. The light scattering is compensated and the pigments of intact cells are readily discerned by this method (Figure 13)[25]. While Keilin's microspectroscopy was limited to the visible region, the dual wavelength method covers the near ultraviolet region where it was difficult for Keilin and Warburg to identify the CO-cytochrome a_3 band at 430 nm. In addition, the cytochrome oxidase bands due to copper are also clearly delineated in the near infrared region in purified cytochrome oxidase[27] and in yeast cells[28]. Thus the kinetics of O_2 and of electron transport are readily followed in intact cells (ascites tumors, etc.[29]).

MYOGLOBIN

Myoglobin has been studied by dual wavelength optical spectoscopy also according to the ideas of Tyndall (Fugre 12)[26]. Tyndall employed a reference sample and a measurement sample, and a differential galvanometer that foreshadowed the basic principles of our optical technology of today. Tyndall's differential system was exploited (Figure 14) by Glenn Millikan[30] (R.A.'s son) who is a special hero in my life as my Thesis Supervisor at Trinity College (Cambs). He illuminated the cat leg with white light and picked up the transmitted light with a differential photovoltaic cell with red and green filters. The output was coupled to a differential galvanometer and fed back to a balnacing vane. The application of this instrument to living muscles gave very clear results (Figure 15) in response to clamping the circulation or stimulating the soleus muscle, the myoglobin (and hemoglobin) immediately gave up oxygen to the cytochrome system and rapidly recovered thereafter — a pioneer observation. Subsequently, Millikan applied this technique to the observation of hemoglobin in the ear lobe[31], and more recently many investigators have exploited the greater transparency of the tissues to near infrared light[32–35] with "pulse oximeters" and "niroscopes" etc., but nevertheless, the basic dual wavelength techniques is employed (see below).

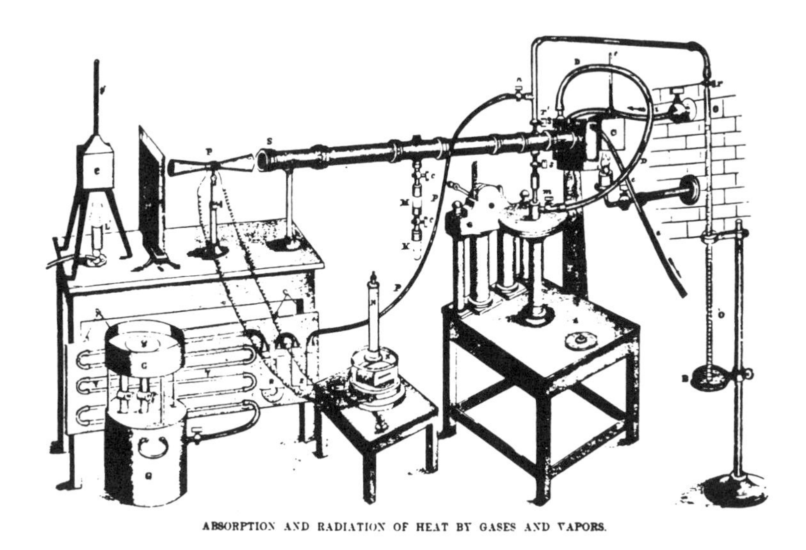

Figure 12. Helmholtz's original "dual wavelength" infrared spectrophotometer.

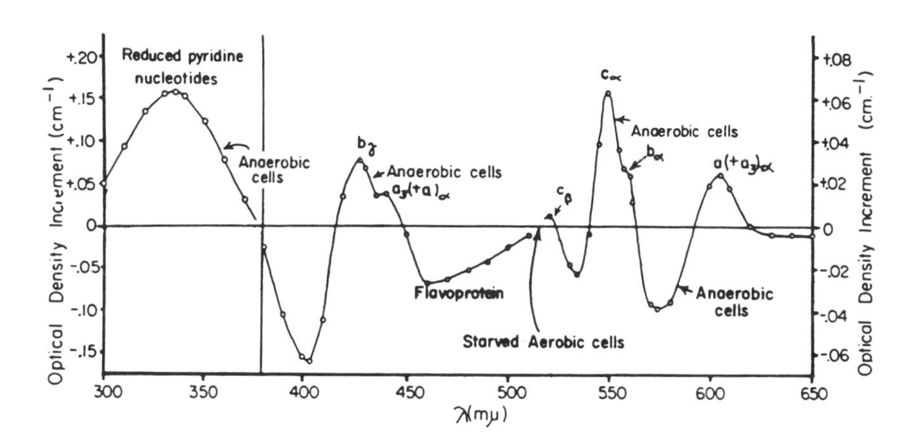

Figure 13. Absorption difference spectrum between the reduced anaerobic baker's yeast cell suspension and the starved aerobic cell suspension[25].

CRYOKINETICS

An important phase of myoglobin studies was initiated by Yonetani who first measured low temperature kinetics of combination of Mb and CO (Figure

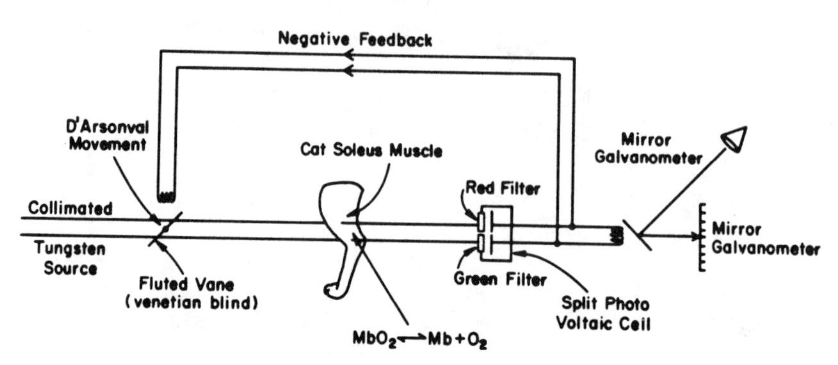

Figure 14. Schematic diagram of Millikan's experimental set up for measuring myoglobin and hemoglobin in cat muscle[30].

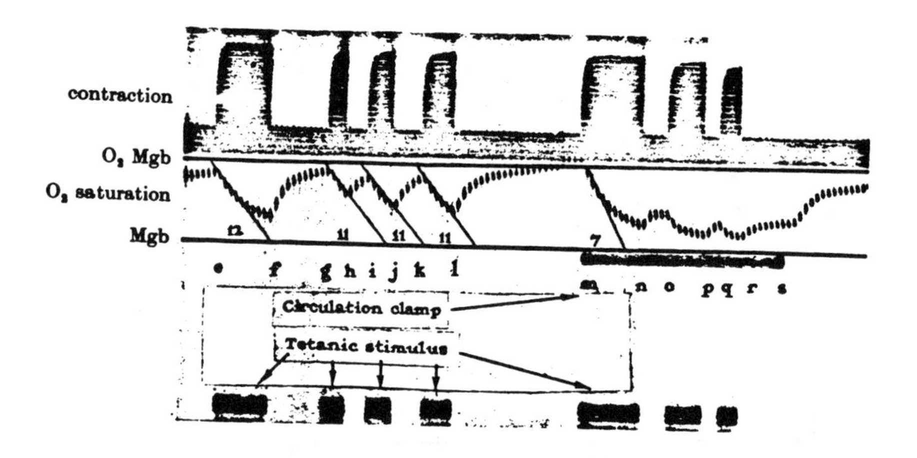

Figure 15. Reproduction of the original record of Millikan's studies of hemoglobin and myoglobin deoxygenation in the cat muscle during stimulation and ischemia[30].

16A) and found biphasic kinetics[36]. The next phase of myoglobin kinetics was initiated by Frauenfelder and his collaborators[37–40] who followed up on Yonetani's observation that myoglobin was quite different from cytochrome in that it continued to combine at low temperatures, for example even at 40 K (see Figure 16B). While optical methods are suitable for rapid recording of

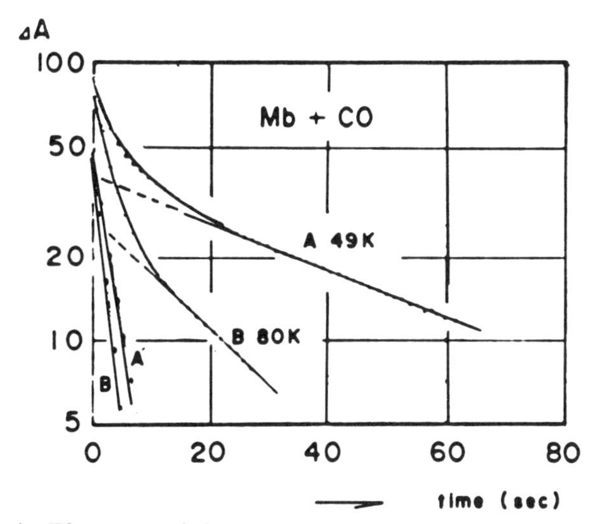

Figure 16A. The original data of Yonetani et.al. indicate the recombination of myoglobin and CO ar very low temperatures, together with the non-exponential nature of the recombination (see ref. 36).

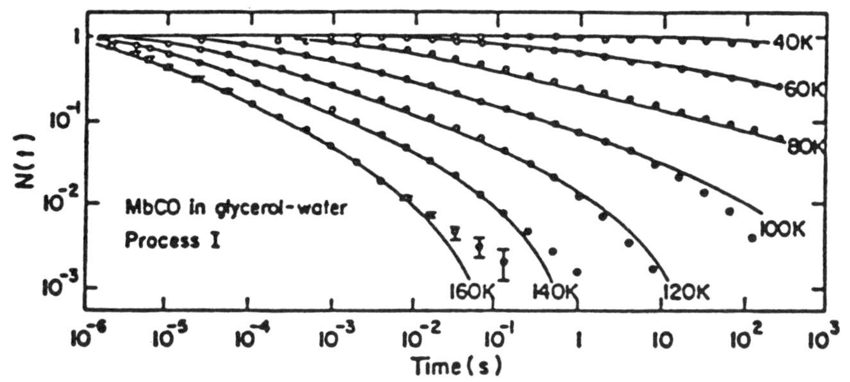

Figure 16B. Illustrating the "power law" recombination of myoglobin with CO in glycerol water glasses at low temperatures (see refs. 37-40).

transient kinetics, several hours may be required for structural studies as in the EXAFS method.

Figure 17 shows how the photoproduct may be maintained at a constant concentration over long times as monitored in the near infreared (NIR) region in connection with the EXAFS studies[41,42]. A time sharing dual wavelength instrument using 760 and 790 nm wavelengths is appropriate since high transparency is obtained for the 1.2 mM 1 mm path samples of MbCO used in the EXAFS studies. It is further possible to time share the optical signals with the white light-illumination of the sample while it is being irradiated by the 7 to 8 KeV electrons. The EXAFS data are acquired over an interval of about four hours. Thus the sample is continuously illuminated and monitored so that a constant population is maintained. When MbCO is photolyzed at the rate, k_4, it recombines at rate $k_1[CO]$ or k_3 as appears to be the case at 7-40 K (Figure 18)[43]. The photons that are effective in photolysis are those that follow the wavelength profile of the extinction coefficient ($\mathcal{E}$). Thus, in the steady state, k_4 is a "clock" setting the recombination rate (k_3). For a given temperature, T, and a given intensity, I, there will be equal values of k_3 and k_4 which give a steady state, and thus the concentration is stabilized over the four hour interval. Figure 19A illustrates optical pumping in the continuously illuminated steady state. As the temperature is lowered, k_3 is diminished until it is negligible compared to k_4 and the sample is completely photolyzed. Ramping the temperature to a higher temperature maintains a portion of the population photolyzed. Activation of ligand motion at the higher temperatures occurs and the structural relationship of the ligand and the iron is determined by EXAFS at various temperatures. The mean Fe-CO distance at 40 K, is increased over that at 7 K by 0.7 Å[42]; the carbon atom of CO is no longer detected in the first shell of the X-ray data (Figure 19b)[42,44,45]. However, we can find evidence for the ligand in the second and third shells. The experiment that we were doing last night was at 80 K with a very intense pump light, i.e., high values of k_3 and k_4. This is the way we are exploring the pathway for ligand motion near the heme site and the trajectory for its recombination.

CRYOSTAT AND TIME SHARED
SPECTROSCOPY AND PHOTOLYSIS

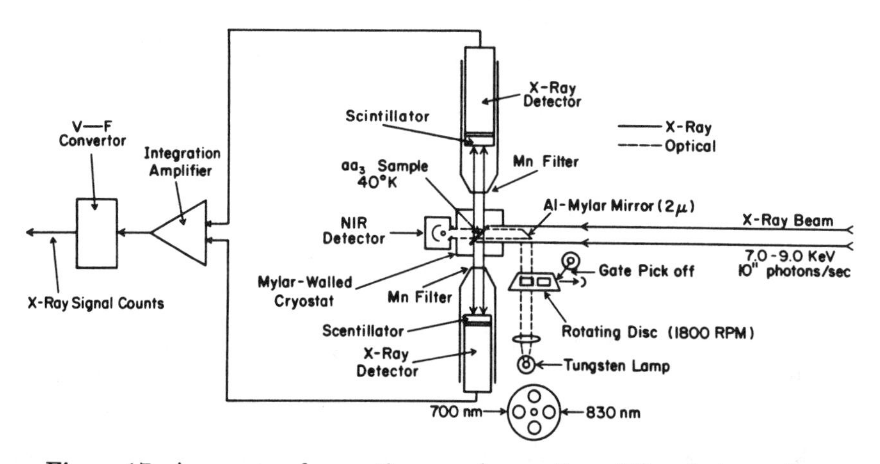

Figure 17. Apparatus for continuous observation of the photoproduct of myoglobin, and MbCO photolysis at various low temperatures as obtained by time sharing photolysis and measuring lights. The apparatus is also capable of X-ray absorption spectroscopy of the well characterized photoproduct.

PHOTOCHEMICAL KINETICS

$$\underset{\substack{\chi \\ e-p_1-p_2}}{Mb + CO} \underset{k_2}{\overset{k_1}{\rightleftharpoons}} \underset{p_1}{Mb^*CO} \underset{k_4}{\overset{k_3}{\rightleftharpoons}} \underset{p_2}{MbCO}$$

$$\frac{dp_1}{dt} = k_1\chi(e-p_1-p_2) - k_2p_1 - k_3p_1 + k_4p_2$$

$$\frac{dp_2}{dt} = k_3p_1 - k_4p_2$$

for $\frac{dp_1}{dt}, \frac{dp_2}{dt} = 0$ steady state pumping

$$p_1 = p_2\frac{k_4}{k_3} \qquad p_1 = \frac{e-p_2}{1+\frac{k_2}{k_1 x}}$$

$$k_4 = \epsilon\phi I$$

$$p_1 = p_2\frac{\epsilon\phi I^{\dagger}}{k_3}$$

ϵ = the molecular extinction coefficient
ϕ = the quantum yield
I = the light intensity

for $p_1 >> p_2$ $\quad \epsilon\phi I >> k_3$

$k_3 \sim 10^6\,sec^{-1}$ $\quad \epsilon\phi I >> 10^6\,sec^{-1}$

Figure 18. Photochemical kinetics.

PERTURBATION OF THE STEADY STATE

Figure 20 shows how flash perturbation of the steady state increases the population of dissociated molecules, giving a display of k_3 with respect to the background clock rate (k_4). k_3 is clearly multiphasic and the restitution of the chemical recombination rate, k_3, to equal the clock rate, k_4, requires about 30 min[46,47]. This is the time required for the system to relax from a perturbation by light (a "photon jump"). Since a temperature perturbation (a "T jump") may also be involved, the temperature rise is measured by a small diode (Figure 21) buried within the sample. The initial 5 minutes of the recombination involves a significant T jump especially at higher concentrations of Mb as required for infrared Mössbauer and EXAFS studies. The deconvolution of the light and temperature effects is currently under study. The temperature and photo jump relaxations are separately recorded in Figure 21 which is obtained by near infrared (NIR) water heating with a 500 Watt lamp and a NIR filter ($> 1.2\ \mu m$) (bottom trace) resulting in a 5° T jump or with a white light flash with a NIR blocking filter, i.e., negligible T jump and a significant photolysis, both at 20 K. The time course of recovery to the steady state is 9-fold different at 15 sec of recombination[46].

RELATIONSHIP TO ATP FORMATIONS

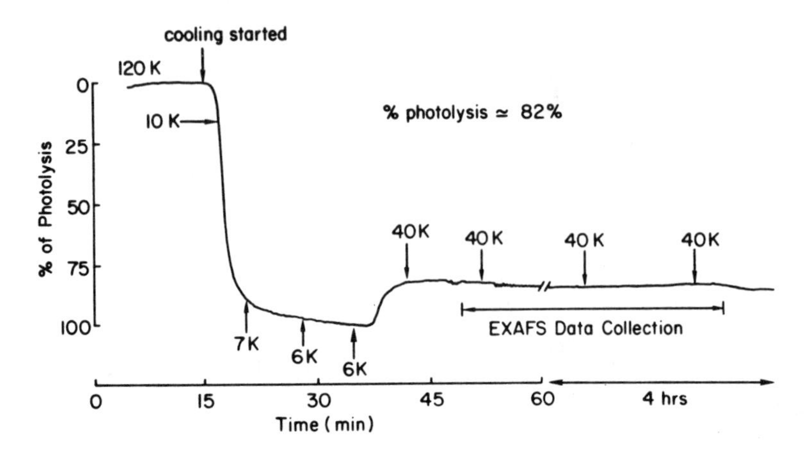

Figure 19A. Illustrating the establishment of steady state optical pumping of the photoproduct of Mb and CO at various low temperatures. At the lowest temperature, MbCO is completely photolyzed and recombines at higher temperatures. The photoproduct can then be stabilized for the necessary four hours of EXAFS data collection.

RESULTS OF FITTING PARAMETERS FOR PHOTOPRODUCTS

	Mb*CO (4°)			Mb** (40°K)		
Parameters	r(Å)	N	$\Delta r^2 \times 10^3$ ($Å^{-2}$)	r(Å)	N	$\Delta r^2 \times 10^3$ ($Å^{-2}$)
Fe-Np	2.03 ± 0.01	4	5.0 ± 1	2.04 ± 0.007	4	5.3 ± 1.5
Fe-Ne	2.22 ± 0.02	1	4.3 ± 1	2.20 ± 0.01	1	6.6 ± 2.0
Fe-C	1.97 ± 0.02	1	3.0 ± 2	—	—	—

Figure 19B. Results of the fitting parameters for photoproducts from Fig. 19A.

At this point I will address further questions of O_2 delivery to tissue, in this case, to the brain and heart. A principal question of physiology, biochemistry and medicine is the adequacy of oxygen delivery and utilization, especially in the heart and the brain, to produce ATP for cell function. In the brain, the cortical neurons (Figure 22) contain many mitochondria, synaptic junctions, axons, dendrites and cell bodies[48]. Continuous supplies of oxygen to the neurons are critical to its viability, principally because the brain does not have the capability of turning off its metabolic function as established and maintained by the Na^+/K^+, ATPase, neurotransmitter synthesis and release. Thus, even

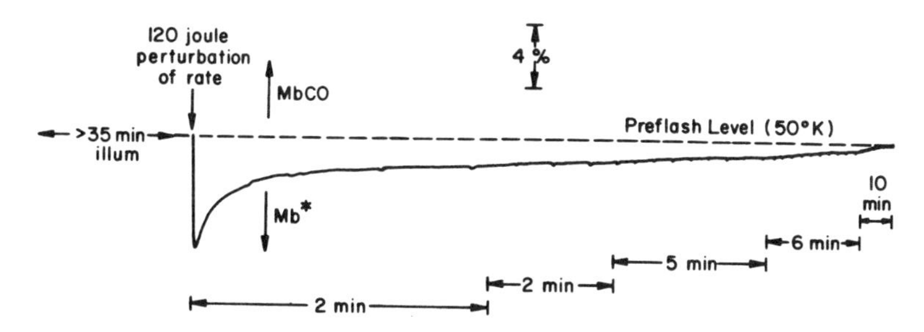

Figure 20. Illustrating the prolonged recovery from photolysis perturbation of the population of MbCO and Mb*. It should be noted that over 10 minutes are required for the relaxation of the population to the steady state composition prior to elimination.

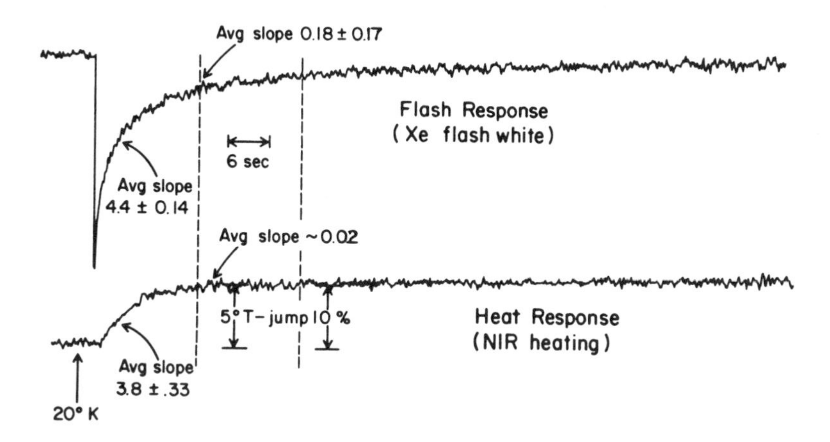

Figure 21. An experimental determination of the response of the steady state optically pumped population to a temperature (T) jump (lower trace) and a photolysis (P) jump (upper trace). At the time marker, the average slope of the T jump is 1/9 that of the P jump. The population change due to the T jump is different from that of P jump.

in the resting brain, an idling neuron must meet the energy requirement for the maintenance of ion gradients. Without energy, the potassium in the cell effuses. Then, sodium and water enter, maximally activating the Na^+/K^+ ATPase, and will enhance the oxygen demand of the tissue even when there is an inadequate O_2 delivery[49]. In addition, some of the neurotransmitters such as glutamate will effuse into the space around the neuron creating a further osmotic stress and as such hence are termed exocytoxins[50]. Thus O_2 lack is to be avoided at all costs and hemoglobin, cytochrome, and myoglobin, and actually the components of the respiratory chain, have an essential role providing extra oxidizing equivalents for cell function in ATP generation.

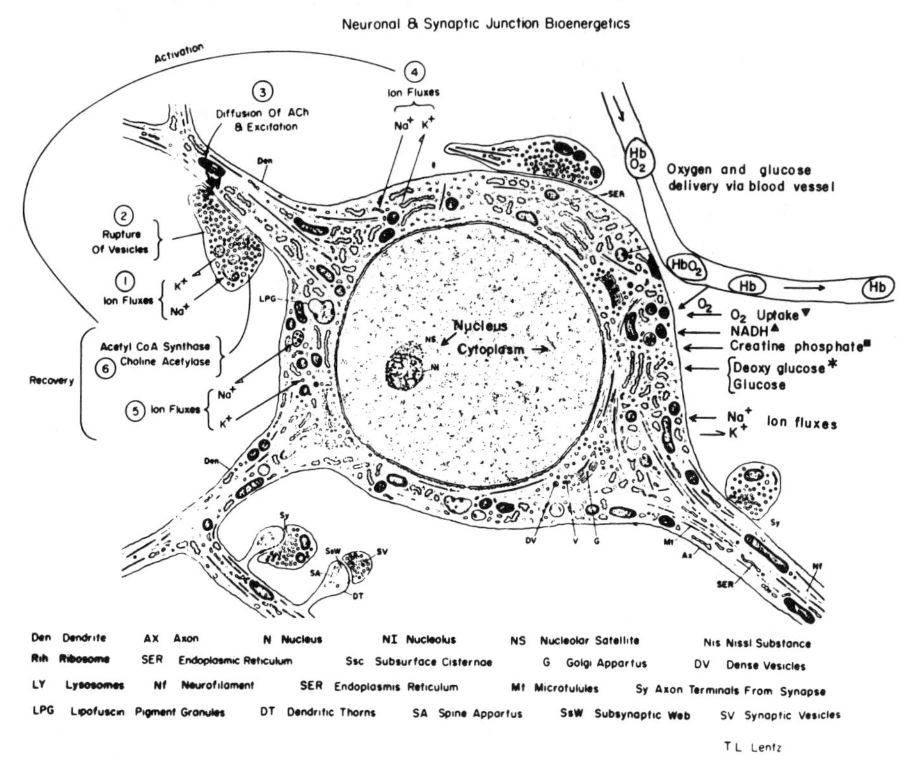

Figure 22. Lentz's diagram of the neuron and its associated processes with annotations of the various parameters that can be measured non-invasively[48].

MAGNETIC RESONANCE SPECTROSCOPY (MRS)

A principal need for evaluating tissue hypoxia is the direct evaluation of intracellular ATP, and MRS uniquely provides this information. The images of the cell that are obtained by Magnetic Resonance Imaging (MRI) do not show ATP but do show that the loss of ATP causes cell rupture due to metabolic and osmotic stress. However, the primary event is the derrangement of the chemistry of the cell by O_2 lack and this is quantified by Magnetic Resonance Spectroscopy (MRS). The MRS detection of the early changes of O_2 delivery that warn of hypoxic cell disaster several hours before MRI can detect changes.

The MRS requires a 1.5 to 2.0 Tesla cryogenic magnet of 7 to 12 inch bore and a 2-4 cm diameter surface coil. This arrangement is suitable for muscle[51], heart[52,53] and brain[54] in the neonate.

MRS detects the biochemicals that are in the cytoplasm of the cell and hence are freely tumbling and remote from paramagnetic disturbances[55]. The biochemicals inside the mitochondria are not recorded (Figure 23A) because they accumulate the Mn^{2+} from the plants we eat. Mn^{2+} is a strong paramagnet and broadens MR lines. Thus, MRS records biochemicals in the cytoplasm: phosphocreatine (PCr), inorganic phosphate (P_i), and adenosine triphosphate (ATP), and pH by the chemical shift of the P_i resonance[50,56,57]. The thermodynamics of oxidative metabolism are readily calculated for the cytoplasm in terms of the ATP/ADP $\times$ P_i ratio or, as the phosphocreatine to inorganic phosphate ratio ($PCr/(P_i)^{2+}$) (Figure 23B)[58,59].

The phosphate potential (PP) (the ability to do work), will increase as the ATP is high and the ADP and P_i are low. The creatine kinase equilibrium enables one to substitute for ADP its value in terms of PCr and Cr and ATP, and assuming Cr and P_i change similarly, PP equals $PCr/(P_i)^2$ to a good approximation at constant pH. Thus, MRS will indicate that O_2 lack has depleted the phosphate potential[59]. In fact, PP depletion causes electrical activity of the brain to cease; ATP need not fall.

But what has turned out to be much more important in the study of living tissues is the velocity (V) of oxidative metabolism (V_m) (Figure 23C) which must match the rate of breakdown of ATP in cell function in order to maintain a steady state. The Michaelis and Menten equation can then be applied to the overall reaction for the synthesis of ATP which is formed from the reaction of oxygen and NADH as catalyzed by the carriers of the respiratory chain and their charge separations[59]:

$$3ADP + 3P_i + NADH + H^+ + 1/2O_2 = 3ADP + NAD^+ + H_2O. \quad (1)$$

A simple Michaelis-Menten formulation suggests that

$$\frac{V}{V_{max}} = \frac{1}{1 + \frac{K_1}{ADP} + \frac{K_2}{P_i} + \frac{K_3}{NADH} + \frac{K_4}{O_2}} \quad (2)$$

where V/V_{max} is the velocity of oxidative metabolism, V is calculated in relation to its maximum (V_m) and K_1 and to K_4 are the apparent Michaelis affinities for the various substrates. As in the case of PP, ADP is approximated by P_i/PCr ($\sim 33\ P_i/PCr$) at constant pH. This equation, while an approximation, tells us the essentials of cell metabolic control, namely, that mass action regulates metabolism *in vivo* and that multiple controls of metabolism are possible.

The key to the stability of the living system is its feedback control as implies by Equation (2). The heart activity or brain activity breaks down ATP and feedback regulation through Equation (2) restores ATP by operation of control, for example, by the raised ADP concentration leading to increased synthesis of ATP. The feedback system is stable if the loop gain is high; i.e.

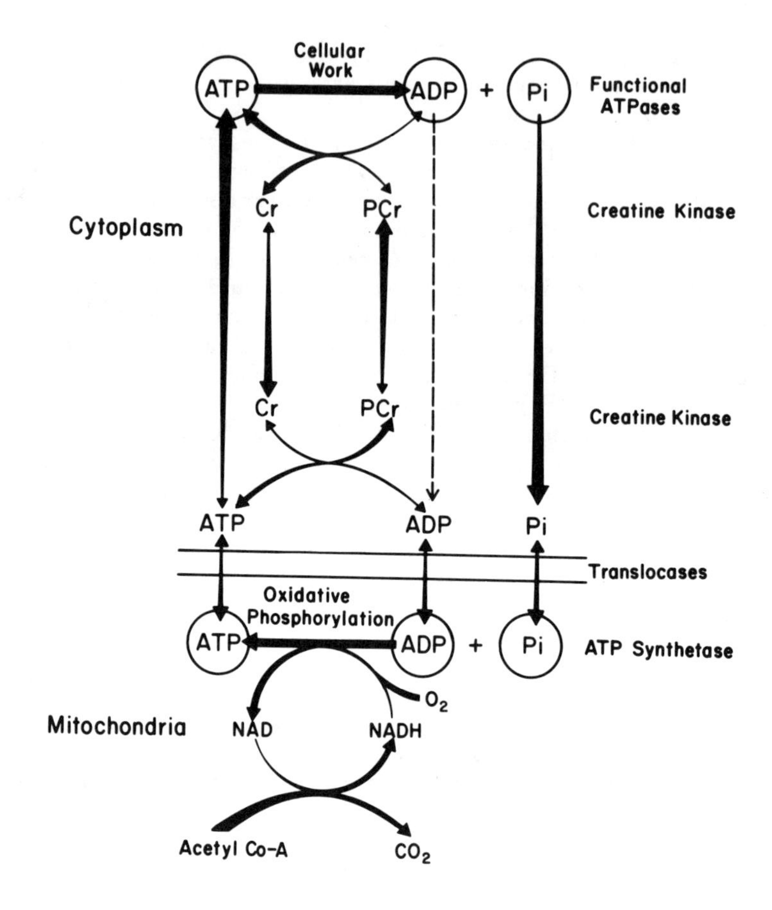

Figure 23A. An illustration of the interrelations between the cytoplasm and mitochrondria as studied by PMRS. Note that the cytoplasm of these components are readily determined by magnetic resonance spectroscopy.

ADP exerts a large effect upon V/V_m. The transfer characteristic of Equation (2) is a hyperbola and, when the operating point is down at the base of the hyperbola, then the input/output transducer relationship is almost linear with respect to ADP and the loop gain is maximal. ADP is usually the controller of rate but when $O_2 \sim k_4$, O_2 can be is a supplementary controller. Thus a rise of ADP can compensate for a decrease of O_2 so that V/V_m is maintained constant. At rest V/V_m is small (10%), the phosphate potential is high, the feedback gain is high; and the loop is highly stable.

PMRS CAN MEASURE CELL THERMODYNAMICS :

The "Equilibrium" of ATP Synthesis

$$ADP + Pi \longleftrightarrow ATP \qquad (1)$$

The Phosphate Potential (P.P.) is defined:

$$[ATP] / [ADP] \times [Pi] \qquad (2)$$

From Cr Kinase Equilibrium:

$$H^+ + ADP + PCr \xleftrightarrow{K_{CrK}} ATP + Cr \qquad (3)$$

$$[ADP] = \frac{[ATP][Cr]}{K_{CrK}[PCr][H^+]} \qquad (4)$$

Substituting 4 into 2:

$$P.P. = K_{CrK} \cdot \frac{PCr}{Cr} \frac{H^+}{Pi} \qquad (5)$$

The PCr Splitting Reaction:

$$PCr \longrightarrow Cr + Pi \qquad (6)$$

$$P.P. \approx K_{CrK} \cdot \frac{PCr\ H^+}{P_i P_i} \qquad (7)$$

Figure 23B. Phosphorus magnetic resonance spectroscopy (PMRS) can measure cell thermodynamics.

The heart is the most "conservative" organ of the body, and it operates at a 10-15% V/V_m and appears to employ manifold feedback loops to achieve high gain – until hear failure occurs[53]. The brain operates higher on the transfer characteristic, about 40-50% in adults and 70% in neonate and thus the brain is more vulnerable to hypoxia. Stresses of high metabolic work (V) that cause V to approach V_m drop the loop gain nearly to zero and the risk of metabolic death is high. While anaerobic glycolysis prevents the loop gain from falling to zero, the price is high and is paid by production of high levels of lactic acid. The phosphate potential achievable by glycolysis is low; no steady state has been observed.

STEADY STATE RELATIONSHIP

1) Michaelis-Menten equation for regulation of cell respiration

$$\frac{V}{V_{max}} = \frac{1}{1+\frac{K_m}{[ADP]}}$$

2) The Creatine-kinase equilibrium

$$PCr + ADP + mH^{+} \overset{K_1}{\rightleftharpoons} ATP + Cr$$

$$[ADP] = \frac{1}{K_1}\frac{[Cr]\times[ATP]}{[PCr]\times[H^{+}]^{m}}$$

$K_1[H^{+}] = 160$ at pH = 7.2, 37°, 3.6 mM Mg^{2+} and 5×10^{-3} M ATP and m = 1

(Lawson and Veech, JBC *254*, 6528-6537 (1979))

3)

$$\frac{V}{V_{max}} = \frac{1}{1+\frac{K_m}{\frac{[Cr]}{[PCr]}\times\frac{[ATP]}{160}}}$$

(Chance and Williams, JBC *217*, 383-393 (1955))

4) $K_m = 2\times10^{-5}$ M

ATP = 5×10^{-3} M in skeletal tissue.

$$\frac{V}{V_{max}} = \frac{1}{1+\frac{0.6}{\frac{[Cr]}{[PCr]}}}$$

5) We can approximate Cr by Cr_0 + Pi or simply by Pi since PCr is converted into equal amounts of Pi and Cr in the 4 to 3 transition.

6)

$$\frac{V}{V_{max}} \approx \frac{1}{1+\frac{0.6}{\frac{[Pi]}{[PCr]}}}$$

Figure 23C. Steady state relationship based on the Michaelis-Menten equation for regulation of cell respiration.

Figure 24 is a typical PMRS spectrum of the brain of a dog which shows that phosphocreatine is high and inorganic phosphate is low in normal function. The spectrum also shows lipid components, phosphomonoester (PME) and phosphodiester (PDE). Resolution of the splittings of the ATP peaks observed in solution by high field high resolution spectroscopy is also obtained in the dog brain at 1.5 to 2.0 Tesla magnetic field[60].

Figure 25 show how these principles have been applied to a model where we stress the brain with bleeding and an occlusion of blood vessels (carotid arteries) to stimulate vascular disease. The thermodynamic potential declines as shown here and acidosis is measured by simultaneously tuning in on the

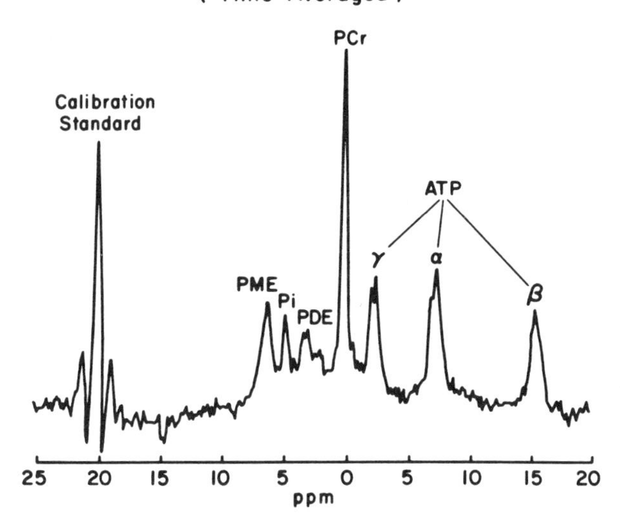

Figure 24. An illustration of PMRS of the adult dog brain. Spectrum was obtained with the surface coil attached to the skull of an anesthetized dog.

protons of the methyl group of lactate during this stress interval. The stress is relieved after two hours, but by that time the feedback loop has a low gain and the system has gone out of feedback control; the metabolism does not recover within the time frame of these studies. Thus, MRS provides an indication of the possibility of metaboolic brain death[61].

NEAR INFRARED SPECTROSCOPY OF TISSUES

Since therapies for hypoxia are mainly preventive, an early warning of hypoxia can be obtained by measuring the deoxyhemoglobin/oxyhemoglobin ratio (Hb/HbO_2) in the near infrared region[32–34]. The NIR wavelength region is desirable because of the better penetration of light and is also used in the EXAFS study of the concentrated myoglobin. In this wavelength region, as my colleague Jobsis[32] has shown that the transillumnination of the neonate cranium is possible. Furthermore, studies in animal models show that PCr/P_i begins to fall at a Hb/HbO_2 value of 80% (Figure 26)[34]. Then the V/V_m may rise to critical values, ATP may be lost and brain damage may arise. In fact, as the hemoglobin is nearly fully deoxygenated, the copper component of cytochrome oxidase becomes detectable at 830 nm as it is reduced as shown in the spectrum of Figure 27. In the blood perfused rat head, the Hb band is shown in Figure 27B. The absorption of the copper component of cytochrome oxidase is shown in the hemoglobin-free head (see Figure 27C)[34].

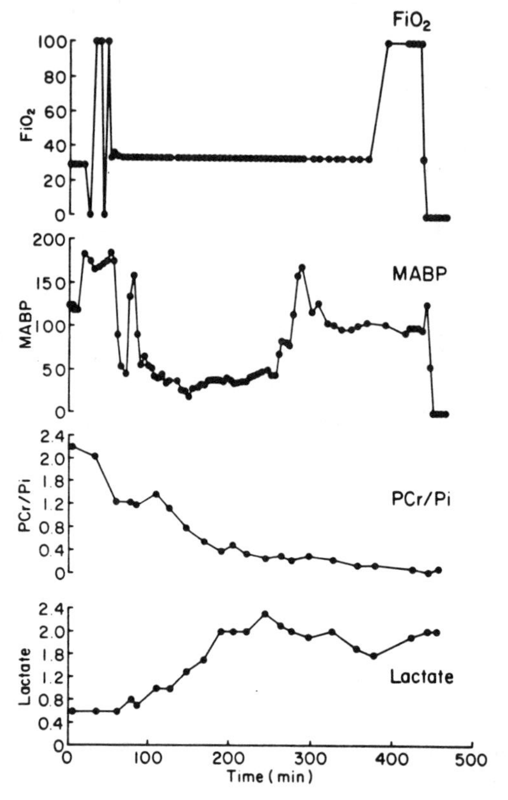

Figure 25. Time course of physiological parameters in a dog brain model stressed with hypovolumia and bilateral carotid artery occlusion. The lactate is measured by time shared proton spectroscopy. Phosphates are measured by PMRS.

The application of this method to follow the loss of HbO_2 and the increase of Hb in the neonate head has been shown by Jobsis[32], Delpy[33], and recently simplified dual wavelength technology permits its use in the adult foreheads[62]. Thus, NIR spectroscopy widens the very narrow "near death" MRS window and complements the MRS biochemical studies in a way that can be very important for long-term non-obtrusive, non-invasive monitoring of O_2 delivery to the brain and eventually other organs.

SUMMARY

This paper traces the versatile role of the optical method in studies of heme pigments in the visible and near infrared regions for the study of the state of MbCO under X-ray examination to that of the brain HbO_2 under conditions of MRS examination.

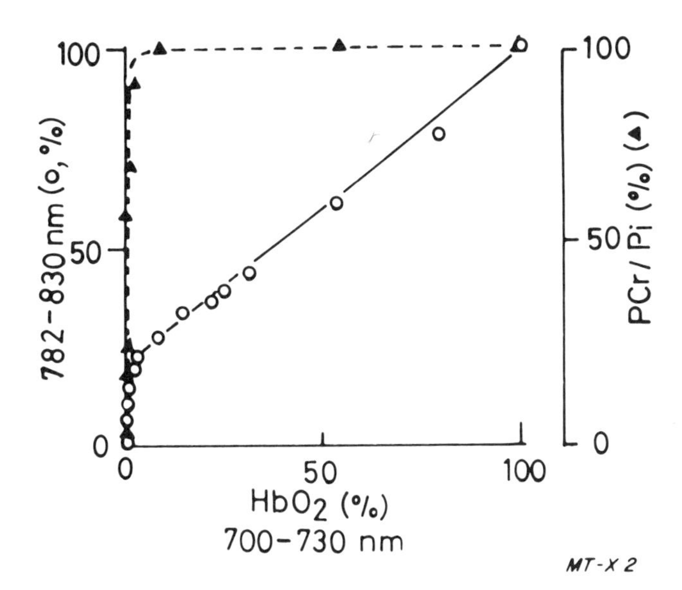

The relationship between PCr/Pi, HbO_2 and absorption change at 780-830 nm . Normal cat head.

Figure 26. A deconvolution of the absorption of hemoglobin and cytochrome in the near infrared spectroscopy by an oxygen titration of the hemoglobin (abscissa) and hemoglobin + cytochrome (ordinate) together with the NMR to simultaneous NMR determination of the PCr/Pi ratio[34].

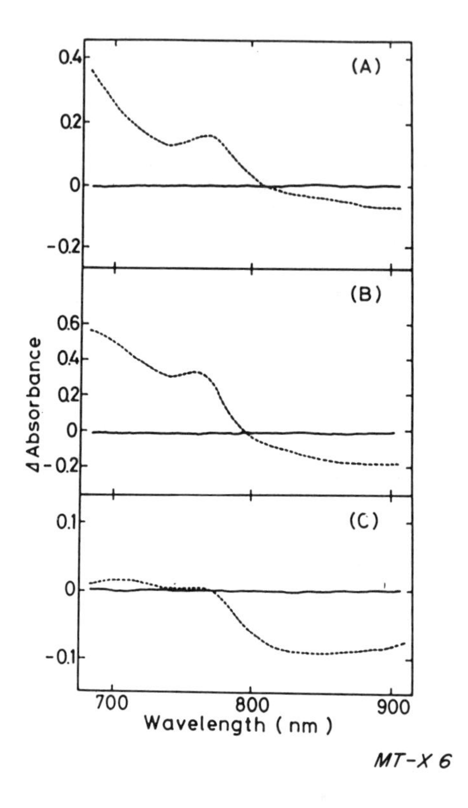

Difference absorption spectra of hemoglobin solution(A), normal rat head (B), and FC-treated rat head (C).

Figure 27. Spectral scans illustrating the contribution of copper to the absorption difference spectrum of the rat head, (B) in which hemoglobin is present and (C) from which Hb has been removed[34].

ACKNOWLEDGEMENTS

This work was supported in part by National Institutes of Health Grants NS-22881, HL 18708, RR 01633, GM 31992, GM 33165 and HL 31909. Also, support was provided by National Science Foundation, DMB 85-18433, and Department of Energy, DOE FG02-86-12476 for PLD, MRG.

REFERENCES

1. D. Keilin, Proc. Roy. Soc. London B98, 312 (1925); D. Keilin, *The History of Cell Respiration and Cytochrome*, (Cambridge University Press, Cambridge 1966).
2. O. Warburg, *Heavy Metal Prosthetic Groups and Enzyme Action,* translation by A. Lawson, (Oxford University Press, London, England 1949).
3. D.C. Wallace, Hospital Practice 21, 77 (1986).
4. P.L. Dutton, in *Encyclopedia of Plant Physiology*, New Series, Vol. 19, L.A. Staechelein and C.J. Arntzen, eds. (Springer Verlag, NY, Berlin, 1986), p. 197.
5. H. Michel, O. Epp and J. Deisenhofer, EMBO Journal 5, 2445 (1986).
6. C.H. Chang, D. Tiede, J. Tang, U. Smith, J. Norris and M. Schiffer, FEBS Lett. 205, 82 (1986).
7. J.P. Allen, G. Feher, T.O. Yeates, H. Komiya and D.C. Rees, Proc. Natl. Acad. Sci. USA 84, 5730 (1987).
8. D. DeVault and B. Chance, Biophys. J. 6, 825 (1966).
9. M.R. Gunner and P.L. Dutton, in *Structure of the Reaction Center X-ray Crystallography and Optical Spectroscopy with Polarized Light* J. Breton and A. Vermeglio, eds. (Plenum Publ. Co., NY, London) in press.
10. M.R. Gunner, D.E. Robertson and P.L. Dutton, J. Phys. Chem. 90, 3783 (1986).
11. M.R. Gunner and P.L. Dutton, Biophys. J., submitted.
12. D. DeVault, Q. Rev. Biophys. 13, 387 (1980).
13. L. Powers, B. Chance, Y. Ching and P. Angiolillo, Biophys. J. 34, 465 (1981).
14. R.A. Scott, J.R. Swartz and S.P. Cramer, Biochem. 25, 5546 (1986).
15. Y. Anaraku, Intl. Symp. on Cytochrome Systems: Molecular Biology and Bioenergetics (IUB Symp. n. 159) Frasano, Italy (April 7-11, 1987).
16. B. Chance and L. Powers, In *Current Topics in Bioenergetics*, Vol. 14, C.P. Lee, ed. (Academic Press, FL, 1985), p. 1.
17. Ehrenberg, this symposium.
18. P. Li, J. Gelles, S. Chan, R. Sullivan and R.A. Scott, Biochem. 26, 2091 (1987).
19. B. Chance, S. Khalid, L. Powers, R. Hettel and T. Toxel, Am. Phys. Sco., submitted.
20. B. Chance, L. Powers, R.O. Poyton and R. Waterland, Biophys. J. 53, 373a (1988).
21. T.E. Patterson, C.E. Trueblood, R.M. Wright and R.O. Poyton, in *Cytochrome Systems: Molecular Biology and Bioenergetics*, S. Papa, ed. (Plenum, New York, 1987), p. 253.

22. Y. Anraku and R.B. Gennis, TIBS 12, 262 (1987).
23. M. Raitio, T. Jalli and M. Saraste, EMBO J. 6, 2825 (1987).
24. M. Saraste, M. Raitio, T. Jalli and A. Peramaa, FEBS Lett. 206, 154 (1986).
25. B. Chance, The Harvey Lecture Series 49, 145 (1955).
26. J. Tyndall, *Contributions to Molecular Physics in the Domain of Radiant Heat*, (Appleton & Co., New York, 1873).
27. B. Chance, in *Biochemistry of Copper* J. Peisach, ed. (Academic Press, New York, 1966), p. 293.
28. B. Chance, J. Biol. Chem. 234, 3036 (1959).
29. B. Chance and B. Hess, J. Biol. Chem. 234, 2416 (1959).
30. G.A. Millikan, Proc. Roy. Soc. London *B*129, 218 (1937).
31. Millikan, ear lobe.
32. F.F. Jobsis-VanderVliet, Adv. Exp. Med. Biol. 191, 833-842 (1985).
33. P. vanderZee and D.T. Delphy, Intl. Soc. on Oxygen Transport to Tissue, Sapporo, Japan (July 22-25, 1987), p. 82.
34. M.H. Tamura, O. Hazeki, S. Nioka, B. Chance and D.S. Smith, in *Chemoreceptors and Reflexes in Breathing*, S. Lahrir, ed. (Oxford Press, New York, 1987), in press.
35. W.E. Blumberg, Biophys. J. 51, 288a (1987).
36. T. Iizuka, H. Yamato, M. Kotani and T. Yonetani, Biochim. Biophys. Acta 731, 126 (1974).
37. N.A. Alberding, R.H. Austin, S.S. Chan, L. Eisenstein, H. Frauenfelder, I.C. Gunsalus and T.M. Nordlund, Science 192, 1002 (1976).
38. N.A. Alberding, R.H. Austin, S.S. Chan, L. Eisenstein, H. Frauenfelder, I.C. Grunsalus and T.M. Nordlund, J. Chem. Phys. 65, 4701 (1976).
39. N.A. Alberding, R.H. Austin, S.S. Chan, L. Eisenstein, H. Frauenfelder, D. Good, K. Kauffman, M.C. Marden, T.M. Nordlund, L. Reinisch, A.H. Reynolds, L.B. Sorenson, T.G. Wagner and K.T. Yue, Biophys. J. 24, 319 (1978).
40. H. Frauenfelder, in *Structure and Motion: Membranes, Nucleic Acids and Proteins*, E. Clementi, G. Corongiu, M.H. Sarma and R.H. Sarma, eds. (Adenine Press, Guilderland, NY, 1985), p. 205.
41. B. Chance, L. Powers, M. Chance, Y. Zhou and K.S. Reddy, in *Adv. Membrane Biochemistry and Bioenergetics* (Symp. Honoring T. King) C. Kim, ed. (Plenum Publ. Co., New York), in press.
42. L. Powers, B. Chance, M.R. Chance, J. Friedman, S. Khalid, C. Kumar, A. Naqui, K.S. Reddy and Y. Zhou, Biochem. 26, 4785 (1987).
43. B. Chance, J. Biol. Chem. 202, 407 (1953).
44. B. Chance, R. Fischetti and L. Powers, Biochemistry 22, 3820 (1983).
45. L. Powers, J.L. Sessler, G.L. Woolery and B. Chance, Biochem. 23, 5519 (1984).
46. K.S. Reddy and B. Chance, Am. Phys. Soc., New Orleans (March 21-25, 1988).
47. B.F. Campbell, M.R. Chance and J.M. Friedman, Science 238, 373 (1987).
48. T.L. Lentz, *Cell Fine Structure, An Atlas of Drawings of Whole Cell Structure*, (W.B. Sanders, Philadelphia, PA, 1971), p. 357.

49. B. Chance, S. Nioka and J.S. Leigh, in *Oxygen Transport and Utilization*, C.W. Bryan-Brown and S.M. Ayres, eds. (Society of Critical Care Medicine, Fullerton, CA, 1987), p. 215.
50. B. Chance, Am. Philos. Soc. (special publ. APS #44) 1, 55 (1987).
51. K.K. McCully and B. Chance, in *Respiratory Muscles and their Neuromotor control* (R. Alan Liss, 1987), p. 317.
52. G.J. Whitman, B. Chance, H. Bode, J. Maris, J. Haselgrove, R. Kelley, B.J. Clark and A.H. Harken, JACC 5, 745 (1985).
53. B. Chance, B.J. Clark, S. Nioka, H.V. Subramanian, J.M. Maris, H. Bode, Phosphorus NMR spectroscopy *in vivo*, circulation 72 (Suppl IV) IV-103-IV-110 (1985).
54. M. Delivoria-Papadopoulos and B. Chance, in *Recent Advances in Neonatal Intensive Care* R.D. Guthrie, ed. (Churchill Livingstone, 1988), p. 153.
55. G.K. Radda, Biochem. Soc. Trans. 14, 517 (1986).
56. J.H. Park, R.L. Brown, C.R. Park, K. McCully, M. Cohn, J. Haselgrove and B. Chance, Proc. Natl. Acad. Sci. USA 84 (1987).
57. D.G. Gadian, *Nuclear Magnetic Resonance and its Application to Living Systems* (Clarendon, Oxford, 1982).
58. B. Chance, J.S. Leigh, B.J. Clark, J. Maris, J. Kent, S. Nioka and D. Smith, Proc. Natl. Acad. Sci. USA 82, 8384 (1985).
59. B. Chance, J.S. Leigh, J. Kent, K. McCully, S. Nioka, B.J. Clark, J.M. Maris and T. Graham, Proc. Natl. Acad. Sci. USA 83, 9458 (1986).
60. S. Nioka, A. Mayevsky, B. Chance, H.V. Subramanian, C. Alter, D. Gilbert, T. Sinnwell, M. Ghosh, E. Donlon and S. Butler, Soc. Mag. Res. in Med. Fifth Ann. Mtg. 3, 674 (1986).
61. B. Chance, N. Eng. J. Med., in preparation.
62. B. Chance, R. Greenfeld, H. Miyake, D. Smith, S. Nioka, G. Holtom, K. Kaufmann and J.S. Leigh, Time resolved infrared spectroscopy of the adult human brain, in preparation.

ADVENTURES AT TWO FRONTIERS IN SCIENCE WITH HANS FRAUENFELDER

Harry J. Lipkin
Department of Physics, Weizmann Institute of Science
Rehovot, Israel

In Israel we have a very nice expression for occasions like this – "ad mea ve esrim" "until 120."

I had the good fortune to be closely associated with Hans in two adventures at the frontiers of science – parity nonconservation and the Mössbauer effect. As soon as we heard the news about the experiment proving that parity was not conserved in beta decay – both Hans and I realized independently that beta rays must be polarized and that very simply experiments could test this. Both in Urbana and Rehovot, experiments were done which measured the beta ray polarization and helped to clarify the confused picture of weak interactions and provide the basis for the universal V-A interaction.

Therefore, when I had a year's leaving coming in 1958-9, it was natural for me to contact Hans and to inquire about the possibility of working with him in Urbana. Instead, I received an answer from Fred Seitz, Head of the Department, informing me that Hans would be away on sabbatical at CERN that year, and offering me the job of looking after his students. Thus I began my fruitful and enjoyable association with the Frauenfelder's group and the University of Illinois which produced much good physics, good times, many friendships and eventually three books.

Our paths crossed in New York, where we met Hans and Vreneli and their two children and Hans told me all about Champaign-Urbana and his group. Of course Hans would not sit still and explain this – we talked on the way to the garage where his car was being repaired and where he talked to the local mechanic – in Swiss German.

I then realized that Hans belonged to an exclusive fraternity whose members spoke this exotic language and could be found everywhere. Only many years later did I learn that this was a very oversimplified picture. Felix Villars and Konrad Bleuler were visiting us in Rehovot and chattering away in Swiss German. Felix then remarked that this was the first time in many years that he had had the opportunity to speak in his native language. I remarked that in Urbana Hans always had Swiss visitors and you heard the language all the time. They looked at me with great disdain and said "Hans is from the North! It's a completely different language!"

In Urbana we immediately received a very warm reception from the department and from the Frauenfelder's group. Although Hans was not there, his spirit and influence were very much evident in the group that he had built, with their warmth and enthusiasm for physics and the way they all worked together. I am very pleased to see Norman and Alma Peacock, Jack Ullman, Renato Bobone and Sy and Cecile Margulies all here again for this symposium. I miss "Peps" de Pasquali, whose name appears on the list of participants, but whom I do not see.

It was during this year that the group began to work on the Mössbauer effect. Hans at CERN, always alert for new exciting frontiers in science had met Rudolf Mössbauer and sent a copy of his paper to Norman Peacock with the suggestion that we look into it. Since we were all nuclear physicists and did not know the solid-state physics necessary to understand the paper, we decided that Urbana was the place where we could get the help we needed. We asked for a meeting with Fred Seitz, who looked at the paper and said, "Who is this fellow Mössbauer? Does anybody know him? Is he reliable?"

Fred asked for a few days to think about it. Later he said that it was perfectly alright, but he admitted that he had first thought it was completely crazy. This was a typical response to Mössbauer's first paper. Anyone who really understood the physics – and there was no new physics in it, everything was in Willis Lamb's old paper on neutron absorption in crystals – knew it was alright. But practically nobody understood this physics because it involved elementary pieces of both nuclear and solid-state physics. People in these two areas did not talk to one another in those days, had completely different intuitions, and were easily misled into false and irrelevant arguments suggesting that Mössbauer was wrong. In Fred's case, it was the 130 keV gamma ray energy – so much larger than lattice energies that there was no possibility of such an energy being deposited in the lattice without totally disrupting it. It took time, thought and elementary kinematics of the nuclear process to realize that the 130 keV energy was irrelevant.

We immediately began plans to start a Mössbauer experiment, and I attempted to find out what other physicists thought about the effect. When I visited Columbia to give a seminar, I learned from Rudolf Peierls, who was then visiting, that a group at Los Alamos headed by Darragh Nagle was working on the Mössbauer effect. I told the members of our group going to the APS Washington Meeting that they should contact the Los Alamos people. But there was no one from Los Alamos at the meeting who knew about their Mössbauer experiment. Instead they heard that someone at Argonne was planning a Mössbauer experiment.

I was planning to spend the summer at Argonne, and learned during my next visit that David Inglis and Maria Mayer had been investigating the theory of the effect and John Schiffer was doing the experiment. But everywhere the communication gap between nuclear and particle physics confused nearly everyone. There were only a few physicists like Peierls who know enough in both areas to understand the effect. At Urbana, I had managed to learn enough elementary solid-state physics to become a "solid-state expert" who could talk to nuclear physicists and vice versa. I then wrote a paper "Some Simple Features of the Mössbauer Effect" explaining the effect and deriving some general results. This was the first publication which called the phenomenon by its present name, the Mössbauer Effect, thereby acknowledging that Mössbauer's work was sufficiently significant to justify naming an effect for him.

In the Argonne cafeteria I met an old friend, Kundan Singwi, an expert on the theory of neutron scattering in crystals, having lunch with the Head of the Argonne Solid State Division. Neither had heard of Mössbauer's work. However, I knew that Kundan would be interested in and could easily extend his previous work to treat the Mössbauer effect. I told them about it, and in the following year Singwi and his colleague Alf Sjolander wrote some of the first papers on the theory of the Mössbauer spectrum. But at that meeting the Director of the Argonne Solid State Division found my description of the effect very fascinating and asked whether anyone had done the experiment. I told him that it had been done by John Schiffer in the next building at Argonne.

The response to Mössbauer's work at that time seems incredible today. The results were explained perfectly by Lamb's paper, which was referred to in Mössbauer's paper. Yet enough serious physicists were so sure that Mössbauer's work was wrong that they repeated his experiment, obtained exactly the same results without adding any new or original ingredient, and published their work in refereed journals as original research. Once the existence of the effect was accepted, it was dismissed as a trivial and useless exhibition of elementary quantum mechanics from which we would learn nothing new. It was no use for nuclear physics, and anything in solid-state physics that could be investigated with the Mössbauer effect would be treated much better by NMR, X-ray electron, neutron diffraction and neutron scattering.

Meanwhile, Hans returned to Urbana and the group continued its first experiments with ^{57}Fe. By the next year, the summer of 1960, the effect had finally come into its own recognition, and Hans organized the first International Mössbauer Conference at Allerton Park of the University of Illinois.

When I returned to Urbana for a second year in 1962, I was fortunate to overlap for a half year with Hans and got to know him better. Mössbauer had already received the Nobel Prize, and Hans and I were invited together by him to a small meeting at Caltech. I remember him taking us on a trip through the countryside and talking about many things. I then learned that you did not have to be a Jew interested in Israel to be a Zionist. Rudolf told us how he was having his arm twisted by Heisenberg with what I would call "pure Zionist propaganda." He should leave the good life in America and return as a pioneer to his ancestral homeland. Heisenberg's answer to all Rudolf's criticisms of academic life and research in Germany was to say that his people needed him because he had the possibility of changing this situation and building a new framework. It sounded so familiar. In the end, Rudolf responded to Heisenberg and became a Zionist pioneer in his homeland.

Rudolf Mössbauer has always been very modest about his effect. His talk at this meeting on neutrino oscillations did not mention the role of the Mössbauer effect in ensuring the coherence of the waves describing neutrinos with different masses.

For those in the audience unfamiliar with the Mössbauer effect, I provide this simple explanation. A downhill skier uses the Mössbauer effect without realizing it. Newton says that the force of gravity on the skier is directed downward. But the skier moves forward as well as downward. Newton says that if a skier moves forward when there is no external force in the forward direction, something else must recoil backward in order to conserve momentum. The motion of the object recoiling backwards has kinetic energy, and this energy must be supplied from somewhere. Mössbauer points out that you can forget about recoil if the recoiling object is very heavy, much heavier than the object going forward. When the skier goes forward, the earth goes backward.

But the earth is so heavy that it easily takes up the skiers' momentum with an imperceptible tiny recoil. This "recoil free" motion is the Mössbauer effect.

After 1963 Hans and I took different paths to different new frontiers. Hans moved into biology, I moved into particle physics. It was actually my friends at Urbana who pushed me to this frontier. The new theory of SU(3) symmetry of Gell-Mann and Ne'eman had aroused great interest in the particle physics community. The particle physicists at Urbana urged me to give a series of lectures on SU(3). The result was my book "Lie Groups for Pedestrians" and my becoming a particle physicist.

At 65, Hans is still at the frontiers of science: Until 120!

APPENDIX

Simple Features of the Mössbauer Effect

Suppose our downhill skier is on an artificial hill weighing only a million times more than the skier, and mounted on wheels so that it is free to recoil. When the skier moves forward, the hill recoils backward, with one-millionth of the skier's velocity. The kinetic energy of the recoil would be only one-millionth of the skier's kinetic energy, and nobody would notice it.

Mössbauer's original experiment studied the emission of a gamma ray by a nucleus a million times heavier than the gamma ray. But even though the energy of the recoiling nucleus was only one-millionth of the gamma ray energy, it could not be ignored. Mössbauer's experiment involved a high precision measurement of the gamma ray energy to one part in a thousand-million. The energy needed for the nuclear recoil had to be supplied by the gamma ray, and a loss of one part per million was very serious in such a precision experiment.

It turns out that the gamma ray energy is just about a million times larger than the energy of the normal thermal motion of the nucleus in thermal equilibrium at room temperature. The thermal energy is thus of the same order as the energy loss by nuclear recoil. This effect was used in designing Mössbauer's original experiment in which a gamma ray emitted from a radioactive source was absorbed by another nucleus in an absorber placed at some distance from the source. This "resonance absorption" can take place only when the gamma ray has the right energy to one part in a thousand-million. Both the nucleus in the source that emits the gamma ray and the nucleus that absorbs the gamma ray must recoil to conserve momentum, with energies about one millionth of the gamma ray energy. There is a small probability that the energy of thermal motion can compensate for the recoil energy loss. This effect would be reduced by cooling the source and absorber, thus reducing the thermal energy.

By studying the temperature dependence of this effect, one can obtain information about the nucleus. This was the original motivation of Mössbauer's experiment. He found, indeed, that when he cooled either the source or the absorber, his absorption signal decreased, as expected, and he actually used this decrease to calculate a property of the nucleus, namely the line width of the transition. But when he cooled both the source and the absorber, instead of a further decrease in the absorption signal, Mössbauer observed an *increase* in the absorption.

Mössbauer learned from Lamb's paper that at sufficiently low temperatures the nucleus emitting or absorbing the gamma ray has a probability of being "frozen" into the crystal, so that the whole crystal recoils together with the nucleus. The crystal is many million times heavier than the gamma ray, and its recoil energy is well below the one part in a thousand-million precision required in this experiment. So when both source and absorber are cooled, there can be a "recoil free" transition, like the skier on the earth.

But Mössbauer's effect was very small, less than 1%. When a student finds a 1% effect in the opposite direction from what convertional wisdom expects, it is natural to disbelieve the student, rather than to read the theoretical paper which explains why the conventional wisdom and conventional intuition are wrong. These enormous factors like one million and one thousand-million tended to cause confusion. Fred Seitz was confused at first because he did not realize that the relevant energy for the crystal was only one millionth of the gamma ray energy.

DISCUSSION
THE BIOMOLECULAR FRONTIER: STRUCTURE, DYNAMICS AND FUNCTION

Moderator: Hans Frauenfelder

Hans Frauenfelder: I hope the discussion will bring about an exchange of ideas between different groups and, in particular, an interchange between experiment and theory. The central issue is, what can we learn about the structure-function relation taking into account that it is a dynamic rather than a static relation.

Britton Chance: I would like to discuss optical and thermal perturbations in photolysis experiments on the myoglobin-CO complex and show that steady state conditions are often useful for the study of slow processes, Figure 1. In the steady state the photolysis rate must equal the recombination rate. The photolysis rate is set exactly by the light intensity and serves as your chronometer; k_3 is set implicitly by the temperature and other parameters. At constant temperature and constant photon input rate you therefore have a clock to measure perturbations of the recombination rate. As illustrated in Figure 2, you see an increase of photolysis if you go with a moderate pump light from 120 K to 5 K. Figure 3 shows profiles of the MbCO fraction versus temperature for a 50 W lamp and a 1 kw lamp. With the 1 kW lamp we now use at Stanford we obtain 50% photolysis near 80 K for prolonged EXAFS studies. We measure the temperature with a diode in the sample, but the temperature of the cold finger, which is usually measured, may be something else. Although we use water filters to absorb all but the visible radiation the illumination produces substantial thermal perturbations of the sample because of the poor heat transfer to the cold finger. When we measured the sample temperature after a 100 J flash from a Xe lamp as it is used in most laboratories we found the temperature to rise from 50 to 55 K with the time dependence shown in Figure 4. With more flashes the perturbation will be even larger, Figure 5. In one of our protocols we waited 10-20 s to let the sample cool after each flash, but in the example above it took 5 minutes for the temperature excursion to drop to 10% of it speak. Figure 6 shows the optical perturbation, that is photolysis. In the process the sample was heated by 5 K after having been under steady state illumination for over 30 minutes near 60 K. The thermal spike is shown in Figure 7, and we can look at the experiment as a T-jump causing more recombination because the recombination rate goes up with temperature. We calculate a 10% photolysis and what we observe is 12%. Thus we have dual perturbations, and what starts out as hot recombination turns into cool recombination after about 5 minutes. Figure 8 shows the same process on a longer time scale. After about we are back to the original temperature within 0.1 K, and the interesting region for analysis is the recovery of the system from the combined effects of the optical and thermal perturbation. In repetitive flash experiments you should wait about 30 minutes between flashes if you want to get back to the initial set of states. On the other hand, if you want to investigate the fact that the recombination rate is getting slower with time you can measure it from these data. The chronometer idea works very well because we have a state that is precisely determined in terms of the recombination rate. The time course shown in Figure 9 suggests that we are measuring in this region a recovery to the pumped or continuously illuminated state. I feel that this is a useful method and would like to open it for discussion. *(No figures were submitted.)*

Jim Alben: I would like to raise the question of how one determines the local temperature of the excited state of the iron-porphyrin complex and the surrounding heme pocket as opposed to the temperature of the bulk matrix which can be measured directly. These two temperatures are quite different and have different time scales, and I think it is the temperature of the iron porphyrin and its surroundings which is critical to the photodissociation process.

Britton Chance: I suspect that Robin Hochstrasser could best answer this question. One has good reason to believe that thermal equilibrium around the heme is established quite fast. Don DeVault and I did an experiment when we first got a Q-switched laser to see if an intense pulse would damage the chlorophyll in leaves. The shock wave destroyed the membranes but the chlorophyll was not altered, suggesting that the energy absorbed by the chlorophyll was dissipated very quickly into the environment. Analogously, in the case of the heme the energy is dissipated extremely fast.

Hans Frauenfelder: There are quite a few additional problems here that could keep us busy for the next two days, but let me stop the discussion here so we can explore other questions as well.

Pierre Douzou: *(Manuscript submitted.)*

Hans Frauenfelder: I know the problem Pierre Douzou is talking about. We have been working on the pressure dependence of CO rebinding to myoblogin for 8 years and yet have to write a first paper. The phenomena are so complex that we need many tools and studies on model systems in order to understand the problem. In order to explore other topics let me now turn to an x-ray crystallographer, then to a theoretician and to an experimentalist again.

Greg Petsko: There are many things to be said about the maximum information one can extract from protein crystallography and the limitations of the methodology, and some of you may know that I am perhaps more concerned with the limitations than some people who commonly employ the technique. Rather than discussing this subject I would like to turn your attention to the possibilities for the future. Conventional protein crystallography requires monochromatic X-ray sources and single counters or area detectors to count the X-rays. Even with the most potent monochromatic source from synchrotron radiation and the most efficient detectors the collection of a complete protein data set still takes days. Figure P, in contract, shows a complete data set of a protein crystal data taken on a single film by using white synchrotron radiation covering the wavelength range from roughly 0.3 Å to 2 Å. This photograph contains all the data necessary to calculate an image to the structure at 2 Å resolution, and the total time required for the exposure of the film was 100 ms. In the future it will be possible to collect complete x-ray data sets on a millisecond time scale. The processing of those data is quite complex, but we can now interpret these photographs and produce reasonable maps of the protein and of protein-substrate complexes formed on a millisecond time scale. This work is part of a collaborative project involving Janos Hajdu and Louise Johnson of the University of Oxford, and the late Pella Machin of the Daresbury Synchrotron in England. The technique is called Laue diffraction. Your should think about how to apply this technology to questions of interest to you.

Peter Wolynes: It is Sunday morning and I will rise to talk about biomolecular metaphysics. Hans has been promoting the area of 4-dimensional molecular biology. Both 1-dimensional molecular biology, dealing with rates of molecular processes, and 3-dimensional molecular biology, dealing with spatial structures, are highly developed, but the combination of these into a 4-dimensional picture has not yet been achieved. We should therefore think about what is needed to achieve such a synthesis, and we may start by asking how 3-dimensional molecular biology got so far along. To some extent it was a question of personalities, and here is a story of Bernal, who was one of the heroes of 3-dimensional molecular biology. The story has a couple of lessons for this morning. Bernal was, although a communist, a consultant to field marshal Montgomery in the war. When they were planning the invasion an officer approached Bernal with the question "Should we use sonar buoys to find the depth of the water?" When Bernal asked why he wanted to know the depth the officer apologized he could not tell him. "Well, I can't give you an answer then", Bernal said, "because I don't know if you are asking me the right question." When the officer reported back, Montgomery said he should tell Bernal they needed to know the depth of the water along the beach to find out where boats could land in an invasion. Given this reason, Bernal said "Now you asked me a good question and I can tell you how to solve your problem. You should fly a plane over there to take pictures of the beach. We will interpret these pictures and tell you the depth anywhere and where to send our troops." Now this is exactly Bernal's attitude in 3-dimensional molecular biology. With the advent of X-ray crystallography you get all information at once instead of chipping away at it. The story has an important implication for 4-dimensional molecular biology as well. The crucial thing Bernal did was to get the baud rate up. A sonar buoy has a very small baud rate, not enough for the purpose. To explore systems as complex as biomolecules obviously requires a higher baud rate. On the theoretical side the problem is solved to some extent by supercomputers. On the experimental side one has to look for approaches with a higher baud rate as well. Chemical dynamics, i.e. the measurement of rate coefficients or distributions thereof is probably not sufficient. The X-ray technique that Greg Petsko described is one possibility. NMR has a high baud rate and perhaps also scanning tunneling microscopy. The problem in 4-dimensional molecular biology is much harder than in 3-dimensions because proteins remain fairly close to a particular structure in spite of all kicking and screaming about motions and disorder in proteins. As a result, the structural information contained in one photograph is hard to match in the time domain. It is not clear if all aspects of 4-dimensional motion are ordered as in a single motion picture. Some degree of order is evident in allostery and all interesting functions of proteins. Hans' results show that there are certain aspects of disorder in protein motion, and the question arises to what extent proteins are random systems. This is a crucial question since random systems have a very complex phase space, and if protein motion were truly random we would be dealing with molecular biology in infinitely many dimensions rather than in four dimensions. If that is the case, it is not sufficient to get the baud rate up, since the number of possible states is so high that one would have to explore them at an exponential pace. So here the problem is how to reduce the search space. Hans' approach to the problem is to start from simple models. The question of how to reduce the search space will persist however, and we must search for other possibilities as well. One is to use the technique of site-directed mutagenesis and perhaps evolutionary site-directed mutagenesis according to Professor Eigen. On the theoretical side there is hope that artificial intelligence can be applied in addition to natural intelligence as we see it already happening in the area of protein folding. In conclusion, I hope that this congregation may find other ways to solve the problems I mentioned, namely how to get the baud rate up and how to reduce the search space so that we can explore it at an exponential pace.

Paul Champion: *(Manuscript submitted.)*

Hans Frauenfelder: Let me add a few remarks and then summarize what was said. First, you all realize that for the purpose of this symposium "Hans" is an abbreviation for a large number of collaborators who have worked extremely well and who always tell me what to do. It starts probably with Bob Austin and goes all the way to Bob Young. Without these collaborators it would not have been possible to carry our work so far.

The second remark concerns hole burning and relates to Paul's talk. Myoglobin has a spectral line at 760 nm that is characteristic for rebinding. The line exists in the deoxy state only, and its disappearance therefore measures how many CO molecules rebind. We find that the line not only decreases with time but its center shifts in the process. Agmon, I think, was the first to suggest that the shift was due to hole burning. In the last few weeks Joel Friedman has shown experimentally that below 60 K the shift of the line is indeed due to kinetic hole burning. We have verified Friedman's result. Above 60 K we believe there is relaxation, but we are not certain yet. The important point about this shift is that it gives us a chance to connect spectroscopic data to structural information.

Let me now summarize the key points of the speakers we heard so far. Britton Chance talked on flash photolysis or photodissociation with very important questions on the role of temperature and of sample heating. Pierre

Douzou's contribution was on the relation between gels and proteins, on the use of gels as model systems and on the role of external parameters. Greg Petsko painted a bright future of X-ray crystallography. It is clear from the developments of the last eight years that much more information can come from X-ray data than was thought before. By the same token, the amount of work required to extract that information is many times larger. Before, one could take one X-ray structure and say, that is it, but now we have to measure all the way from 10 K to 300 K and analyze the data much more carefully. Peter Wolynes asked key questions about molecular dynamics in four dimensions which can keep us busy for the next 30 years and can be approached by models that not only include structure but specific time-dependent aspects. Paul Champion treated a particular problem in this field, namely binding of CO and O_2 in heme protein with a model that connects spectroscopic data from X-ray to Raman scattering to IR absorption. So here are five different topics and I suggest that we do not go systematically through each one but rather throw questions at the panelists.

Marianne Grunberg-Manago: I have some comments and questions concerning the future of our techniques Site-specific mutagenesis was already mentioned, but the recombinant DNA techniques really give you enormous possibilities. In the future you can change a protein as much as you want, you can increase the affinity for a substrate, you can change the transition state and the reaction rate, so there is the potential for tremendous progress in understanding how a protein works. I understand that X-ray diffraction has made great progress and there is much more to come, but there is one limiting factor where I have seen no progress whatsoever. This limiting factor is how to crystallize a protein . Do the physicists have any answers? In our work we have many mutants both in the protein and the substrate, but if we do not know the tertiary structure we cannot go on to 4-dimensional molecular biology or whatever. This is really a serious limitation, and I would like to know if there has been any progress in protein crystallization. Also, if we can not get crystals, what other models are there to determine the 3-dimensional structure?

Greg Petsko: There are three answers to these questions. First, some progress in crystallization techniques has been made, mainly by accident, and our chances of getting crystals from a pure sample of biological macromolecules are higher now than they were 5 or 10 years ago. The increased use of polymer percipitants, such as polyethylene glycol and other systems, have given us a wider range of things to try with any given protein. However, progress has not been systematic, we do not know how to approach the problem most efficiently, and there are enormous gaps in our knowledge of what is really going on. Basically, two approaches have been taken. A couple of groups in the US, in particular George Feher's, and one is Europe are looking at the crystallization process in a systematic way. In contrast to this scientific approach several laboratories, in particular the ones at DuPont and Lilly, are pursuing a more industrial approach. They have developed robots that are capable of setting up hundreds of crystallization set-ups a day automatically in an attempt to do by brute force what we are unable to do by cunning.

If you cannot crystallize a protein you may still get a solution structure by 2D-NMR that is as good as a 4 Å or 3 Å resolution crystal structure, provided that the number of amino acids in the protein does not exceed about 80. That number will probably increase to 120—140 in the next few years with the advent of 600 MHz instruments and other improvements.

The third prospect of getting structural information is, of course, the ability to compute the structure de novo. If your protein has at least 50% homology to a protein of known structure, then you can obtain a model that is equivalent to a 4 Å resolution crystal structure by a variety of techniques. If the homology is less than 50%, you have little chance to obtain a realistic model with the kind of detail that you need.

Peter Wolynes: It is amazing how little thought has gone into the process of protein crystallization with the notable exception of Feher's work. One of the general ideas is to investigate the phase diagram of the protein system, and Pierre Douzou's work is a good example of that approach.

I am more optimistic than Greg about the prospects of NMR. With the advent of high-temperature superconductors, which have much higher critical fields as well, it might be possible to increase the field by a factor of ten with proton resonances at 6 GHz. This is bound to have an impact on structural studies.

Britton Chance: Marianne Grunberg-Manago asked very pertinent questions, and let me make two remarks. One is that point mutations generally cause small changes of the active site and, working in the vineyard of large membrane proteins where crystallization is way down the line, we use EXAFS to look for these small changes at the active site down to ± 0.05 Å. We have used this admittedly very limited approach on several point mutants of cytochrome *c* oxidase where the cytoplasmic gene alters the conformation of the whole protein which is coded by both nuclear and cytosolic genes.

The second point concerns the need for a prokaryote myoglobin to carry out genetic manipulations. Yeast has a hemoglobin which is expressed under appropriate conditions and could thus be used for such studies.

Pierre Douzou: I noticed a striking difference between biochemists and polymer chemists. The latter struggle to avoid crystallization of long-chain polymers whereas the former struggle to induce crystallization of their biopolymers. Maybe there is something to be learned if the two groups talked to each other.

Paul Champion: Lately, we have been looking at resonance Raman scattering from myoglobin crystals, and Tim Sage pointed out that we are really probing the crystal surface since the light does not penetrate very far into the crystal at resonance. It has been suggested that one reason for the termination of crystal growth may be the build-up of a disordered layer on the surface.

Raman scattering might be a good probe for this effect, and we plan to check whether we can see it.

Larry Smarr: I have some comments concerning the computational side of Peter's and Greg's remarks. It appears that one can anticipate an explosion in computational needs, whether it is for millisecond X-ray crystallography, for 2D-NMR, for molecular dynamics calculations, or for protein folding. What one really needs is a protein computer, a protein work station coupled with a supercomputer, a massive data base, probably on laser disks, and so forth. In your field the data rate is extremely high, you are dealing with complex systems, and rather than using simplifying models you ought to have the computer do the real thing and present what you are interested in in intelligible form. This involves a lot of algorithms, it involves software and, as Peter was saying, a lot of "Artificial Intelligence" techniques. What you should envisage is a computational lab, probably on a national scale, so that everyone has access to the new protien structures as they come out. The question is how we best make this happen.

Peter Wolynes: To paraphrase Larry's comments, the question is how to get computer power in the hands of the people who will generate a huge amount of data from theory or from experiment. The need for more computer power is obvious. Imagine the big difference when we get a thousand crystal structures a year from Greg Petsko's group rather than ten. For people like myself who have difficulties visualizing 3-dimensional structures good computer graphics are very important. I understand that NIH is planning some biological computation center.

Hans Frauenfelder: We should really have a computer in each of our labs that has access to Greg's data set so we do not have to call him up and ask for, say, the Fe-C-O angle in a particular state of myoglobin. To a certain extent it is intellectual laziness, apart from the lack of money, that we still do it the traditional way. Things will eventually change for the better if Larry Smarr keeps walking around with a 2×4 hitting us over the head whenever he can.

Greg Petsko: The idea of Hans having direct access to my computer is slightly troubling but I suppose progress has its prices. I simply agree with the last three speakers. Free computing is a constitutional right.

Kaspar Winterhalter: I think the question raised by Marianne Grunberg-Manago is a very pertinent one, but there are even harder questions. You physicists talk about isolated proteins in solutions. From the biological point of view these are no longer in the center of interest. Biologists now work on proteins in membranes, a rather ill defined solvent, and here X-ray crystallography is still in its infancy. But apart from membrane proteins, even more interesting, important and more difficult to solve are those proteins that function in conjunction with other proteins. Some effort has been made to characterize these by electron microscope techniques at rather low resolution.

There is a lot to be done in these fields if you physicists should ever run out of problems.

Hans Frauenfelder: Basically, Kaspar Winterhalter enquires when we are going to do real work rather than look at the simplest system. Each of us here, whether physicists, chemist, or biologist, works in a different layer of complexity, and we certainly see a beginning of contacts. We should always be told to make the connection, but unless we start at the simplest point and solve one problem completely, the models we develop will be like a card house that may collapse at some point as we build it up more and more. What we are trying, instead, is to build a solid structure at each level, but we hope to make more and more contacts.

Britton Chance: With NMR we are now able to measure velocity constants, genetic deletions and genetic alterations *in vivo*, so this a field where physics and biology converge.

Stephen Sligar: I want to make a comment about microbial or bacterial myoglobin that Britton Chance alluded to. Figure S1 shows a totally synthetic gene for sperm whale myoglobin that can be over-expressed in *E. coli*. We constructed the entire gene in our laboratory using codons that would be expected to be found if myoglobin were normally synthesized in *E. coli*. This is not the place to discuss the details, but we managed to put it together with a lot of hard work in overlapping all the oligonucleotides. This synthetic gene can be used for site-directed mutagenesis experiments to alter the amino acids at the active site. Our lab is involved in a more general program of the synthesis of genes for proteins whose structure and biophysical properties are well known. We have synthesized the complete genes for poplar plastocyanin, a plant protein, for horse heart cytochrome c and for rate liver cytochrome b_5. The genes for these eucariotic proteins are genetically engineered for expression in high yield in bacteria. Figure S2 illustrates the spectacular results for the synthesis of rate liver cytochrome b_5 in *E. coli*. The dark color of the beaker on the right indicates that the apoprotein is expressed in high yield and that the heme group is correctly incorporated in the apoprotein. Figure S3 shows the expression of the synthetic myoglobin gene in *E. coli* in a more quantitative way. What is shown is the difference spectrum of CO-treated minus untreated whole cells, and the peak at 423 nm indicates an intact myoglobin being expressed in high yield in the bacterial environment. An overnight fermentation of 28 liters of ferment yields gram quantities of holomyoglobin after purification. This technology, using a microbial expression system, opens up new avenues in altering protein structure.

Hans Frauenfelder: You all realize, of course, that Steve just had complicated our problem enormously by making modified proteins available. Myoglobin alone is very complex, although we were told it was simple when we started. In 18 years of work we have steadily progressed backwards and the more we learn the less we understand. The only way to move forward is to concentrate on the crucial questions, but the problem is to find them. Mies van der Rohe once said "God is in the details", but there is also the saying "The devil is in the details", and it really matters what details you have in mind.

Anders Ehrenberg: I would like to comment on the possibilities of measuring larger structures by NMR than is practical today. The improved resolution expected at higher fields and frequencies was already mentioned. Another way to get sharper lines in slowly tumbling molecules is to apply the techniques developed for solid state NMR. A third way is to reduce the number of protons in your protein and to focus on the few remaining ones. If you grow your bacteria in a deuterated medium to which you add an amino acid of interest in its protonated form, chances are that some of this amino acid shows up in your protein. Obviously this method works best with mutants that cannot synthesize the amino acid of interest. In this way you can assign proton NMR signals in rather large structures. Harden McConnell has applied this technique successfully to the large proteins of an antibody binding site.

Rudolf Mössbauer: I would like to amplify on the point Hans just raised. We have the possibility of studying structures with ever improving means; synchrotron radiation, in particular, will bring a big break-through in the next few years. We also have the possibility of studying protein dynamics by several techniques, and NMR, X-ray and γ-ray scattering will improve dramatically in the near future. If you want to measure all the vibrations of all the atoms in a protein, the measurement is very tedious because the problem is so complex, but it can be solved. However, the crucial questions then are the relations between structure and dynamics on one hand and protein function on the other hand. It is very difficult even to define that question, and before we can do that it is impossible to say what aspects of structure and dynamics are important for function.

Hans Frauenfelder: I have a partial answer to that question. I have introduced the concept of FIM's, namely functionally important motions, and Tony Crofts immediately pointed out that there must by BUM's as well, namely biologically unimportant motions. Anyhow, the question of importance can be answered only if the conditions are defined. Suppose you have a car and ask what is functionally important on a car. Is the rear view mirror, for instance, functionally important? Obviously, you can drive a car without the mirror, but if you are stopped by the police for that reason, then you have a functionally important problem. One question that has always worried me is why essentially all mammalian myoglobins have 153 amino acids. You can certainly have reversible oxygen binding even if you take away a few amino acids, so they presumably are there for some other purpose. It is hard to imagine that evolution would not have led to a spread in the number of amino acids otherwise. Not knowing what else to look for we focus on oxygen binding and on the motions that are important in this context. With the techniques that Marianne and Steve described we now have an excellent handle to check the relationship between structure and function because we can change individual amino acids and check what is happening. Of course it takes much more work to do this systematically, and it is very important to ask directed questions.

Peter Wolynes: I want to respond to the question of Professor Mössbauer because it is a problem that comes up at every meeting but is never dealt with properly. On one hand Hans' physical studies suggest that proteins behave like random systems in many respects, and analogies to spin glasses have been pointed out. This may indicate that we are looking at things that are biologically important in some sense but that we are looking with the wrong probe. If a system has a very complex phase space the reason is normally the presence of external constraints; maybe this even relates to Professor Winterhalter's comments that we should study proteins in their real environment with the molecules they are meant to interact with. At some point we have to begin to do that. I wonder if NMR experiments *in vivo* might help, say, Paul Lauterbur's technique or the measurements Britton Chance talked about. We would have to do very incisive, highly molecular experiments *in vivo*. Maybe Brit has some thoughts on that.

Britton Chance: There are clearly severe limitations. We can measure velocities of specific systems *in vivo* by use of saturation inversion transfer, and there are powerful techniques for the detection of an altered enzyme, for instance. High-frequency NMR with high-field superconductors will bring us closer to the sensitivity needed.

Ernesto DiIorio: I have some comments relating to the questions of Professor Mössbauer. We made a systematic study of the differences between myoglobins from different species, hemoglobin chains and modified myoglobins. Systematic differences clearly exist. The peak activation energies for geminate recombination of CO turn out to be about 10 kJ/mole in all the myoglobins we studied while the four human hemoglobin chains, α, β, γ and δ all have peak activation energies of about 4 kJ/mole. In collaboration with Professor Maurizio Brunori we also looked at "minimyoglobin" prepared from horse heart myoglobin by proteolytic treatment. About one third of the myoglobin is removed, and the shortened chain, which corresponds almost exactly to the polypeptide encoded by the central exon of the myoglobin gene, has a molecular weight of $\sim$ 11 kD. According to our preliminary data the CO-rebinding kinetics from 40 K to 300 K in this "minimyoglobin" are indistinguishable from the kinetics of the whole protein. It thus appears that the myoglobin core coded by the central exon has all the features important for ligand binding, and we do not know the function of the added piece.

Anthony Crofts: I want to mention three possible points of interaction between biologists and physicists. Biologists tend to trust biology, and biology has worked rather hard to find the best mechanism for enzymes by natural selection. In the process it has produced many different enzymes that perform similar function and a great deal can be learned by comparing these enzymes. On the whole biology has sought to conserve essential functions. This is in response to a question that has frequently cropped up: How do we select things to look at? Does biology help us in our selection?

Another way in which biology can help is by being stressed. Take photosynthesis, for instance. There are more and more weeds that become resisitant to the conventional herbicides by modifying their catalytic sites in rather specific ways. It is possible to stress other enzymes in similar fashions to get modifications of the enzymes in response to the stress in the environment. These modifications are useful in telling you which portions of the enzyme or the catalytic site are functionally important.

Thirdly, Hans has stressed the need to lay a foundation in physics by using simple systems to study biological function. It is also worth considering more complex systems, in particular those with several chromophores. A lesson can be learned here from bacteriorhodopsin and from photosynthetic reaction centers. When bacteriorhodopsin was discovered it was thought to become a key in the study of proton pumping. Unfortunately, bacteriorhodopsin has only a single (visible) chromophore, although the range of chromophores has been extended by the use of other spectroscopic techniques. As a consequence, we have learned very little about proton pumping. On the other hand, in more complicated systems like the photosynthetic reaction center, which has several chromophores that can be studied by a number of probes, we have learned a great deal more about its function.

Hans Frauenfelder: Let me add just two remarks: When we got interested in biomolecules the first thing I learned from Gunny is that the best model system is the real biological system because evolution has selected it well. Secondly, we are also studying bacteriorhodopsin, but it is very complex.

Vitalii Goldanskii: First, I would like to ask whether there is general consensus now concerning two of the long standing problems in ligand binding to myoglobin, namely the spin state of the heme iron after ligand dissociation and at the different stages of rebinding, and on the polychromatic kinetics or the distributed parameter picture. I remember discussing with Hans the attempts to describe geminate recombination by just two rate coefficients, an attempt that finally failed.

Secondly, I note that the data obtained from Rayleigh scattering of Mössbauer radiation may shed some light on the role of water in protein dynamics and the structure-function relation. Figure G shows a plot of the f'-value along the z-axis as function of the degree of hydration and temperature along the x- and y- axis, respectively; from the near symmetry of the diagram it is clear that hydration and increasing temperature have similar effects. The data show that the protein gets less and less rigid as water is added in much the same way as increasing tempertaure softens the protein structure. This general picture applies to all proteins we studied, but different proteins also show characteristic differences.

Hans Frauenfelder: Vitalii, can you stay here for another week? The role of hydration is protien dynamics is crucial and Enrico, who has long worked on this topic, may want to talk about it. Let me just say that the first two problems you mentioned are still open as Brit and I disagree on some aspects which are quite technical.

Enrico Gratton: Protein hydration and its influence on proten dynamics is clearly very important. We have been studying this problem for several years and concluded that if there is no hydration there is essentially no dynamics for a protein at room temperature.

I would like to pursue the suggestion of Peter Wolynes that we should aim for a 4-dimensional molecular biology, that we should get the baud rate up and find ways to reduce the search space. A 4-dimensional description of proteins clearly requires an exponential number of experiments, and we have to restrict ourselves to just a few because we are limited in time and techniques. We therefore have to select the experiments we want ot do very carefully. If we are to embark on a new program we need some leading ideas. Is the concept of distributed parameters and states crucial to a better understanding of proteins? Should we study the effects of hydration instead or do we want to study the interaction of proteins with other molecules of the cell?

Peter Wolynes: I let the crystallographers talk about distributions because in their field this concept is both revolutionary and relevant because it allows to solve many important problems. One comment I want to make concerns a point Hans has been making for years. In a biological process like an allosterism it makes a big difference if the thermodynamics is energetic or entropic. If it is entropic, then the distribution concept is really crucial. I would like to see a theoretical model that relates these distributions to questions of allostery and related phenomena. It may turn out that it is not important how I go between states but that the presence of many states is important.

One of the real problems we have to address is the question whether there is any structural correlate of any dynamical process. We have many X-ray structures and we have motions on all time scales but we have no correlation of the two at all. I would like to see that in at least one case.

Greg Petsko: Each of us does what he can, and that's the way this kind of science works. We tend to be faily technology-driven and we use the tools we know. If they are the best tools for the problem, that's fine, and if they are not the best tools they still provide information that is usually relevant. I guess that everything in a protein is important in one way or another because eventually you must know how the thing behaves at every level. That's how understanding works. What is related to a particular function is a secondary aspect for finding of everything about a molecule. Hans compared the study of a protein with the study of a car. We all study a car in a different way. I try to figure out how a car works by looking in detail at a picture of all its parts when the car is not moving. Hans is freezing the car in a block of ice and then tries to start the engine, and Steve Sligar is going to change the round wheels to square ones. We all do different things and the important point is that we keep talking to each other because eventually we will figure out that this stupid thing is a car!

George Phillips: What crystallography ususally tells us is the average structure, but we know that there are all kinds of dynamic processes going on. I want to comment on some new development that will reveal the distribution of states as well as the average structure. The technique I am going to describe gives no information about the time constant of the motions, but it lets us guess the distribution of states about the average. In Figure P1 on the left is a simulated crystal of a fibrous protein called tropomyosin and on the top right is an actual diffraction pattern along a particular direction of the crystal. In addition to the usual Bragg spots that allow the crystallographer to get the average structure you can see some additional streaks. Most crystallographers ignore the incoherent scattering that gives rise to these diffuse streaks, but we were able to predict it, after having solved the average structure, by assuming a certain distribution of states. The tropomyosin molecule is designed to move in the muscle filament by as much as 10 Å, and we see 5 Å to 7 Å mode variations in the crystal. In response to an earlier question, here is an example where the dynamic aspect is an inherent part of the molecular design and we can recognize it in the crystal. On the bottom of the figure is a successful simulation of a number of aspects of this diffuse scattering.

We are trying to apply this approach to globular proteins as well. We are simulating myoglobin crystals with various structures that were slightly perturbed from the average X-ray structure by a Monte-Carlo algorithm. Figure P2 shows different patterns of diffuse scattering, all of which arise form the same average structure but assume different distributions of states. In one model each atom of the myoglobin moves independently (Figure P2a), in another model each molecule in the lattice moves as a rigid unit (Figure P2b). We know that none of these limiting cases is realistic, but the point is that the diffuse scattering allows us to discriminate between different sizes of the moving parts. Figure P2c shows a perhaps more realistic model in which each of the helices moves independently. These diffuse scattering patterns of myoglobin are all theoretical, and we still have to collect data and analyze them. I hope I made the point, however, that there is information in the diffuse background of X-ray patterns that tells us something about the distribution of states.

Heiner Roder: NMR has been brought up repeatedly as a promising method of the future, in particular for the determination of solution structures in moderately large molecules. The application of high-resolution NMR to problems of dynamics has not been stressed so far, and I would like to illustrate two possibilities.

One example is from our work on cytochrome *c*. We have been studying the exchange of hydrogen, mainly the amide hydrogens, against solvent deuterium, using 2D NMR methods. We have complete sets of exchange rates for about 50 individual protons in reduced and oxidized cytochrom *c* as well as in two complexes with external ligands. We are just beginning to analyze all this rich information. Ultimately, we hope to characterize the local mobility and stability of the protein on a detailed atomic level. According to Takano and Dickerson the crystal structure of oxidized and reduced cytochrome *c* are very similar, but we find that the mobilities are dramatically different. This difference may be a hint of the functional importance of such motions and, as Professor Mössbauer mentioned, the functional importance is really what we should go after.

Another experiment that provides dynamical information is something we just started two weeks ago, and we are quite excited about the preliminary results which indicate far-reaching possibilities. We use the paramagnetism of O_2 dissolved in the NMR sample as a relaxation probe. The idea is very simple but turns out to be quite powerful. Paramagnetic centers are long known to relax protons, and fluorescence quenching by O_2 in solution has been studied in detail. In several small proteins we can monitor hundreds of assigned proton resonances by 2D NMR and measure local O_2 concentration throughout the structure. We started with the bovine pancreatic trypsin inhibitor, and our first results indicate that some O_2 is found essentially everywhere inside the protein although to varying extent.

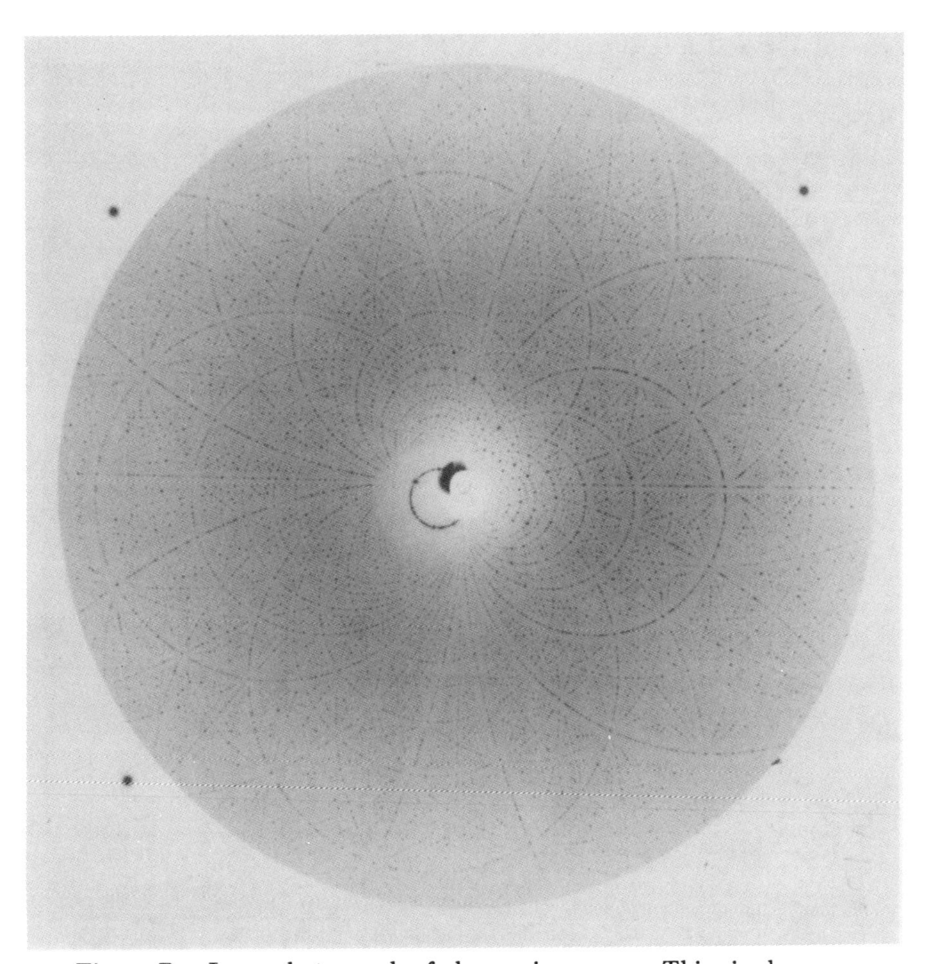

Figure P. Laue photograph of glucose isomerase. This single exposure of less than 0.1s contains more than 70% of a complete data set to 2 Å resolution. The photograph was taken at the Daresbury Synchrotron by Gregory K. Farber, Steve Almo and Janos Hajdu. The 3-dimensional structure of the enzyme has been solved by Greg Farber and Dagmar Ringe of M.I.T. using conventional diffractometry; the collection of each data set took more than a week.

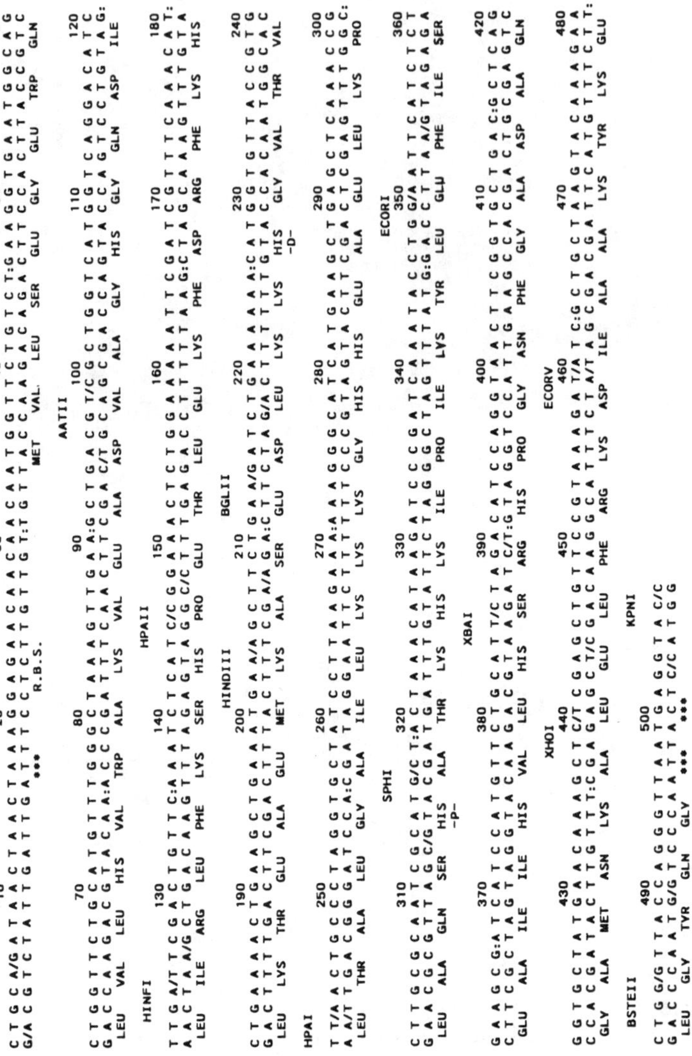

```
PSTI
                  10                  20                  30                  40                  50                  60
C T G C A/G A T A A C T A A C T A A A G G A G A A C A A C A A C A A T G G T T C T G T C T:G A A G G T G A A T G G C A G
G/A C G T C T A T T G A T T G A T T T C C T C T T G T T G T:T G T T A C C A A G A C A G A C T T C C A C T T A C C G T C
                     •••         R.B.S.                    MET   VAL   LEU   SER   GLU   GLY   GLU   TRP   GLN

                                                          AATII
                  70                  80                  90                 100                 110                 120
C T G G T T C T G C A T G T T T G G G C T A A A G T T G A A:G C T G A C G T/C G C T G G T C A T G G T C A G G A C A T C
G A C C A A G A C G T A C A A:A C C C G A T T T C A A C T T C G A C/T G C A G C G A C C A G T A C C A G T C C T G T A G:
LEU   VAL   LEU   HIS   VAL   TRP   ALA   LYS   VAL   GLU   ALA   ASP   VAL   ALA   GLY   HIS   GLY   GLN   ASP   ILE

HINFI                                       HPAII
                 130                 140                 150                 160                 170                 180
T T G A/T T C G A C T G T T C:A A A T C T C A T C/C G G A A A C T C T G G A A A A A T T C G A T C G T T T C A A A C A T:
A A C T A A/G C T G A C A A G T T T A G A G T A G G C/C T T T G A G A C C T T T T T A A G:C T A G C A A A G T T T G T A
LEU   ILE   ARG   LEU   PHE   LYS   SER   HIS   PRO   GLU   THR   LEU   GLU   LYS   PHE   ASP   ARG   PHE   LYS   HIS

                                    HINDIII               BGLII
                 190                 200                 210                 220                 230                 240
C T G A A A A C T G A A G C T G A A A T G A A/A G C T T C T G A A/G A T C T G A A A A A A:C A T G G T G T T A C C G T G
G A C T T T T G A C T T C G A C T T T A C T T T C G A/A G A:C T T C T A G/A C T T T T T T G T A C C A C A A T G G C A C
LEU   LYS   THR   GLU   ALA   GLU   MET   LYS   ALA   SER   GLU   ASP   LEU   LYS   LYS   HIS   GLY   VAL   THR   VAL
                                                                                    -D-
HPAI
                 250                 260                 270                 280                 290                 300
T T/A A C T G C C C T A G G T G C T A T C C T T A A G A A A:A A A G G G C A T C A T G A A G C T G A G C T C A A A C C G
A A/T T G A C G G G A T C C A:C G A T A G G A A T T C T T T T T T C C C G T A G T A C T T C G A C T C G A G T T T G G C:
LEU   THR   ALA   LEU   GLY   ALA   ILE   LEU   LYS   LYS   LYS   GLY   HIS   HIS   GLU   ALA   GLU   LEU   LYS   PRO

                              SPHI                                                      ECORI
                 310                 320                 330                 340                 350                 360
C T T G C G C A A T C G C A T G/C T:A C T A A A C A T A A G A T C C C G A T C A A A T A C C T G G/A A T T C A T C T C T
G A A C G C G T T A G C/G T A C G A T G A T T T G T A T T C T A G G G C T A G T T T A T G:G A C C T T A A/G T A G A G A
LEU   ALA   GLN   SER   HIS   ALA   THR   LYS   HIS   LYS   ILE   PRO   ILE   LYS   TYR   LEU   GLU   PHE   ILE   SER
                        -P-
                                                XBAI
                 370                 380                 390                 400                 410                 420
G A A G C G:A T C A T C C A T G T T C T G C A T T/C T A G A C A T C C A G G T A A C T T C G G T G C T G A C:G C T C A G
C T T C G C T A G T A G G T A C A A G A C G T A A G A T C/T:G T A G G T C C A T T G A A G C C A C G A C T G C G A G T C
GLU   ALA   ILE   ILE   HIS   VAL   LEU   HIS   SER   ARG   HIS   PRO   GLY   ASN   PHE   GLY   ALA   ASP   ALA   GLN

                                XHOI                                ECORV
                 430                 440                 450                 460                 470                 480
G G T G C T A T G A A C A A A G C T C/T C G A G C T G T T C C G T A A A G A T/A T C:G C T G C T A A G T A C A A A G A A
C C A C G A T A C T T G T T T:C G A G A G C T/C G A C A A G G C A T T T C T A/T A G C G A C G A T T C A T G T T T C T T:
GLY   ALA   MET   ASN   LYS   ALA   LEU   GLU   LEU   PHE   ARG   LYS   ASP   ILE   ALA   ALA   LYS   TYR   LYS   GLU

BSTEII                                          KPNI
                 490                 500
C T G G/G T T A C C A G G G T T A A T G A G G T A C/C
G A C C C A A T G/G T C C C A A T T A C T C C/C A T G G
LEU   GLY   TYR   GLN   GLY   •••   •••
```

Figure S1. Nucleotide sequence of the synthesized sperm whale myoglobin gene as designed from the known amino acid sequence. In this construction we used frequently used codons for highly expressed *E. coli* genes, and we incorporated a ribosome binding site (R.B.S.) known to efficiently express proteins in *E. coli*. This synthetic gene construction expresses native sperm whale myoglobin to 10% of the total *E. coli* soluble cell protein when placed in the vector pUC19. The restriction enzyme sites designed into the synthetic gene will allow for convenient mutagenesis of the myoglobin active site.

Figure S2. *E. coli* cultures without (left) and with (right) the synthetic gene for rat liver cytochrome b_5.

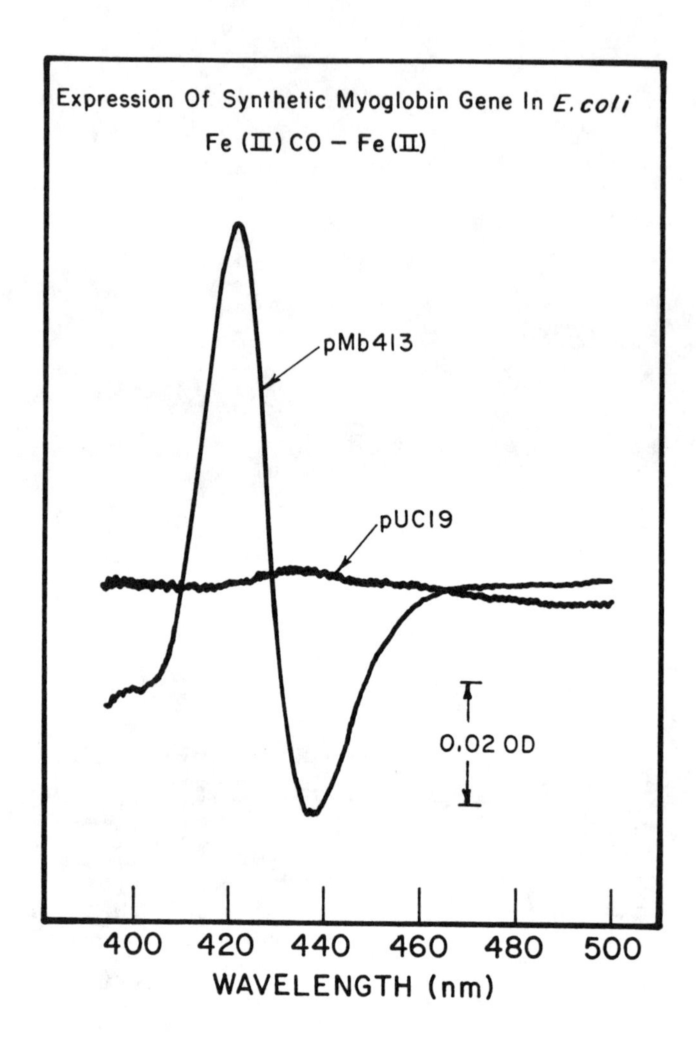

Figure S3. Whole cell UV-visible carbon monoxide difference spectrum of the *E. coli* TB-1 strain harboring the pUC19 plasmid alone and the synthetic sperm whale myoglobin gene (pMb413). The difference spectrum represents the Fe(II)CO - Fe(II) heme protein states and clearly shows that the expressed myoglobin is functional.

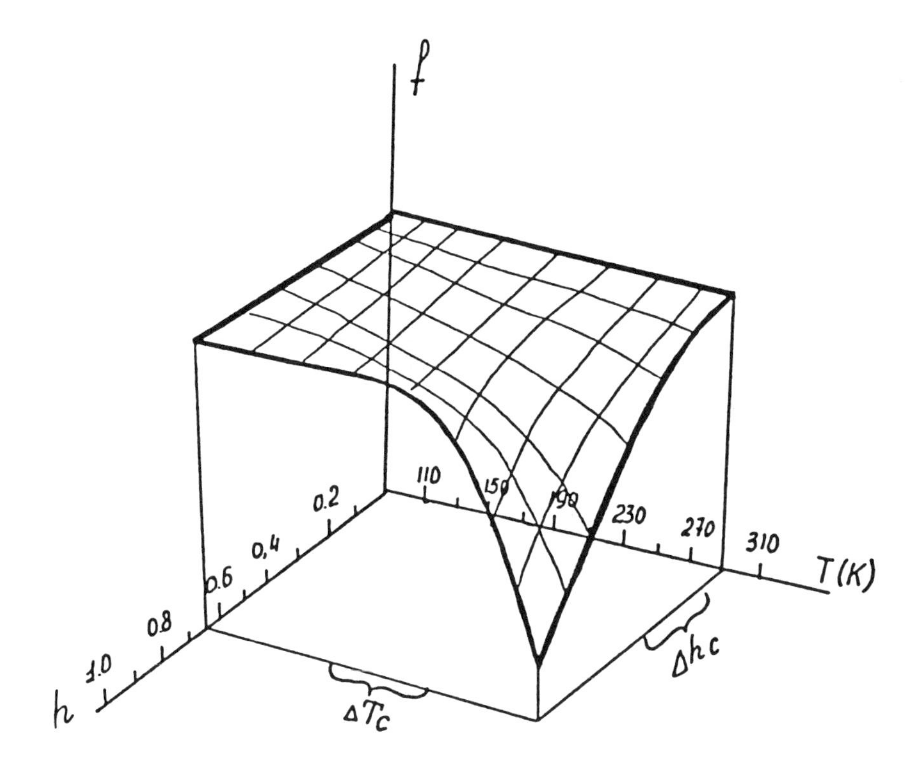

Figure G. The f-value from Rayleigh scattering as a function of the degree of hydration (h) and temperature (T).

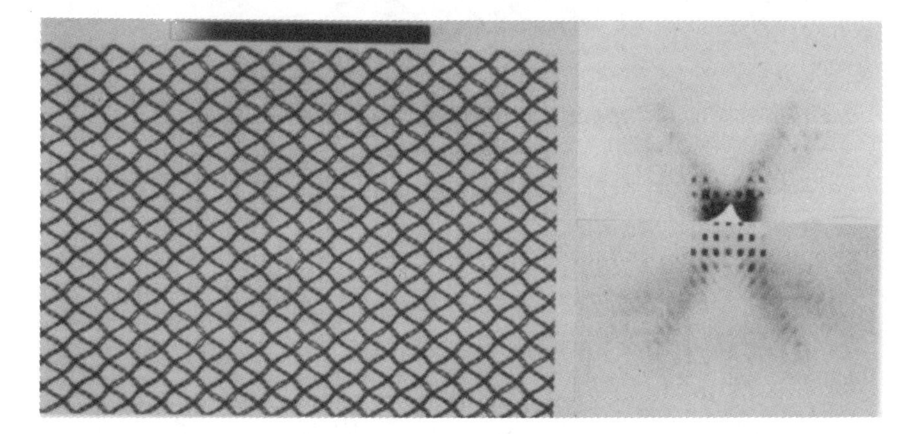

Figure P1. Comparison of observed and calculated diffuse X-ray scattering pattern for tropomyosin. The positions of the tropomyosin filaments are displaced from their average positions by including both random and wave-coupled perturbations (left). The array is Fourier-transformed, scaled, convoluted with a square to represent the X-ray beam size, corrected for the curvature of Ewald's sphere and displayed (lower right). The resulting theoretical pattern can be compared with the experimentally observed X-ray diffraction pattern (upper right). Aspects of the continuous scatter and the streaks near the Bragg peaks are accounted for by the simulation.

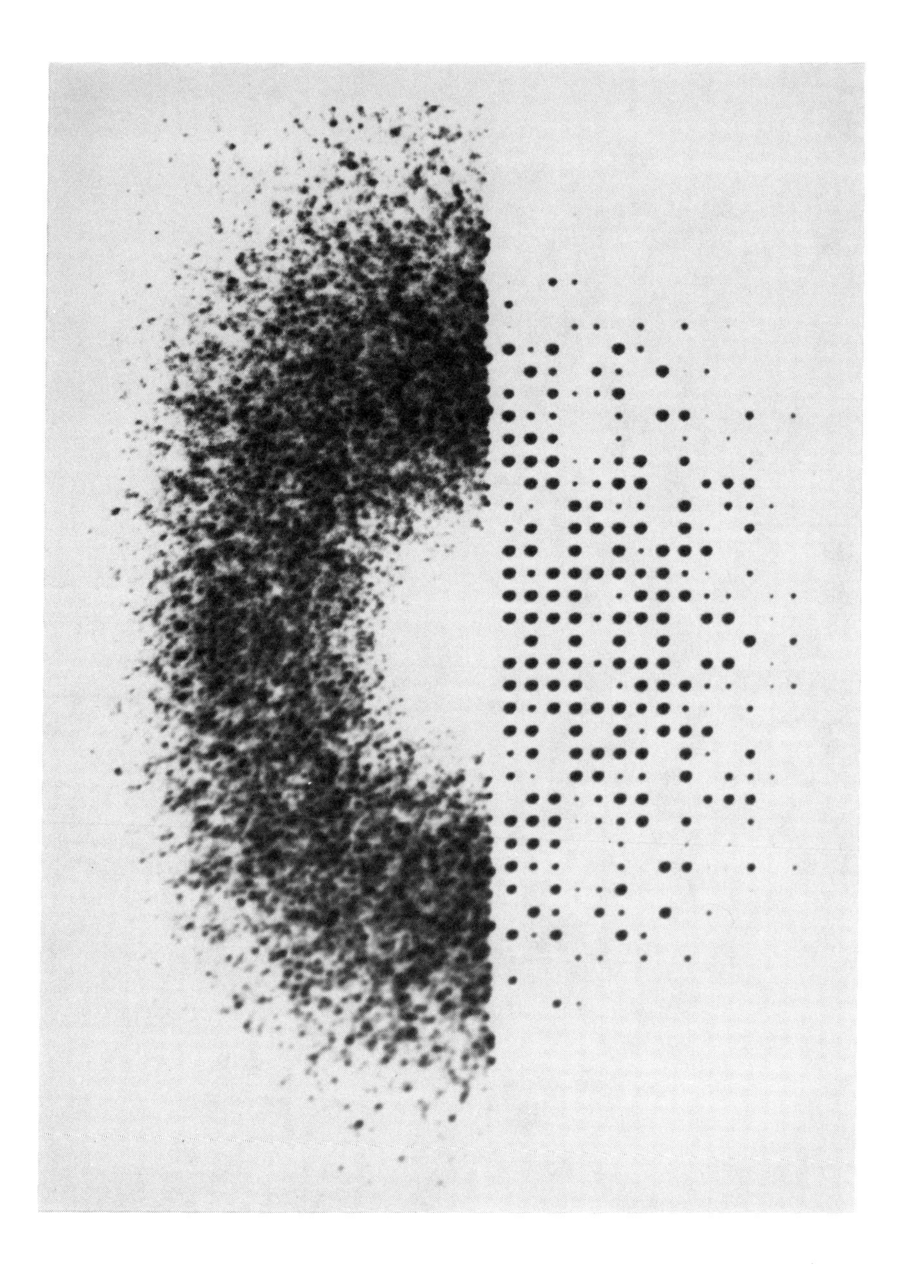

Figure P2a. Calculated continuous diffuse diffraction from a $P2_1$ myoglobin crystal with every atom moving independently. The resulting diffuse pattern is simply the intensity transform of the individual atoms.

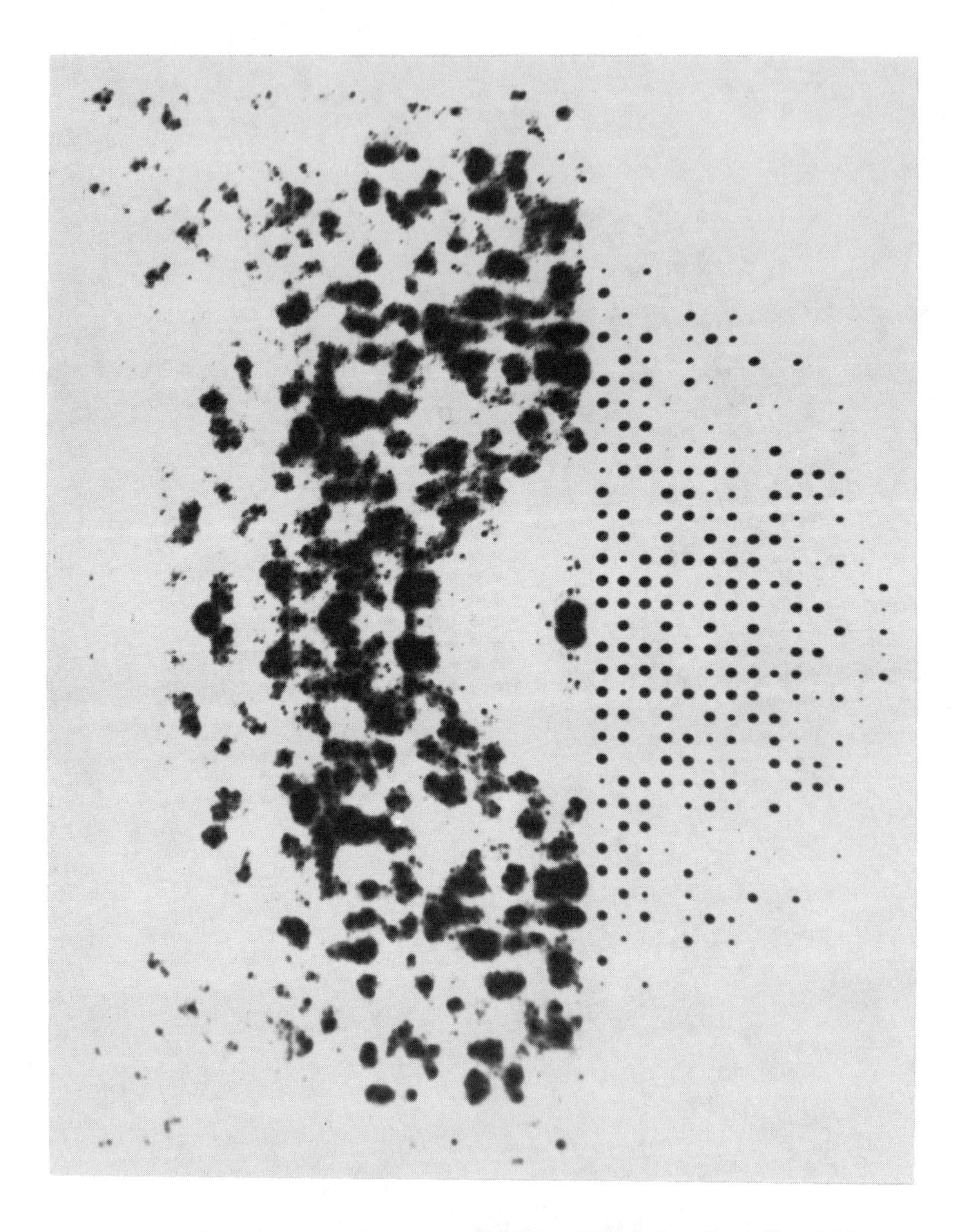

Figure P2b. Calculated continuous diffuse diffraction from myoglobin crystals with each complete myoglobin molecule moving as a whole, but independently of other myoglobin molecules.

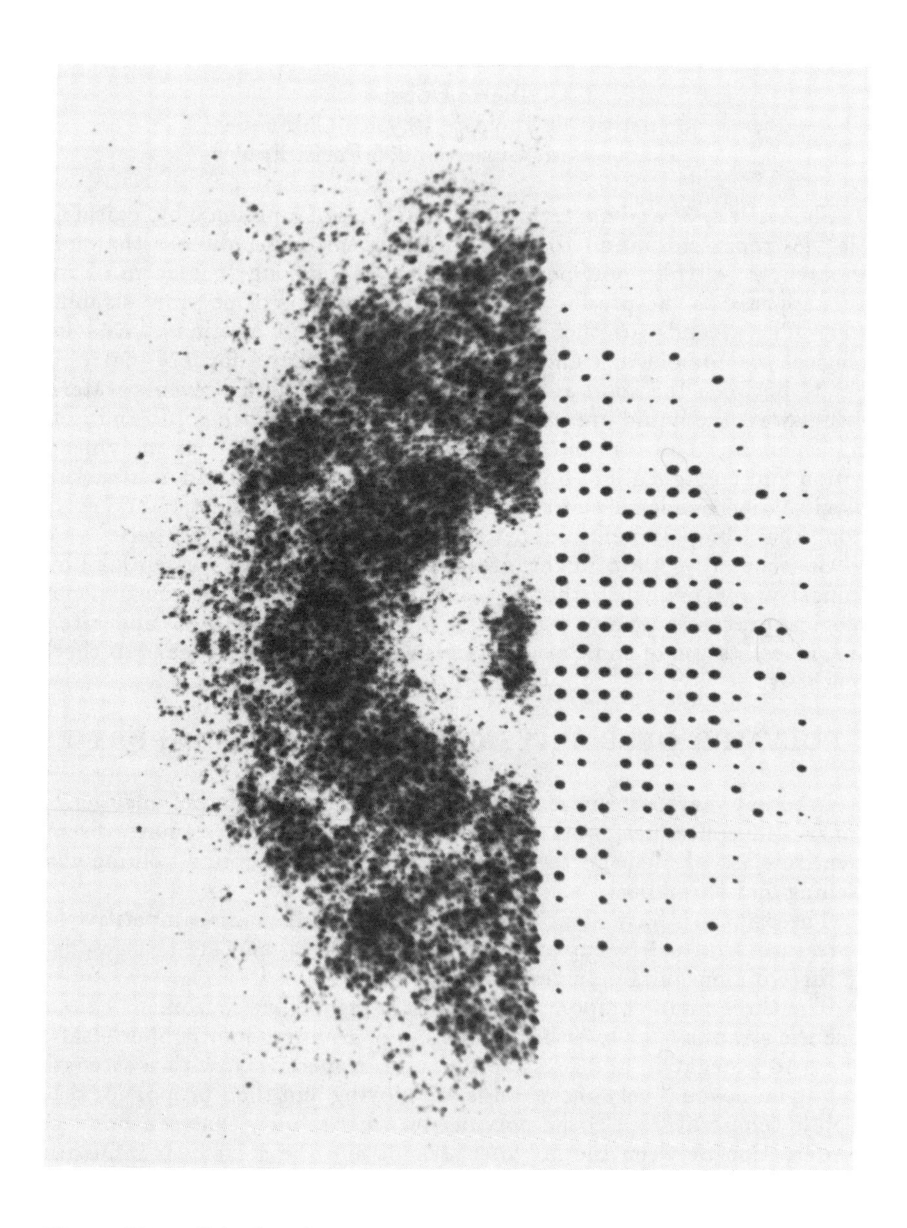

Figure P2c. Calculated pattern from a model of myoglobin dynamic behavior in which each helix of each molecule moves as an independent unit. In this model there is no correlation between the movements of helices, but each helix moves as a unit.

PRESUMPTIVE EVIDENCE FOR SIMILAR MECHANICS IN BIOLOGICAL MACROMOLECULES AND GELS

Pierre Douzou
Institut de Biologie Physico-Chimique
13, rue P.M. Curie - 75005 Paris, France

The tertiary 3° and quaternary 4° structures of functional biological macromolecules represent submicroscopic three-dimensional networks that may be compared to jellylike polymer networks. Both belong to a form of matter which appears as the product of viscous interactions of polymer strands and water. The strands are flexible and show a certain elasticity. And finally, biological macromolecules and polymer gels represent a form of matter intermediate between the solid and the liquid. The network prevents water from flowing away; the liquid prevents the network from collapsing into an insoluble, compact mass. Thus in spite of important differences in size, polymer composition and configuration, bonding and solvation, the 3° and 4° structures of biological macromolecules arising from folding, cross-connection and solvation of long chain biopolymers support a comparison with polymer gels.

Consequently, they might present some of the gel's essential physical chemical properties when they reveal a process. And since most, if not all, biological processes involve a sequence of elementary equilibria and rate reactions, investigation of some of them in isolation provides a means to check the hypothesis.

POLYMER GELS: PHYSICAL CHEMICAL PROPERTIES

In recent years, extensive studies have been devoted to polymer gels[1], and Tanaka and colleagues, working on polyacrylamide gels[2-5], have discovered an entire class of unsuspected phenomena centering around volume changes (swelling and shrinking).

These authors have identified and analyzed three main competitive forces, or pressures that act to expand or contract the gels, and are characteristic of any three-dimensional polymer network.

The three main component forces, or pressures, arise from: (1) the resistance the strands offer to either stretching or compression (rubber elasticity), leading to a negative and positive pressure, respectively, with a strength depending on how actively the strands are moving and then proportional to the absolute temperature; (2) the polymer-polymer affinity that can be traced to an interaction between the polymer strands and the solvent, is influenced by the solubility of the strands, and develops a negative force, the magnitude of which depends on the volume of the network; and (3) the hydrogen-ions concentration in polyanionic networks. Hydrogen atoms can be assimilated to molecules of gas in a confining vessel, and they exert a positive pressure which is proportional to the absolute temperature. Each of these component forces can be selectively influenced by external conditions (temperature, solvent composition, pH and ionic strength), and their changing balance gives rise to volume changes. An essential principle in interpreting these changes is that, whenever possible, a gel will adjust its volume so that the total osmotic pressure is zero. If the pressure is initially positive, the gel takes up water and expands. If the pressure is negative the gel expels water and contracts.

The expansion or contraction continues until an equilibrium is reached, where the various positive and negative forces exactly cancel one another. This behavior is schematized (Figure 1) in a simple graph of pressure versus volume at a given (ambient) temperature.

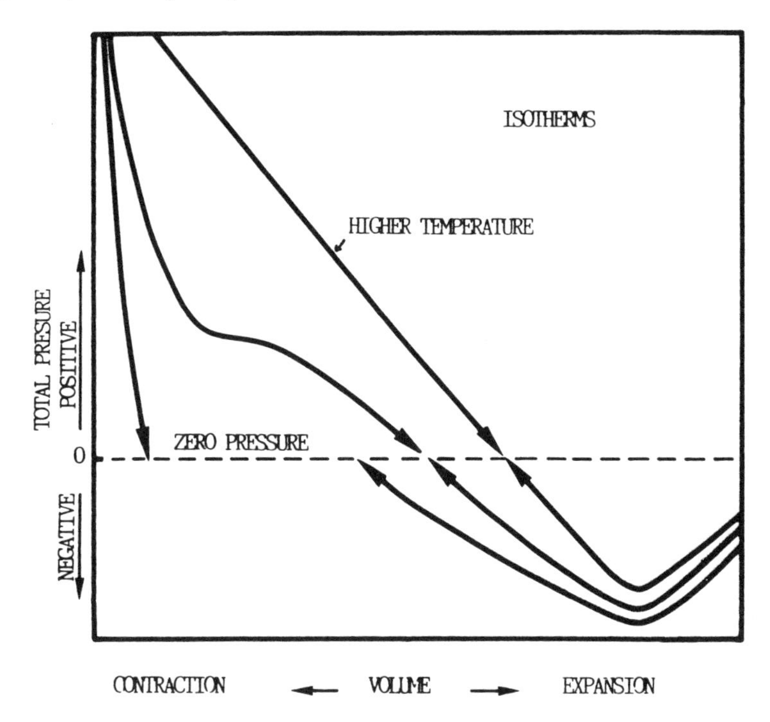

Figure 1. Isotherms of the osmotic pressure vs. volume. The gel adjusts its volume so that the total pressure is zero. If the pressure is initially positive, the gel expands. If the pressure is negative, the gel contracts.

The curve shows how a gel "migrates" to zero pressure from regions of either positive and negative pressures, with expansion or contraction, respectively. It can be seen that, as temperature is reduced or the concentration in a cosolvent is increased, adding isotherms to such diagrams does not result in tracing a set of parallel curves.

Using many combinations of volume plus solvent composition, temperature, or effective ionization of polymer strands, Tanaka[5] constructed phase diagrams revealing three regions as shown on Figure 2.

The first region is obtained at high temperature or low concentration in a cosolvent. In this region the response of the gel is continuous and moderate. A second region is attained when the temperature falls or the concentration in

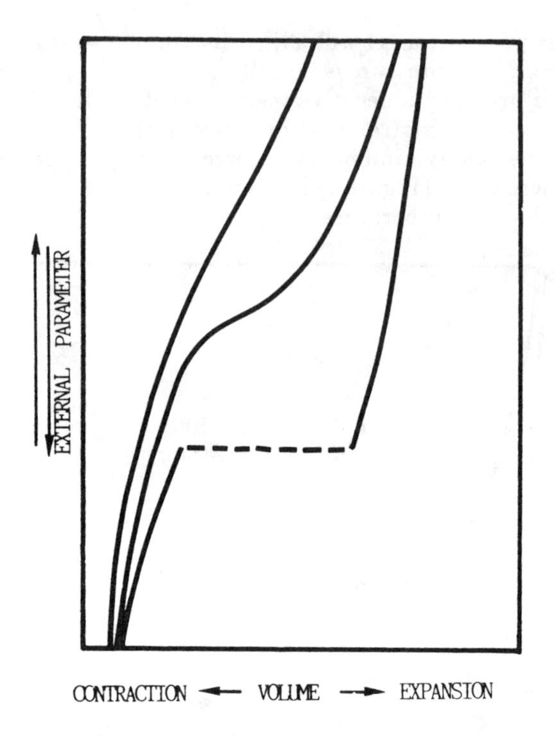

Figure 2. Simplified phase diagram showing the volume variations corresponding to three regions. Above the critical point, the response of a gel to external parameters is continuous. At the critical point, the response is of the "cooperative" type. Below this point, the gel undergoes a discrete phase transition.

cosolvent increases. The curve has an inflection point corresponding to "critical" conditions. The third region, below the critical point, is characterized by "broken" curves. The phenomenon of two segments being separated by a horizontal (dashed) line represents a discrete "phase transition." Thus, depending on the internal characteristics of the gel (particularly its effective ionization) as well as on initial external conditions, any variation of one of these conditions can determine three main types of responses. Above and near the critical point of the gel, the response will be monotonic and sigmoidal, and the gel wil be able to switch from one to another by changing the external conditions.

FUNCTIONAL SIMILARITIES IN BIOLOGICAL MOLECULES

The varied responses of polymer gels, that are due to significant changes in sensitivity depending on initial external conditions, are familiar to biochemists. However, explanation of such response from biological molecules for instance the fact that they may be similar under the influence of parameters bearing no physical chemical resemblance (say, cations, organic solvents, temperature and solvent composition) is not possible except in the context of the gel's properties.

For years, we analyzed kinetically (isolaing) a number of elementary equilibria and rate processes and found data uninterpretable in a different context. Introducing the polymer gel model and its underlying physical mechanisms, the task becomes much easier.

Let us consider E. Coli ribosomes as the representative of nucleoprotein systems. Most sequential equilibria and rate processes stepping the initiation and elongation stages of polypeptide synthesis are influenced and modulated either by physiological agents and effectors or cosolvents. These coslovents often "mimic" the effects of Mg^{2+} cations. This is the case for the association equilibrium of $30S$ and $50S$ subunits giving the $70S$ particles:

$$30S + 50S \overset{Mg^{2+}}{\rightleftharpoons} 70S$$

As shown in Figure 3, association equilibrium varies with Mg^{2+} concentration, and the value $[Mg^{2+}]_{1/2}$ for half association is readily recorded by light scattering. Such concentration decreases significantly in the presence of selected amounts of various cosolvents, and it is obvious that these solvents show additive effects with Mg^{2+} ions[6].

According to the polymer gel model, such effects can be explained by introducing an additional negative pressure through the solvent, then opposing the positive pressure due to the hydrogen-ion concentration and decreasing the amount of Mg^{2+} which is necessary to reduce this pressure.

The model also permits to understand the effect of lowering temperature on $[Mg^{2+}]_{1/2}$. The decrease observed may be explained by a reduction of the positive pressure of the hydrogen-ions.

Perturbations of other elementary equilibria and rate processes involved in protein synthesis, and also in DNA and RNA syntheses, may be understood in the context of the component conspiring to adjust new configurations and then to modulate activity.

These mechanisms seem to be also involved in functional (enzymatic) proteins. Intermediate complexes engaged in equilibria and rate processes may respond differently to changes in one parameter, or similarly to different parameters, depending on initial external conditions influencing their state. We checked such behavior on several enzymes systems, and especially on the camphor-bound bacterial cytochrome P_{450} complex[7] by using cosolvents and temperature variations at different ionic strength values[8].

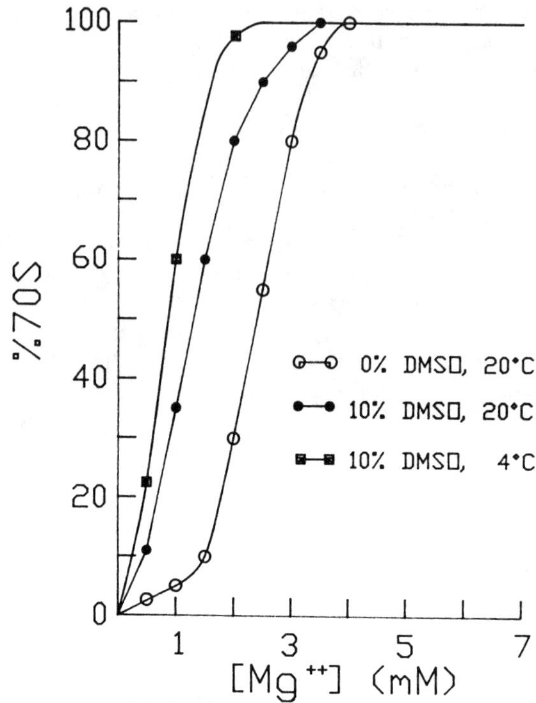

Figure 3. Association equilibrium curves of E.Coli ribosomes as a function of Mg^{2+} concentration: in water, in a water dimethylsulfoxide (DMSO) mixture at 20°C and 4°C. The additive effects of the cosolvent and of lowering temperatures are shown.

CONCLUSION

Functional similarities between polymer gels, nucleoprotein systems and enzymes provide presumptive evidence for similar underlying mechanisms. Consequently one may assume that biological macromolecules may be characterized as gels, an assumption recently postulated by Austin, Stein and Wang[9].

The polymer gel model suggests new investigations and explains some of the most challenging responses of reacting biomolecular system. This model, in its present and highly perfectible form, also provides one mechanistic explanation for the transmission of information in biological macromolecule, since any local perturbations of one of the component forces gives rise to a change in pressure, and conformation, in the molecule as a whole.

No doubt that further investigation and refinements of the model applied to biological macromolecules and systems would provide new explanations about their most intimate mechanism.

ACKNOWLEDGEMENTS

This work was supported by the INSERM, INRA, and the MRES.
It is dedicated to the memory of Laura Eisenstein, in remembrance of her sabbatical year in the Laboratory and of years of friendship and cooperation.

REFERENCES

1. P.G. deGennes, *Scaling Concepts in Polymer Physics*, (Cornell University Press, Ithaca, NY, London, 1979).
2. T.Tanaka, L.P. Hocker and G.B. Benedek, Chem. Phys. 59, 5151 (1973).
3. T. Tanaka, Phys. Rev. Lett. 40, 820 (1978).
4. T. Tanaka, D.J. Fillmore, I. Nishio, S.T. Sun, G. Swislov and A. Shah, Phys. Rev. Lett. 45, 1636 (1980).
5. T. Tanaka, Sci. Amer. 240, 110 (1980).
6. B.H.G. Hui, E. Begard, P. Beaudry, P. Maurel, M. Grunberg-Manago and P. Douzou, Biochem. 19, 3080 (1980).
7. M. Marden and B.H.G. Hui, Eur. J. Biochem. 129, 111 (1982).
8. P. Douzou, Compte-rendus Acad. Sci. Paris, in press.
9. R.H. Austin, D.L. Stein and J. Wang, Proc. Natl. Acad. Sci. USA 84, 1541 (1987).

COMMENTS ON Mb·CO REBINDING KINETICS

Paul M. Champion
Department of Physics, Northeastern University, Boston, MA 02115

1. THE PROBLEM

In 1975 Austin et al.[1,2] measured the rebinding of CO to myoglobin (Mb) over a wide range of time and temperature. I will confine my comments to the simplest of the several rebinding regimes that are observed; namely, the geminate recombination that takes place between $T \cong 60 - 160\text{K}$:

$$\text{Mb} \cdot \text{CO} \underset{\{k\}}{\overset{\gamma}{\rightleftarrows}} \text{Mb} + \text{CO}. \tag{1}$$

The curly brackets around the $\{k\}$, indicate that a single rate is not sufficient to describe the rebinding process. In fact, over the last ten years Hans Frauenfelder and his group have presented convincing evidence[3] that these rates are distributed and arise from conformational substates of the myoglobin which are frozen into the ensemble below the phase transition of the solvent (quenched disorder). The goal of this short presentation is to discuss some recent work which tries to quantitatively account for these results using a simple mathematical model. The model is set up in an intuitive way and allows the results from other independent measurements (e.g. X-ray, Mössbauer, electronic absorption, Raman and infrared spectroscopies) to be incorporated into a general framework that also describes the rebinding kinetics.

2. DISTRIBUTED RELAXATION

The general concept of distributed relaxation pervades most subfields of physics that involve glassy or amorphous systems[4]. We have recently been drawn into this area through resonant light scattering studies of heme proteins[5]. In particular, we believe that distributions in the nonradiative decay of the $\pi - \pi^*$ excitations of the heme chromophore in cytochrome c may lead to anomalous resonance enhancement of the Rayleigh scattering[5,6]. In general, one can write the relaxation funciton for an ensemble of simply relaxing systems as:

$$\phi(t) = \int P(\Gamma) e^{-\Gamma t} d\Gamma \tag{2}$$

where Γ represents the exponential decay constant (i.e. damping factor or rate constant) of a correlation and $P(\Gamma)d\Gamma$ is the probability of finding a molecule with a particular Γ. In the case of non-radiative decay of an electronic excitation of the heme, the time scale for decay can be extremely fast ($10^{-13} - 10^{-14}s$). Quite often it is *assumed* that $P(\Gamma) = \delta(\Gamma - \Gamma_0)$ and single exponential decay dominates the ensemble.

The same basic approach holds for the rebinding of CO to Mb in Eq. 1. If we identify $N(t)$ with the normalized concentration of Mb (unbound) following the photolysis (γ), we can write by analogy:

$$N(t) = \int P(k)e^{-kt}dk. \tag{3}$$

The equivalent formulation used by Hans and his group over the years is:

$$N(t) = \int g(H)e^{-k(H)t}dH \tag{4}$$

with

$$k(H) = k_0 e^{-H/\mathbf{k}_B T}. \tag{5}$$

At temperatures above the tunneling regime ($T \geq 60$K), the rate constant $k(H)$ varies in the usual Arrhenius way with a barrier height (H) and $g(H)dH$ describes the probability of finding a Mb molecule in the frozen matrix with a barrier height between H and $H + dH$. Hans and his group have shown that $g(H)$ is *not* a delta function, and by using inverse Laplace transform techniques they have used Eq. 4 to explicitly document the $g(H)$'s for a wide variety of heme protein systems.

3. COORDINATE SPACE AS A PARAMETRIC

The present model rests upon our ability to identify a single configurational coordinate that plays a dominant role in determining the rebinding barrier height. This special coordinate is one of many protein coordinates that fluctuate above the freezing point (T_f) and are found in a static distribution below T_f. We parameterize the problem with respect to this coordinate and its distribution in coordinate space. This approach has worked well in other contexts[4-7] and we proceed following the same logic. (Section 8 briefly discusses the relationship between this simple view and more formal theories of multiphonon group transfer.)

We first consider the work needed to bring the Mb into a "transition" state for binding the CO. By analogy to various other chemical systems (e.g. the inversion of ammonia), it seems probable that the key coordinate involves the iron displacement towards the porphyrin center (distance *ca.* 0.45 Å). We assume that the work involved in bringing the iron-porphyrin system into the planar transition state configuration can be written as:

$$H_p = \frac{1}{2}KQ^2, \tag{6}$$

where the "force constant" K is yet to be determined, but involves all linear restoring forces between iron-protein and iron-porphyrin. We shall refer to this term as the "proximal" work.

In addition to Eq. 6, there is another term which contributes to the total barrier height. We refer to this term as the "distal pocket" work. It may involve a variety of effects (e.g. the work needed to tilt the CO molecule off the preferred linear binding geometry or the work needed to "unstick" the CO form the pocket). All distal pocket-CO energy that needs to be expended in order to reach the transition state should be included in this term. We denote this term as H_D and write for the total barrier height:

$$H(Q) = \frac{1}{2}KQ^2 + H_D. \tag{7}$$

The distribution of the coordinate, Q, is taken to be Gaussian

$$P(Q) = \frac{1}{\sigma\sqrt{2\pi}} e^{-(Q-Q_0)^2/2\sigma^2} \tag{8}$$

and a schematic diagram of the situation is shown in Figure 1.

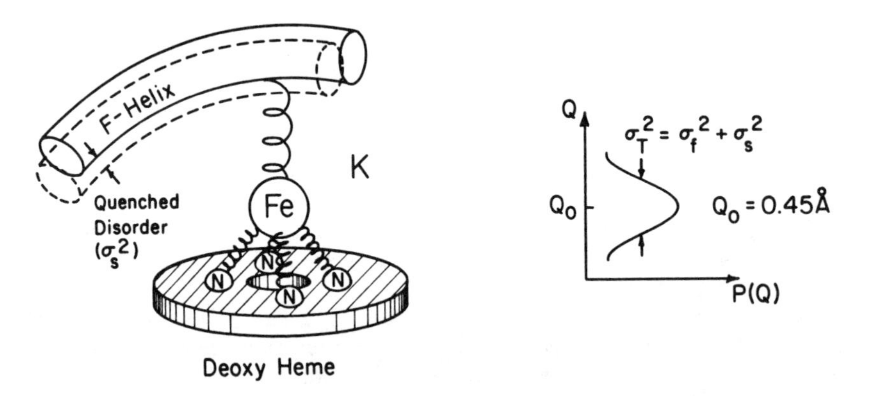

Figure 1. A schematic diagram showing the source of the "proximal pocket work," H_p. The constant, K, represents all linear restoring forces involved in displacing the iron-porphyrin system toward the planar transition state.

4. DISTRIBUTION OF BARRIER HEIGHTS

We note that the width of the coordinate distribution has been partitioned into two regimes in Figure 1[8]. A "fast" regime (σ_f) of time scales, $\tau < 10^{-8}s$, involves local vibrational modes and fast protein fluctuations which occur in the high temperature ($T > T_f$) limit. A "slow" regime (σ_s) of time scales, $\tau > 10^{-8}s$, involves the protein fluctuations which have been frozen out leading to the "static" distribution of conformations (low temperature quenched disorder). This latter regime is of direct interest to the rebinding kinetics, since the typical time scales for rebinding are on the order of $\langle k \rangle^{-1} \geq 10^{-9}s$.

The delineation of the two time regimes at $10^{-8}s$ is particularly convenient from an experimental point of view, since mean square displacements of the iron atom can be obtained from Mössbauer spectroscopy for $\tau < 10^{-8}s$ and thus yield σ_f^2 directly[8,9]. The static disorder σ_s of the iron atom at low temperature can be then found from:

$$\sigma_s^2 = \sigma_T^2 - \sigma_f^2 \tag{9}$$

where σ_T^2 is obtained from the x-ray data after removal of the lattice disorder[8] (see Figure 2). However, since Q is a relative coordinate, one must also extimate and remove contributions due to iron- porphyrin correlation. Use of EXAFS may be helpful in this regard.

It is worth noting that the description of inhomogeneous broadening in the electronic absorption band[7] utilizes the full σ_T^2, since the electronic processes are fast compared to the nuclear motion and the nuclei can be considered as "frozen" in place. For the sake of notational simplicity, we make the identification $\sigma \equiv \sigma_s$ and remember to consider only the "static" distributions of iron-porphyrin displacements which are frozen into the ensemble below T_f. In the frozen state the local motion in the fast regime (e.g. Fe-ligand vibrations, $10^{-12}s$) is small compared to σ and can be averaged, (i.e. $\langle Q^2 \rangle_f = \langle Q \rangle_f^2 + \sigma_f^2$). We thus make another notational simplification, $\langle Q \rangle_f \equiv Q_s \equiv Q$, and remember that Q represents the mean position of the iron atom averaged over the fast motion. The σ_f can be neglected ($\sigma_f^2 \ll \sigma_s^2$) or factored to k_0 in the classical limit ($\sigma_f^2 \cong \mathbf{k}_B T / m\omega_f^2$).

Thus, given the "static" distribution of Eq. 8 along with the parametric Eq. 7, we can find $g(H)$ using:

$$g(H) = P(Q) \left| \frac{dQ}{dH} \right| \tag{10}$$

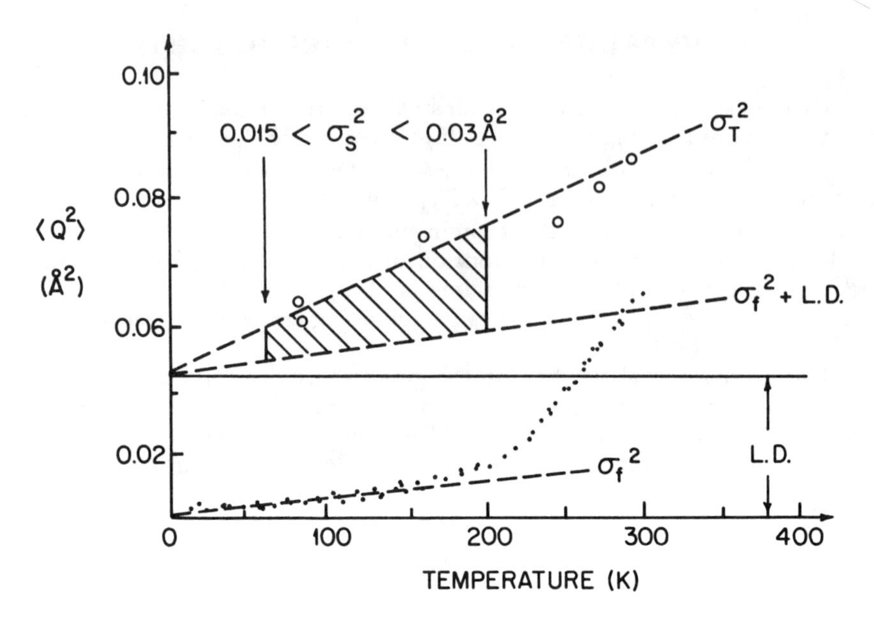

Figure 2. A schematic of the Mössbauer (solid data points) and X-ray results (open circles) for deoxy Mb as discussed in Ref. 9. The slow (static) contribution to the mean square displacement σ_s^2 can be estimated as indicated (see Eq. 9). The lattice disorder (L.D.) contribution to the X-ray results is also shown (L.D. ~ 0.045 Å^2).

This leads to:

$$g(H) = \frac{\left[(H - H_D)\pi K\right]^{-1/2}}{2\sigma}\left[\exp - \frac{\left[\sqrt{H - H_D} - Q_0\sqrt{K/2}\,\right]^2}{K\sigma^2} + \exp - \frac{\left[\sqrt{H - H_D} + Q_0\sqrt{K/2}\,\right]^2}{K\sigma^2}\right], \tag{11}$$

which is essentially the same distribution that was used in fitting the Soret band lineshape of deoxy Mb[7].

5. CALCULATION OF $N(t)$

The calculation of the observable quantity, $N(t)$, is straightforward. We write

$$N(t) = \int_{H_D}^{\infty} g(H) e^{-k_0 t e^{-H/k_B T}} dH \tag{12}$$

where we note that $g(H) = 0$ if $H < H_D$ and we have combined Eqs. 4 and 5. We also see that:

$$e^{-k_0 t e^{-H/\mathbf{k}_B T}} \equiv e^{-e^{(H'-H)/\mathbf{k}_B T}} = \Theta(H - H') \tag{13}$$

where:

$$H' = \mathbf{k}_B T ln(k_0 t). \tag{14}$$

The function $\Theta(H - H')$ closely approximates a step function at H', so that to a good approximation:

$$N(t) \cong \int_{H'}^{\infty} g(H) dH. \tag{15}$$

If $H' < H_D$ we have $N(t) = 1$ via normalization. If $H' > H_D$ we find a closed form expression for $N(t)$:

$$N(t) \cong \frac{1}{2}\left[erfc \frac{\sqrt{H' - H_D} - Q_0\sqrt{K/2}}{\sigma\sqrt{K}} + erfc \frac{\sqrt{H' - H_D} + Q_0\sqrt{K/2}}{\sigma\sqrt{K}} \right] \tag{16}$$

with H' given by eq. 14. Eq. 16 provides a convenient approximation to the observed $N(t)$ and allows for efficient fitting of data via least squares analysis. The approximation can be checked by direct numerical integration of Eq. 12.

6. FITS TO THE Mb • CO DATA

Figure 3 shows the rebinding data for Mb·CO over many orders of magnitude in time and at a number of different temperatures. The dashed curves represent the error function approximation of Eq. 16 and the solid curves are the result of exact integration of Eq. 12. The parameters used in the fits are compared with other experimental information in Table I.

Generally, we find that the model parameters are well determined by the low temperature rebinding data, once Q_0 is fixed. The set of parameters given in the table assumes that the iron atom moves to its full out-of-plane displacement (0.45Å) after photolysis at low temperature. In the event that the heme does not relax to the full displacement seen at room temperature, we have also carried out the fitting procedure using $Q_0 = 0.35$ Å and 0.25 Å. The overall fits to the data are the same quality as shown in Figure 3. However, the quantity σ_s must be reduced along with Q_0 (to 0.11 Å and 0.08 Å, respectively), while the value of K is increased (to 13.8 N/m and 26.9 N/m, respectively). Each of these parameter sets can be considered "reasonable" and the final determination awaits the experimental specification of Q_0 at low temperature. The reduced values for σ_s associated with the smaller Q_0's may actually be more appropriate, since the correlated iron-porphyrin motion must be removed

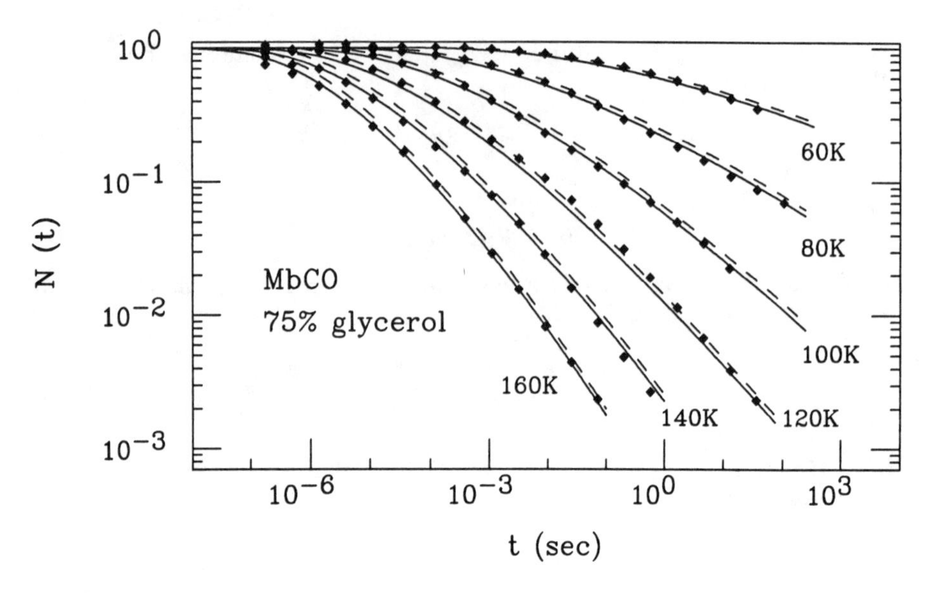

Figure 3. The theoretical fits to the Mb·CO rebinding data are shown. The dashed curves are the closed form approximation and the solid curves are the exact integral (Eq. 12). The parameters are listed in Table I.

from the X-ray and Mössbauer determined mean square displacements. We also note that the value for K, determined from Mössbauer experiments (Ref. 9), does not take into consideration the coupling constants that relate the normal and cartesian coordinates. (P. Debrunner, private communication). When such effects are taken into account, the Mössbauer data can only specify a lower limit $K > 2\text{N/m}$.

Table I: Fitting parameters for Mb•CO Rebinding

Parameter	Rebinding	Independent value	Techniques
Q_0	0.45Å	0.45Å	X-ray diffraction
σ_s	0.146Å	0.12 Å$< \sigma_s <$ 0.17Å	X-ray combined with Mössbauer (see Eq. 9, Ref. 9, Fig. 2)
K	8.38 N/m (8.38 x 10^{-2} mdyne/Å)	5.1 N/m	Soret Absorption (see Ref. 7)
		21 N/m	Mössbauer (Ref. 9) $\sigma_f^2 \cong \mathbf{k}_B T/K_f$
		88 N/m	Raman $\tilde{\nu}_{\mathrm{Fe-N_{His}}}$ =220cm^{-1}
k_0	$2.86 \times 10^9 s^{-1}$	—	—
H_D	7.01 kJ/mole	7.1 kJ/mole	High Temperature rebinding limit (Ref. 10) $\langle H \rangle = 12.2$ kJ/mole $H_D = \langle H \rangle - 1/2KQ_0^2$ (see Ref. 11)

Table II: Fe-C-O Frequencies of Mb·CO

State	$\tilde{\nu}_{\mathrm{Fe-CO}}$	$\tilde{\nu}_{\mathrm{C-O}}$	pH
A_0	490 cm^{-1}	1966 cm^{-1}	≤ 5
A_1	510 cm^{-1}	1945 cm^{-1}	≥ 7

7. SPECULATION ON THE ORIGIN OF H_D

Figures 4 and 5 show various examples of the barrier height distributions for different heme proteins. Figure 4 is the theoretical distribution used in the fits of Figure 3. It should be noted that the effect of varying H_D is to shift the $g(H)$ curve along the abscissa. Figure 5 is the result of inverse Laplace transforming the $N(t)$ data for a variety of compounds[12].

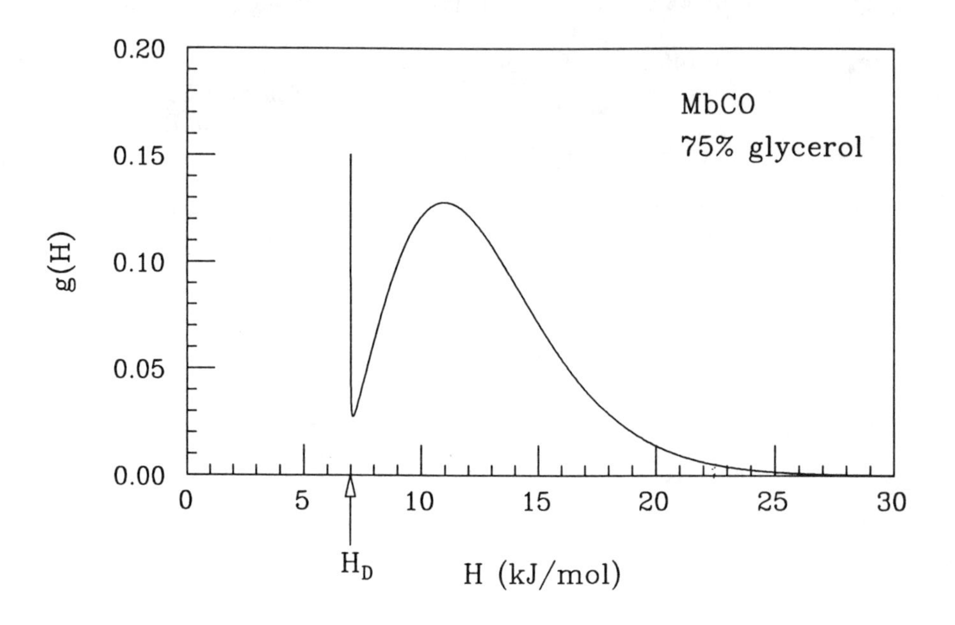

Figure 4. The probability distribution, $g(H)$, used to fit the data in Figure 3. The arrow denotes the value of H_D. The area under the spike vanishes when $g(H)dH$ is calculated.

The pH dependence of the Mb·CO data (Figure 5) is of particular interest in view of recent Raman and infrared experiments which probe the Fe–CO and C–O stretching frequencies as a function of pH. Table II lists the frequencies and the dominant species as a function of pH.

It should be noted that the A_0 state is observed below the freezing point at low pH and the A_1 state is dominant in the liquid phase for pH $\simeq 5$. Thermodynamic analysis of these states at high temperature using the IR technique[13] places the A_0 state at lower enthalpy but this is compensated by a much higher entropy associated with the A_1 state. This probably reflects a modulation of the protein pocket by a pH controlled conformational equilibrium involving a hydrogen bond (e.g. between arginine 45 and the proprionic acid side chain of

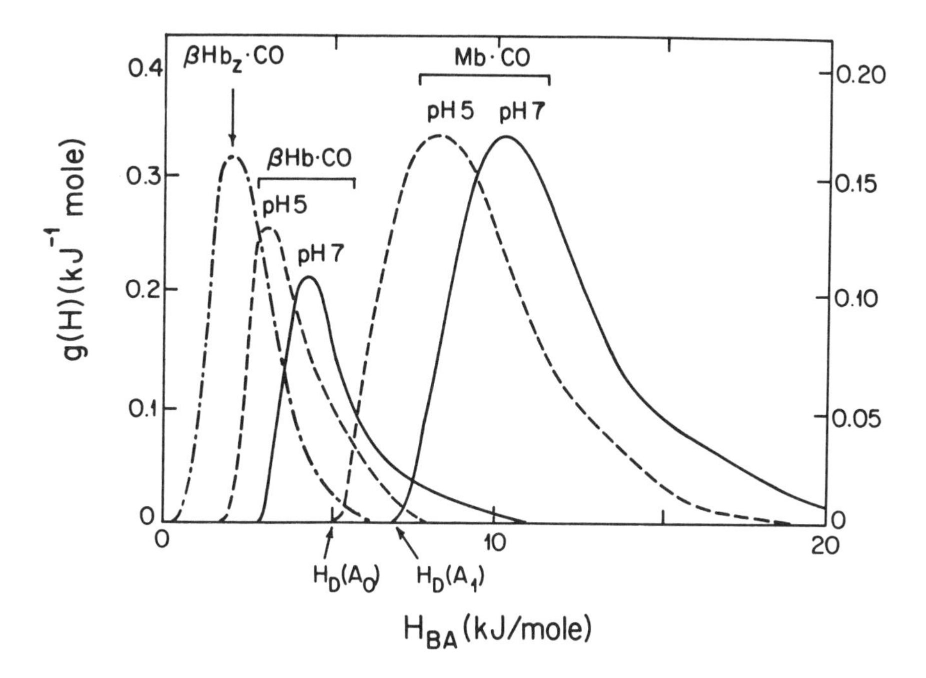

Figure 5. The probability distributions obtained by inverse Laplace transforming the rebinding data. The pH dependence of the rebinding kinetics may reflect distal pocket contributions to the transition state barrier. The states A_0 and A_1 are identified by spectroscopic signature and can be correlated with the Fe–C–O bond angles determined from X-ray diffraction. The arrows denote the positions of $H_D(A_0)$ and $H_D(A_1)$. Note that the vertical scale for the Mb·CO is given in the right side of the figure. Hb_z stands for Hb Zürich.

the heme). As temperature is reduced, the $T\Delta S$ term becomes less important and the A_0 state will be observed spectroscopically if the pH is sufficiently low to accomodate the "open" form of Mb and allow CO "access" to the A_0 geometry.

If we compare the data of Table II with correlations between $\tilde{\nu}_{Fe-CO}$ and $\tilde{\nu}_{CO}$ (Figure 6) that have been previously established[14], it is clear that A_1 and A_0 have different Fe–C–O bond angles. The A_1 state is thought[14] to be more bent (with subsequent Fe($d\pi$) $\rightarrow$ CO(π^*) donation). This leads to a higher Fe–CO and lower C–O frequency for A_1. The A_0 state is apparently the less bent of the two states. We assign the respective angles by referencing to the recent X-ray structures[15]. This leads to off-axis bend angles of *ca.* 40° for the A_0 state and 60° for the A_1 state. If we now refer back to Figure 5 and note that $H_D(A_0) \sim 5$ kJ/mole and $H_D(A_1) \sim 7$ kJ/mole, it suggests that part of the distal pocket energy barrier may involve the work needed to bend the CO molecule off of its preferred linear binding geometry. If more bending is required, the transition state barrier becomes higher. These ideas can be used to predict the off-axis bend angles of the other species in Figure 5:

$$\begin{aligned}\mathrm{Mb\cdot CO,\ pH\ 7\ (60^\circ)} &> \mathrm{Mb\cdot CO,\ pH\ 5\ (40^\circ)}\\ &> \beta\mathrm{Hb\cdot CO,\ pH\ 7} > \beta\mathrm{Hb\cdot CO,\ pH\ 5} > \beta\mathrm{Hb_s\cdot CO}.\end{aligned}$$

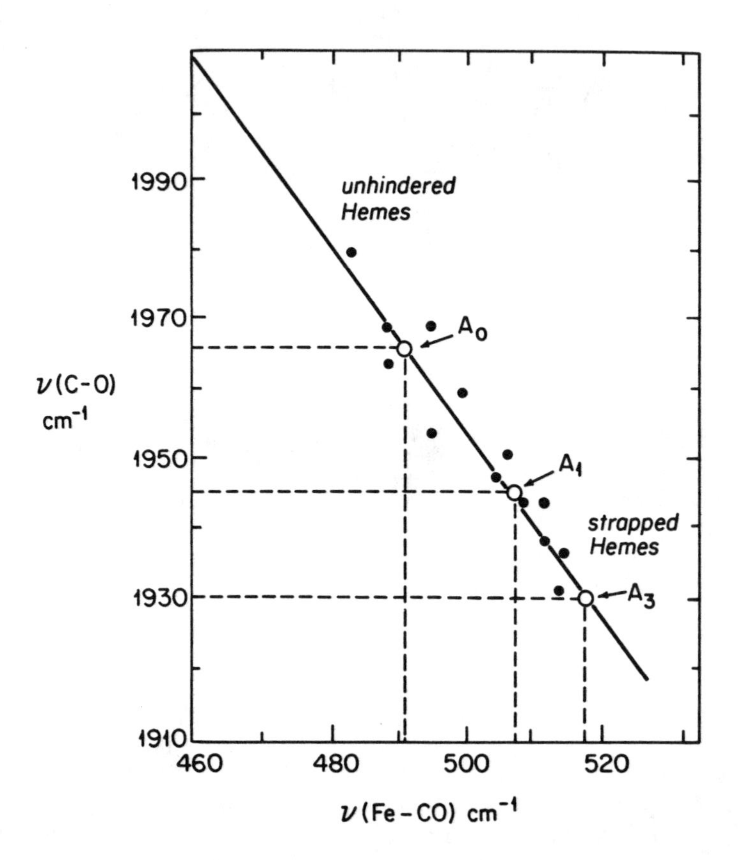

Figure 6. A schematic of the correlation diagram from Ref. 14. The dotted lines denote the positions of the A_0 and A_1 states. The position of a third, minority state, is also shown.

A more quantitative analysis of the energy of the heme system as a function of the Fe–C–O bond angle might be achieved by careful electronic structure calculations. The problem is quite complex due to the interaction of the porphyrin, iron and CO electrons. Nevertheless if such calculations were carried out, the values of H_D as obtained from the low-temperature rebinding studies, might be used to obtain quantitative structural information.

We certainly acknowledge that structural factors other than the Fe–C–O bond angle may play a role in the determination of H_D. For example, the torsional angle (ϕ) of the C–O axis around the heme normal may influence the transition state energy. Similarly, we must consider the weak forces (dipoles, H-bonds, etc.) between the CO and the pocket; such forces will lead to "unsticking" energy or polarization effects that could influence the CO(π^*) electron densities. These interesting topics should provide the basis for further experiments.

8. FORMAL THEORY

I would now like to briefly sketch a more formal extension of the present theory and comment on a number of other theoretical approaches that have been applied to this problem[3c,16–18]. The details of the formal theory will be fully presented elsewhere.

The basic approach involves extensions of the work of Buhks and Jortner[16] and Agmnon and Hopfield[17] to include quadratic coupling (change in force constant upon CO binding) as well as *distributions* of linear coupling (origin shifts). The initial state is taken to be Mb · CO and the vertical transitions (low-spin to high-spin energy difference at fixed in-plane geometry) are held constant throughout the ensemble. This corresponds to the idea of a heme iron atom that is "locked" in place in the Mb·CO state. The "coupled" state corresponds to deoxy, Mb, and the linear electron-nuclear coupling (S) is distributed throughout the ensemble. This distribution of deoxy iron equilibrium positions results from different protein conformations that are fixed into the ensemble as the solvent freezes (i.e. the *protein* still fluctuates in the Mb·CO state, but the CO bound heme is insensitive to the different protein conformation until the CO is flashed off). Thus, the vertical transitions from high-spin to low-spin at the various deoxy equilibrium positions throughout the ensemble are *not* fixed. The different values of S correspond to displacements of the iron atom and F-helix to different equilibrium positions after the CO is flashed off. The various iron cyrstal field parameters that result, lead to different high-spin to low-spin vertical transition energies for the deoxy species. It is interesting to note that the introduction of ensemble distributions of S breaks the symmetry usually associated with the linear harmonic models. One must take extreme care in the definition of the initial (ground) and final (excited) states. In this respect, the rebinding rate of CO to Mb is obtained by evaluating the zero frequency component of a distributed "fluorescence" spectrum rather than an "absorption" spectrum. The quadratic coupling is introduced to account for the "looser" environment of the iron atom in the deoxy system when compared to Mb·CO. Use of recently developed absorption and resonance Raman time correlator models could be employed in non-adiabatic theoretical calculations[19]. The advantage of the full quantum mechanical treatment is that the very low temperature ($T < 60$K) data can be explained with the same theory that is used for the region 60–180K. However. a crossover from non-adiabatic calculations in the tunneling regime[16] to adiabatic at higher temperatures seems likely[20].

The previous work[16] used a simple linear coupling model and has looked separately at only the low temperature (< 60K) data without any attempt at introducing distributions. Work along similar lines by Bialek and Goldstein[18] also holds to linear coupling but introduces damping of the vibrational states via anharmonic coupling. Such (single mode) calculations lead to sharp structure in the transmission function (Franck-Condon spectrum), so that small shifts in mode frequency "detune" the zero frequency component and lead to large changes in rebinding rate. The distributions are inserted into quadratic terms (i.e. force constants) rather than the linear terms and the different frequencies that result throughout the ensemble map into a wide distribution of rates. There appears to be two main problems with this model: (1) The damping is kept extremely small ($\gamma = 0.01$ cm^{-1}) in order to obtain the needed sharp structure in the spectrum. This seems inconsistent with observed vibrational linewidths (10 cm^{-1}) and decay times. (2) The sharp structure is also a result of recurrences in the correlator that will be (self) damped once the presence of more than a single mode frequency is acknowledged. When points 1 and 2 are considered, and the transmission function is broadened out, it seems unlikely that a distribution of force constants will be such an effective means for distributing the rates.

The works of Agmon and Hopfield[17], and Young and Bowne[3c], are in a similar vein. Here, distributions in a generalized protein coordinate are separated from the reaction coordinate and used to modulate the barrier height. Non-equilibrium conformations of the protein which result in a transition state heme (and therefore low barriers) appear to be excluded from the solutions of Young and Bowne[3c]. Agmon and Hopfield let the potential surfaces of the entire protein be frozen at all temperatures. Thus, the iron relaxation out of the heme plane cannot be accounted for at low temperatures and the iron porphyrin displacement energy is not considered. This leads directly to a linear mapping function and a Gaussian $g(H)$. The predictions of the high temperature rates are also flawed as discussed by Young and Bowne[3c]. We would argue that fluctuations and disorder in the globular protein structure affect the high-spin deoxy heme much more than the low-spin Mb · CO and that distributions in conformation lead to distribution in the out of plane iron displacement. The optical spectra attest to this fact[7]. The effects of bounded diffusion on the melting transition and the high temperature limit are discussed in Ref. 17 and will be treated elsewhere in the context of the current model, which correctly predicts the high temperature rates.

9. SUMMARY AND FUTURE PROBLEMS

In summary, it appears that a simple intuitive picture of proximal and distal pocket work can go a long way toward explaining the observed rebinding data. It seems that the spread of the distribution in H is primarily determined by the iron-porphyrin disorder and that the distal pocket along with the Stokes shift due to linear coupling controls the mean barrier height and is therefore a primary source of regulatory control. We should not exclude the possibility of distributions in H_D. We have not treated that problem here, due to the added complexity and additional unknown parameters. One could imagine that, for pH in the range 5–7, a bimodal distribution in H_D might be appropriate.

Our understanding of the O_2 binding reaction is much more limited. It is somewhat surprising to find a wide spread in the values of H_D for the O_2 complexes, since the ligand is naturally bent in these cases. This may implicate the torsional degree of freedom or the "sticking" energy as important contributors to H_D.

Further studies of these issues, as well as the extrapolation of the low temperature data through the melting transition will present us with future challenges. I, for one, am now thoroughly convinced that the low temperature rebinding studies reflect unexpected details about the mechanism of action of proteins in the physiological regime.

ACKNOWLEDGEMENTS

I would like to thank my colleagues Vukica Srajer and Lou Reinisch for their important contributions to this work. Financial support was obtained through NSF (84-17712) and NIH (DK 35090). Hans, of course, deserves credit for inspiring us all.

REFERENCES

1. R.H. Austin, K. Beeson, L. Eisenstein, H. Frauenfelder, I.C. Gunsalus and V.P. Marshall, Phys. Rev. Lett. 32, 403 (1974).
2. R.H. Austin, K. Beeson, L. Eisenstein, H. Frauenfelder and I.C. Gunsalus, Biochem. 14, 5355 (1975).
3. (a) A. Ansari, J. Berendzen, S. Bowne, H. Frauenfelder, I. Iben, T. Sauke, E. Shyamsunder and R. Young, Proc. Natl. Acad. Sci. USA 82, 5000 (1985).
   (b) H. Frauenfelder, in *Structure, Dynamics, Interaction and Evolution of Biological Macromolecules*, C. Helene ed., (D. Reidel, New York, 1983), p. 227.
   (c) R.D. Young and S. Bowne, J. Chem. Phys. 81, 3730 (1984).
4. J. Klafter and M. Shlesinger, Proc. Natl. Acad. Sci. USA 83, 848 (1986).
5. K.T. Schomacker, V. Srajer and P.M. Champion, J. Chem. Phys. 86, 1796 (1987).
6. L. Reinisch, K.T. Schomacker and P.M. Champion, J. Chem. Phys. 87, 150 (1987).
7. V. Srajer, K.T. Schomacker and P.M. Champion, Phys. Rev. Lett. 57, 1267 (1986).
8. H. Frauenfelder, G. Petsko and D. Tsernoglou, Nature (London) 280, 558 (1979).
9. F. Parak and E. Knapp, Proc. Natl. Acad. Sci. USA 81, 7088 (1984).

10. W. Doster, D. Beece, S. Bowne, E. DiIorio, L. Eisenstein, H. Frauenfelder, L. Reinisch, E. Shyamsunder, K. Winterhalter and K. Yu, Biochem. 21, 4831 (1982).
11. The issue of averaging over $\langle H \rangle$ versus averaging over $\langle k \rangle$ will be discussed elsewhere. It is probably an average over k that is required for the high temperature limit and this will lead to slightly different results. The main impact will fall on the evaluation of H_B, the enthalpy of CO in the pocket (B-state) with respect to the gas phase (see Refs. 3c, 10).
12. L. Reinisch, Ph.D. Thesis, University of Illinois (1982).
13. A. Ansari, J. Berendzen, D. Braunstein, B. Cowen, H. Frauenfelder, M. Hong, I. Iben, J. Johnson, P. Ormos, T. Sauke, R. Scholl, A. Schulte, P. Steinbach, J. Vittitow and R. Young, Biophys. Chem. 26, 337 (1987).
14. M. Tsubaki, A. Hiwataski and Y. Ichikawa, Biochem. 25, 3563 (1986).
15. J. Kuriyan, S. Wilz, M. Karplus and G. Petsko, J. Mol. Biol. 192, 133 (1986).
16. E. Buhks and J. Jortner, J. Chem. Phys. 83, 4456 (1985).
17. N. Agmon and J.J. Hopfield, J. Chem. Phys. 79, 2042 (1983).
18. W. Bialek and R. Goldstein, Biophys. J. 48, 1027 (1985).
19. H.M. Lu and J.B. Page, Chem. Phys. Lett. 131, 87 (1986).

STUDENTS OF HANS FRAUENFELDER

1953	J.S. Lawson, Jr.
1957	Sidney Singer
1958	Nathan Levine, H. Ralph Lewis, R. Norman Peacock
1960	Renato Bobone, Jack D. Ullman
1961	William H. Rosenfeld, Eberhard von Goeler
1962	J. Bernard Blake, Seymour Margulies
1964	Frank A. Franz, David W. Hafemeister, David N. Pipkorn
1965	John W. Burton, Rollin J. Morrison
1966	Martin H. Garrell, Robert P. Godwin
1968	Arthur R. Cooke
1969	Dwight D. Cook, Thomas A. Tumolillo
1971	Bruce L. Chrisman, Joel L. Groves, Charles J. Henkin, Eli I. Rosenberg, Allison D. Russell
1974	Mary J. Potasek
1975	Robert H. Austin, Karl W. Beeson, James M. Potter
1976	Shirley S. Chan, Thomas M. Nordlund
1978	Neil A. Alberding
1979	Douglas M. Alde
1980	Albert H. Reynolds, Larry B. Sorensen
1981	Michael C. Marden, David E. Good
1982	Lou Reinisch, Daniel K. Beece, Kwok To Yue
1984	Samuel F. Bowne

1986	Erramilli Shyamsunder, Kimberly A. Bagley
1987	Icko E.T. Iben, Anjum Ansari,
1988	Mi Kyung Hong
Current	Joel Berendzen, David Braunstein, Benjamin Cowen, Judy Mourant, Bruce Johnson, Donald Lamb, Todd Sauke, Reinhard Scholl, Peter Steinbach, Joseph Vittitow

RESEARCH ASSOCIATES AND LONG-TERM VISITORS

1957	Angelo Rossi, Kessar Alexopoulos
1961	Harry Lipkin, Hendrik de Waard
1962	Peter G. Debrunner
1965	Robert Ingalls, Charles F. Perdrisat
1971	Eckard Münck
1972	Elias Greenbaum
1973	Laura Eisenstein, Moshe Pasternak
1977	Vincent Yuan, Peter Hänggi
1979	Wolfgang Doster
1981	Ronald Harper, Heinrich Roder
1982	Peter Langer, Robert D. Young
1986	Alfons Schulte, Pal Ormos, Stan Luck
1987	Aihua Xie

INTERNATIONAL SYMPOSIUM ON FRONTIERS IN SCIENCE

PARTICIPANTS

Hans Frauenfelder
Department of Physics, University of Illinois
1110 West Green Street, Urbana, IL 61801

Alben, J.O.	Department of Physiological Chemistry Ohio State University Columbus, OH 43210	USA
Alde, D.M.	MS D-456, P2 Division Los Alamos National Laboratory Los Alamos, NM 87545	USA
Alpert, D.	Center for Advanced Study University of Illinois Urbana, IL 61801	USA
Anderson, A.C.	Department of Physics University of Illinois Urbana, IL 61801	USA
Ansari, A.	Department of Physics University of Illinois Urbana, IL 61801	USA
Atac, M.	Research Division Fermi National Accelerator Laboratory Batavia, IL 60510	USA
Austin, R.H.	Department of Physics Princeton University Princeton, NJ 08544	USA
Bagley, K.	Department of Physics University of California-San Diego La Jolla, CA 92093	USA
Bardeen, J.	Department of Physics University of Illinois Urbana, IL 61801	USA
Baym, G.	Department of Physics University of Illinois Urbana, IL 61801	USA

Berendzen, J.	Department of Physics University of Illinois Urbana, IL 61801	USA
Bobone, R.	1036 University Place Schenectady, NY 12308	USA
Bohner, H.	Wiesentalstrasse 15 CH-8684 Hombrechtikon	SWITZ.
Braunstein, D.	Department of Physics University of Illinois Urbana, IL 61801	USA
Burton, J.W.	Box 62, Rt. 1 Mountainview Drive New Market, TN 37820	USA
Champion, P.	Department of Physics Northeastern University Boston, MA 02115	USA
Chan, S.S.	Department of Chemistry Rutgers University New Brunswick, NJ 08903	USA
Chance, B.	Department of Biochemistry & Biophysics University of Pennsylvania Philadelphia, PA 19104	USA
Cowen, B.	Department of Physics University of Illinois Urbana, IL 61801	USA
Debrunner, P.	Department of Physics University of Illinois Urbana, IL 61801	USA
DePasquali, G.	R.R. #1 Box 132 Sidney, IL 61877	USA
Di Iorio, E.	Biochemistry I ETH Zentrum Universitätstrasse 16 8092 Zurich	SWITZ.

Dlott, D.	Department of Chemistry University of Illinois Urbana, IL 61801	USA
Douzou, P.	Institut de Biol. Phys.-Chem. 13 rue Pierre et Marie Curie 75005 Paris	FRANCE
Drickamer, H.G.	Department of Chemical Engineering University of Illinois Urbana, IL 61801	USA
Ehrenberg, A.	Biophysics Institute University of Stockholm S106 91 Stockholm	SWEDEN
Eigen, M.	Max Planck Institute for Biophysical Chemistry D-3400 Göttingen	WEST GERMANY
Everhart, T.E.	Chancellor, Swanlund Bldg. University of Illinois 601 East John Street Champaign, IL 61820	USA
Faulkner, L.R.	Department of Chemistry University of Illinois Urbana, IL 61801	USA
Ford, K.W.	American Institute of Physics 335 East 45th Street New York, NY 10017	USA
Franz, F.A.	Academic Affairs and Research West Virginia Univeristy Morgantown, WV 26506	USA
Frauenfelder, U.	Schoof 52 NL 6581 SH Malden	NETHERLANDS
Goldanskii, V.I.	Institute of Chemical Physics USSR Academy of Sciences, Ulitsa Kosygina 4, Moscow 117334	USSR
Good, D.E.	AT&T Laboratory 1100 E. Warrenville Rd. Naperville, IL 60566	USA

Gratton, E.	Department of Physics University of Illinois Urbana, IL 61801	USA
Greenbaum, E.	Oak Ridge National Laboratory P.O. Box X Oak Ridge, TN 37830	USA
Grunberg-Manago, M.	Institute de Biol. Phys.-Chem. 13 Rue Pierre et Marie Curie 75005 Paris	FRANCE
Gunsalus, I.C.	Department of Biochemistry University of Illinois Urbana, IL 61801	USA
Hager, L.P.	Department of Biochemistry University of Illinois Urbana, IL 61801	USA
Henley, E.M.	College of Arts & Sciences University of Washington Seattle, WA 98195	USA
Hong, M.K.	Department of Physics University of Illinois Urbana, IL 61801	USA
Iben, I.	Department of Physics University of Illinois Urbana, IL 61801	USA
Johnson, B.	Department of Physics University of Illinois Urbana, IL 61801	USA
Jonas, J.	School of Chemical Sciences Univeristy of Illinois Urbana, IL 61801	USA
Karplus, M.	Department of Chemistry Harvard University Cambridge, MA 02138	USA
Keszthelyi, L.	Institute of Biophysics Hungarian Academy of Sciences Odesszai Krt. 62 701 Szeged	HUNGARY

Kruse, U. — Department of Physics, University of Illinois, Urbana, IL 61801, USA

LaMar, G. — Department of Chemistry, University of California, Davis, CA 95616, USA

Leggett, A.J. — Department of Physics, University of Illinois, Urbana, IL 61801, USA

Leonard, N. — Department of Chemistry, University of Illinois, Urbana, IL 61801, USA

Lewis, H.R. — Los Alamos National Laboratory, MS-F642, Los Alamos, NM 87545, USA

Lin, S.L. — Department of Physics, University of Illinois, Urbana, IL 61801, USA

Lipkin, H. — Department of Physics, Weizmann Institute of Science, 76100 Rehovot, ISRAEL

Luck, S. — Department of Physics, University of Illinois, Urbana, IL 61801, USA

Lüscher, E. — Department of Physics, Technical University-Munich, D-8046 Garching, W. GERMANY

Lustig, H. — The American Physical Society, 335 East 45th Street, New York, NY 10017, USA

Margulies, S. — Department of Physics, University of Illinois, Chicago, IL 60680, USA

May, M. — Brookhaven National Laboratory, 20 Pennsylvania Ave., Upton, NY 11973, USA

Mössbauer, R.L.	Department of Physics Technical University-Munich D-8046 Garching	W. GERMANY
Münck, E.	Freshwater Biological Institute College of Biological Science Navarre, MN 55392	USA
Nagle. D.E.	Meson Physics Division Los Alamos National Laboratory Los Alamos, NM 87545	USA
Noble, R.W.	Department of Medicine & Biochemistry State University of New York Buffalo, NY 14215	USA
Ormos, P.	Department of Physics University of Illinois Urbana, IL 61801	USA
Ossipyan, Y.A.	Solid State Physics Institute USSR Academy of Sciences 142432 Chernogolovka	USSR
Peacock, R.N.	8845 Elgin Drive Lafayette, CO 80026	USA
Petsko, G.A.	Department of Chemistry Massachusetts Institute of Technology Cambridge, MA 02139	USA
Phillips, G.	Department of Biochemistry Rice University Houston, TX 77005	USA
Pines, D.	Department of Physics University of Illinois Urbana, IL 61801	USA
Raether, M.R.	Department of Physics University of Illinois Urbana, IL 61801	USA
Reimer, P.	Department of Physics University of Illinois Urbana, IL 61801	USA

Reinisch, L.	Department of Physics Northeastern University Boston, MA 02115	USA
Roder, H.	Department of Biochemistry & Biophysics University of Pennsylvania Philadelphia, PA 19104	USA
Sard, R.D.	Department of Physics University of Illinois Urbana, IL 61801	USA
Sauke, T.	Department of Physics University of Illinois Urbana, IL 61801	USA
Scholl, R.	Department of Physics University of Illinois Urbana, IL 61801	USA
Schulte, A.	Department of Physics University of Illinois Urbana, IL 61801	USA
Schulten, K.	Department of Physics Technical University-Munich D-8046 Garching	W. GERMANY
Shapero, D.	Board of Physics & Astronomy National Research Council Washington, DC 20418	USA
Shyamsunder, E.	Department of Physics Princeton University Princeton, NJ 08544	USA
Simmons, R.O.	Department of Physics University of Illinois Urbana, IL 61801	USA
Slichter, C.P.	Department of Physics University of Illinois Urbana, IL 61801	USA
Sligar, S.G.	Department of Biochemistry University of Illinois Urbana, IL 61801	USA

Smarr, L.	National Center for Supercomputing Applic. University of Illinois Champaign, IL 61820	USA
Sorensen, L.B.	Department of Physics University of Washington Seattle, WA 98195	USA
Stapleton, H.J.	Department of Physics University of Illinois Urbana, IL 61801	USA
Steinbach, P.	Department of Physics University of Illinois Urbana, IL 61801	USA
Suslick, K.S.	Department of Chemistry University of Illinois Urbana, IL 61801	USA
Ullman, J.D.	Department of Physics Herbert Lehman College Bronx, NY 10468	USA
Vittitow, J.	Department of Physics University of Illinois Urbana, IL 61801	USA
Wattenberg, A.	Department of Physics University of Illinois Urbana, IL 61801	USA
Winterhalter, K.	Biochemistry I ETH Zentrum Universitatsstrasse 16 CH-8092 Zurich	SWITZ.
Wolynes, P.G.	Department of Chemistry University of Illinois Urbana, IL 61801	USA
Young, R.D.	Department of Physics University of Illinois Urbana, IL 61801	USA
Yuan, V.	AT-2 Division Los Alamos National Laboratory Los Alamos, NM 87545	USA

Yue, K.T. Department of Physics
Emory University
Atlanta, GA 30322 USA

AIP Conference Proceedings

		L.C. Number	ISBN
No. 1	Feedback and Dynamic Control of Plasmas – 1970	70-141596	0-88318-100-2
No. 2	Particles and Fields – 1971 (Rochester)	71-184662	0-88318-101-0
No. 3	Thermal Expansion – 1971 (Corning)	72-76970	0-88318-102-9
No. 4	Superconductivity in *d*- and *f*-Band Metals (Rochester, 1971)	74-18879	0-88318-103-7
No. 5	Magnetism and Magnetic Materials – 1971 (2 parts) (Chicago)	59-2468	0-88318-104-5
No. 6	Particle Physics (Irvine, 1971)	72-81239	0-88318-105-3
No. 7	Exploring the History of Nuclear Physics – 1972	72-81883	0-88318-106-1
No. 8	Experimental Meson Spectroscopy –1972	72-88226	0-88318-107-X
No. 9	Cyclotrons – 1972 (Vancouver)	72-92798	0-88318-108-8
No. 10	Magnetism and Magnetic Materials – 1972	72-623469	0-88318-109-6
No. 11	Transport Phenomena – 1973 (Brown University Conference)	73-80682	0-88318-110-X
No. 12	Experiments on High Energy Particle Collisions – 1973 (Vanderbilt Conference)	73-81705	0-88318-111–8
No. 13	π-π Scattering – 1973 (Tallahassee Conference)	73-81704	0-88318-112-6
No. 14	Particles and Fields – 1973 (APS/DPF Berkeley)	73-91923	0-88318-113-4
No. 15	High Energy Collisions – 1973 (Stony Brook)	73-92324	0-88318-114-2
No. 16	Causality and Physical Theories (Wayne State University, 1973)	73-93420	0-88318-115-0
No. 17	Thermal Expansion – 1973 (Lake of the Ozarks)	73-94415	0-88318-116-9
No. 18	Magnetism and Magnetic Materials – 1973 (2 parts) (Boston)	59-2468	0-88318-117-7
No. 19	Physics and the Energy Problem – 1974 (APS Chicago)	73-94416	0-88318-118-5
No. 20	Tetrahedrally Bonded Amorphous Semiconductors (Yorktown Heights, 1974)	74-80145	0-88318-119-3
No. 21	Experimental Meson Spectroscopy – 1974 (Boston)	74-82628	0-88318-120-7
No. 22	Neutrinos – 1974 (Philadelphia)	74-82413	0-88318-121-5
No. 23	Particles and Fields – 1974 (APS/DPF Williamsburg)	74-27575	0-88318-122-3
No. 24	Magnetism and Magnetic Materials – 1974 (20th Annual Conference, San Francisco)	75-2647	0-88318-123-1
No. 25	Efficient Use of Energy (The APS Studies on the Technical Aspects of the More Efficient Use of Energy)	75-18227	0-88318-124-X

No. 26	High-Energy Physics and Nuclear Structure – 1975 (Santa Fe and Los Alamos)	75-26411	0-88318-125-8
No. 27	Topics in Statistical Mechanics and Biophysics: A Memorial to Julius L. Jackson (Wayne State University, 1975)	75-36309	0-88318-126-6
No. 28	Physics and Our World: A Symposium in Honor of Victor F. Weisskopf (M.I.T., 1974)	76-7207	0-88318-127-4
No. 29	Magnetism and Magnetic Materials – 1975 (21st Annual Conference, Philadelphia)	76-10931	0-88318-128-2
No. 30	Particle Searches and Discoveries – 1976 (Vanderbilt Conference)	76-19949	0-88318-129-0
No. 31	Structure and Excitations of Amorphous Solids (Williamsburg, VA, 1976)	76-22279	0-88318-130-4
No. 32	Materials Technology – 1976 (APS New York Meeting)	76-27967	0-88318-131-2
No. 33	Meson-Nuclear Physics – 1976 (Carnegie-Mellon Conference)	76-26811	0-88318-132-0
No. 34	Magnetism and Magnetic Materials – 1976 (Joint MMM-Intermag Conference, Pittsburgh)	76-47106	0-88318-133-9
No. 35	High Energy Physics with Polarized Beams and Targets (Argonne, 1976)	76-50181	0-88318-134-7
No. 36	Momentum Wave Functions – 1976 (Indiana University)	77-82145	0-88318-135-5
No. 37	Weak Interaction Physics – 1977 (Indiana University)	77-83344	0-88318-136-3
No. 38	Workshop on New Directions in Mossbauer Spectroscopy (Argonne, 1977)	77-90635	0-88318-137-1
No. 39	Physics Careers, Employment and Education (Penn State, 1977)	77-94053	0-88318-138-X
No. 40	Electrical Transport and Optical Properties of Inhomogeneous Media (Ohio State University, 1977)	78-54319	0-88318-139-8
No. 41	Nucleon-Nucleon Interactions – 1977 (Vancouver)	78-54249	0-88318-140-1
No. 42	Higher Energy Polarized Proton Beams (Ann Arbor, 1977)	78-55682	0-88318-141-X
No. 43	Particles and Fields – 1977 (APS/DPF, Argonne)	78-55683	0-88318-142-8
No. 44	Future Trends in Superconductive Electronics (Charlottesville, 1978)	77-9240	0-88318-143-6
No. 45	New Results in High Energy Physics – 1978 (Vanderbilt Conference)	78-67196	0-88318-144-4
No. 46	Topics in Nonlinear Dynamics (La Jolla Institute)	78-57870	0-88318-145-2
No. 47	Clustering Aspects of Nuclear Structure and Nuclear Reactions (Winnepeg, 1978)	78-64942	0-88318-146-0
No. 48	Current Trends in the Theory of Fields (Tallahassee, 1978)	78-72948	0-88318-147-9

No. 49	Cosmic Rays and Particle Physics – 1978 (Bartol Conference)	79-50489	0-88318-148-7
No. 50	Laser-Solid Interactions and Laser Processing – 1978 (Boston)	79-51564	0-88318-149-5
No. 51	High Energy Physics with Polarized Beams and Polarized Targets (Argonne, 1978)	79-64565	0-88318-150-9
No. 52	Long-Distance Neutrino Detection – 1978 (C.L. Cowan Memorial Symposium)	79-52078	0-88318-151-7
No. 53	Modulated Structures – 1979 (Kailua Kona, Hawaii)	79-53846	0-88318-152-5
No. 54	Meson-Nuclear Physics – 1979 (Houston)	79-53978	0-88318-153-3
No. 55	Quantum Chromodynamics (La Jolla, 1978)	79-54969	0-88318-154-1
No. 56	Particle Acceleration Mechanisms in Astrophysics (La Jolla, 1979)	79-55844	0-88318-155-X
No. 57	Nonlinear Dynamics and the Beam-Beam Interaction (Brookhaven, 1979)	79-57341	0-88318-156-8
No. 58	Inhomogeneous Superconductors – 1979 (Berkeley Springs, W.V.)	79-57620	0-88318-157-6
No. 59	Particles and Fields – 1979 (APS/DPF Montreal)	80-66631	0-88318-158-4
No. 60	History of the ZGS (Argonne, 1979)	80-67694	0-88318-159-2
No. 61	Aspects of the Kinetics and Dynamics of Surface Reactions (La Jolla Institute, 1979)	80-68004	0-88318-160-6
No. 62	High Energy e^+e^- Interactions (Vanderbilt, 1980)	80-53377	0-88318-161-4
No. 63	Supernovae Spectra (La Jolla, 1980)	80-70019	0-88318-162-2
No. 64	Laboratory EXAFS Facilities – 1980 (Univ. of Washington)	80-70579	0-88318-163-0
No. 65	Optics in Four Dimensions – 1980 (ICO, Ensenada)	80-70771	0-88318-164-9
No. 66	Physics in the Automotive Industry – 1980 (APS/AAPT Topical Conference)	80-70987	0-88318-165-7
No. 67	Experimental Meson Spectroscopy – 1980 (Sixth International Conference, Brookhaven)	80-71123	0-88318-166-5
No. 68	High Energy Physics – 1980 (XX International Conference, Madison)	81-65032	0-88318-167-3
No. 69	Polarization Phenomena in Nuclear Physics – 1980 (Fifth International Symposium, Santa Fe)	81-65107	0-88318-168-1
No. 70	Chemistry and Physics of Coal Utilization – 1980 (APS, Morgantown)	81-65106	0-88318-169-X
No. 71	Group Theory and its Applications in Physics – 1980 (Latin American School of Physics, Mexico City)	81-66132	0-88318-170-3
No. 72	Weak Interactions as a Probe of Unification (Virginia Polytechnic Institute – 1980)	81-67184	0-88318-171-1
No. 73	Tetrahedrally Bonded Amorphous Semiconductors (Carefree, Arizona, 1981)	81-67419	0-88318-172-X

No. 74	Perturbative Quantum Chromodynamics (Tallahassee, 1981)	81-70372	0-88318-173-8
No. 75	Low Energy X-Ray Diagnostics – 1981 (Monterey)	81-69841	0-88318-174-6
No. 76	Nonlinear Properties of Internal Waves (La Jolla Institute, 1981)	81-71062	0-88318-175-4
No. 77	Gamma Ray Transients and Related Astrophysical Phenomena (La Jolla Institute, 1981)	81-71543	0-88318-176-2
No. 78	Shock Waves in Condensed Matter – 1981 (Menlo Park)	82-70014	0-88318-177-0
No. 79	Pion Production and Absorption in Nuclei – 1981 (Indiana University Cyclotron Facility)	82-70678	0-88318-178-9
No. 80	Polarized Proton Ion Sources (Ann Arbor, 1981)	82-71025	0-88318-179-7
No. 81	Particles and Fields –1981: Testing the Standard Model (APS/DPF, Santa Cruz)	82-71156	0-88318-180-0
No. 82	Interpretation of Climate and Photochemical Models, Ozone and Temperature Measurements (La Jolla Institute, 1981)	82-71345	0-88318-181-9
No. 83	The Galactic Center (Cal. Inst. of Tech., 1982)	82-71635	0-88318-182-7
No. 84	Physics in the Steel Industry (APS/AISI, Lehigh University, 1981)	82-72033	0-88318-183-5
No. 85	Proton-Antiproton Collider Physics –1981 (Madison, Wisconsin)	82-72141	0-88318-184-3
No. 86	Momentum Wave Functions – 1982 (Adelaide, Australia)	82-72375	0-88318-185-1
No. 87	Physics of High Energy Particle Accelerators (Fermilab Summer School, 1981)	82-72421	0-88318-186-X
No. 88	Mathematical Methods in Hydrodynamics and Integrability in Dynamical Systems (La Jolla Institute, 1981)	82-72462	0-88318-187-8
No. 89	Neutron Scattering – 1981 (Argonne National Laboratory)	82-73094	0-88318-188-6
No. 90	Laser Techniques for Extreme Ultraviolt Spectroscopy (Boulder, 1982)	82-73205	0-88318-189-4
No. 91	Laser Acceleration of Particles (Los Alamos, 1982)	82-73361	0-88318-190-8
No. 92	The State of Particle Accelerators and High Energy Physics (Fermilab, 1981)	82-73861	0-88318-191-6
No. 93	Novel Results in Particle Physics (Vanderbilt, 1982)	82-73954	0-88318-192-4
No. 94	X-Ray and Atomic Inner-Shell Physics – 1982 (International Conference, U. of Oregon)	82-74075	0-88318-193-2
No. 95	High Energy Spin Physics – 1982 (Brookhaven National Laboratory)	83-70154	0-88318-194-0
No. 96	Science Underground (Los Alamos, 1982)	83-70377	0-88318-195-9

No. 97	The Interaction Between Medium Energy Nucleons in Nuclei – 1982 (Indiana University)	83-70649	0-88318-196-7
No. 98	Particles and Fields – 1982 (APS/DPF University of Maryland)	83-70807	0-88318-197-5
No. 99	Neutrino Mass and Gauge Structure of Weak Interactions (Telemark, 1982)	83-71072	0-88318-198-3
No. 100	Excimer Lasers – 1983 (OSA, Lake Tahoe, Nevada)	83-71437	0-88318-199-1
No. 101	Positron-Electron Pairs in Astrophysics (Goddard Space Flight Center, 1983)	83-71926	0-88318-200-9
No. 102	Intense Medium Energy Sources of Strangeness (UC-Sant Cruz, 1983)	83-72261	0-88318-201-7
No. 103	Quantum Fluids and Solids – 1983 (Sanibel Island, Florida)	83-72440	0-88318-202-5
No. 104	Physics, Technology and the Nuclear Arms Race (APS Baltimore –1983)	83-72533	0-88318-203-3
No. 105	Physics of High Energy Particle Accelerators (SLAC Summer School, 1982)	83-72986	0-88318-304-8
No. 106	Predictability of Fluid Motions (La Jolla Institute, 1983)	83-73641	0-88318-305-6
No. 107	Physics and Chemistry of Porous Media (Schlumberger-Doll Research, 1983)	83-73640	0-88318-306-4
No. 108	The Time Projection Chamber (TRIUMF, Vancouver, 1983)	83-83445	0-88318-307-2
No. 109	Random Walks and Their Applications in the Physical and Biological Sciences (NBS/La Jolla Institute, 1982)	84-70208	0-88318-308-0
No. 110	Hadron Substructure in Nuclear Physics (Indiana University, 1983)	84-70165	0-88318-309-9
No. 111	Production and Neutralization of Negative Ions and Beams (3rd Int'l Symposium, Brookhaven, 1983)	84-70379	0-88318-310-2
No. 112	Particles and Fields – 1983 (APS/DPF, Blacksburg, VA)	84-70378	0-88318-311-0
No. 113	Experimental Meson Spectroscopy – 1983 (Seventh International Conference, Brookhaven)	84-70910	0-88318-312-9
No. 114	Low Energy Tests of Conservation Laws in Particle Physics (Blacksburg, VA, 1983)	84-71157	0-88318-313-7
No. 115	High Energy Transients in Astrophysics (Santa Cruz, CA, 1983)	84-71205	0-88318-314-5
No. 116	Problems in Unification and Supergravity (La Jolla Institute, 1983)	84-71246	0-88318-315-3
No. 117	Polarized Proton Ion Sources (TRIUMF, Vancouver, 1983)	84-71235	0-88318-316-1

No. 118	Free Electron Generation of Extreme Ultraviolet Coherent Radiation (Brookhaven/OSA, 1983)	84-71539	0-88318-317-X
No. 119	Laser Techniques in the Extreme Ultraviolet (OSA, Boulder, Colorado, 1984)	84-72128	0-88318-318-8
No. 120	Optical Effects in Amorphous Semiconductors (Snowbird, Utah, 1984)	84-72419	0-88318-319-6
No. 121	High Energy e^+e^- Interactions (Vanderbilt, 1984)	84-72632	0-88318-320-X
No. 122	The Physics of VLSI (Xerox, Palo Alto, 1984)	84-72729	0-88318-321-8
No. 123	Intersections Between Particle and Nuclear Physics (Steamboat Springs, 1984)	84-72790	0-88318-322-6
No. 124	Neutron-Nucleus Collisions – A Probe of Nuclear Structure (Burr Oak State Park - 1984)	84-73216	0-88318-323-4
No. 125	Capture Gamma-Ray Spectroscopy and Related Topics – 1984 (Internat. Symposium, Knoxville)	84-73303	0-88318-324-2
No. 126	Solar Neutrinos and Neutrino Astronomy (Homestake, 1984)	84-63143	0-88318-325-0
No. 127	Physics of High Energy Particle Accelerators (BNL/SUNY Summer School, 1983)	85-70057	0-88318-326-9
No. 128	Nuclear Physics with Stored, Cooled Beams (McCormick's Creek State Park, Indiana, 1984)	85-71167	0-88318-327-7
No. 129	Radiofrequency Plasma Heating (Sixth Topical Conference, Callaway Gardens, GA, 1985)	85-48027	0-88318-328-5
No. 130	Laser Acceleration of Particles (Malibu, California, 1985)	85-48028	0-88318-329-3
No. 131	Workshop on Polarized ^{3}He Beams and Targets (Princeton, New Jersey, 1984)	85-48026	0-88318-330-7
No. 132	Hadron Spectroscopy–1985 (International Conference, Univ. of Maryland)	85-72537	0-88318-331-5
No. 133	Hadronic Probes and Nuclear Interactions (Arizona State University, 1985)	85-72638	0-88318-332-3
No. 134	The State of High Energy Physics (BNL/SUNY Summer School, 1983)	85-73170	0-88318-333-1
No. 135	Energy Sources: Conservation and Renewables (APS, Washington, DC, 1985)	85-73019	0-88318-334-X
No. 136	Atomic Theory Workshop on Relativistic and QED Effects in Heavy Atoms	85-73790	0-88318-335-8
No. 137	Polymer-Flow Interaction (La Jolla Institute, 1985)	85-73915	0-88318-336-6
No. 138	Frontiers in Electronic Materials and Processing (Houston, TX, 1985)	86-70108	0-88318-337-4
No. 139	High-Current, High-Brightness, and High-Duty Factor Ion Injectors (La Jolla Institute, 1985)	86-70245	0-88318-338-2

No. 140	Boron-Rich Solids (Albuquerque, NM, 1985)	86-70246	0-88318-339-0
No. 141	Gamma-Ray Bursts (Stanford, CA, 1984)	86-70761	0-88318-340-4
No. 142	Nuclear Structure at High Spin, Excitation, and Momentum Transfer (Indiana University, 1985)	86-70837	0-88318-341-2
No. 143	Mexican School of Particles and Fields (Oaxtepec, México, 1984)	86-81187	0-88318-342-0
No. 144	Magnetospheric Phenomena in Astrophysics (Los Alamos, 1984)	86-71149	0-88318-343-9
No. 145	Polarized Beams at SSC & Polarized Antiprotons (Ann Arbor, MI & Bodega Bay, CA, 1985)	86-71343	0-88318-344-7
No. 146	Advances in Laser Science–I (Dallas, TX, 1985)	86-71536	0-88318-345-5
No. 147	Short Wavelength Coherent Radiation: Generation and Applications (Monterey, CA, 1986)	86-71674	0-88318-346-3
No. 148	Space Colonization: Technology and The Liberal Arts (Geneva, NY, 1985)	86-71675	0-88318-347-1
No. 149	Physics and Chemistry of Protective Coatings (Universal City, CA, 1985)	86-72019	0-88318-348-X
No. 150	Intersections Between Particle and Nuclear Physics (Lake Louise, Canada, 1986)	86-72018	0-88318-349-8
No. 151	Neural Networks for Computing (Snowbird, UT, 1986)	86-72481	0-88318-351-X
No. 152	Heavy Ion Inertial Fusion (Washington, DC, 1986)	86-73185	0-88318-352-8
No. 153	Physics of Particle Accelerators (SLAC Summer School, 1985) (Fermilab Summer School, 1984)	87-70103	0-88318-353-6
No. 154	Physics and Chemistry of Porous Media—II (Ridge Field, CT, 1986)	83-73640	0-88318-354-4
No. 155	The Galactic Center: Proceedings of the Symposium Honoring C. H. Townes (Berkeley, CA, 1986)	86-73186	0-88318-355-2
No. 156	Advanced Accelerator Concepts (Madison, WI, 1986)	87-70635	0-88318-358-0
No. 157	Stability of Amorphous Silicon Alloy Materials and Devices (Palo Alto, CA, 1987)	87-70990	0-88318-359-9
No. 158	Production and Neutralization of Negative Ions and Beams (Brookhaven, NY, 1986)	87-71695	0-88318-358-7

No. 159	Applications of Radio-Frequency Power to Plasma: Seventh Topical Conference (Kissimmee, FL, 1987)	87-71812	0-88318-359-5
No. 160	Advances in Laser Science–II (Seattle, WA, 1986)	87-71962	0-88318-360-9
No. 161	Electron Scattering in Nuclear and Particle Science: In Commemoration of the 35th Anniversary of the Lyman-Hanson-Scott Experiment (Urbana, IL, 1986)	87-72403	0-88318-361-7
No. 162	Few-Body Systems and Multiparticle Dynamics (Crystal City, VA, 1987)	87-72594	0-88318-362-5
No. 163	Pion–Nucleus Physics: Future Directions and New Facilities at LAMPF (Los Alamos, NM, 1987)	87-72961	0-88318-363-3
No. 164	Nuclei Far from Stability: Fifth International Conference (Rosseau Lake, ON, 1987)	87-73214	0-88318-364-1
No. 165	Thin Film Processing and Characterization of High-Temperature Superconductors	87-73420	0-88318-365-X
No. 166	Photovoltaic Safety (Denver, CO, 1988)	88-42854	0-88318-366-8
No. 167	Deposition and Growth: Limits for Microelectronics (Anaheim, CA, 1987)	88-71432	0-88318-367-6
No. 168	Atomic Processes in Plasmas (Santa Fe, NM, 1987)	88-71273	0-88318-368-4
No. 169	Modern Physics in America: A Michelson-Morley Centennial Symposium (Cleveland, OH, 1987)	88-71348	0-88318-369-2
No. 170	Nuclear Spectroscopy of Astrophysical Sources (Washington, D.C., 1987)	88-71625	0-88318-370-6
No. 171	Vacuum Design of Advanced and Compact Synchrotron Light Sources (Upton, NY, 1988)	88-71824	0-88318-371-4
No. 172	Advances in Laser Science–III: Proceedings of the International Laser Science Conference (Atlantic City, NJ, 1987)	88-71879	0-88318-372-2
No. 173	Cooperative Networks in Physics Education (Oaxtepec, Mexico 1987)	88-72091	0-88318-373-0
No. 174	Radio Wave Scattering in the Interstellar Medium (San Diego, CA 1988)	88-72092	0-88318-374-9
No. 175	Non-neutral Plasma Physics (Washington, DC 1988)	88-72275	0-88318-375-7

No. 176	Intersections Between Particle and Nuclear Physics (Third International Conference) (Rockport, ME 1988)	88-62535	0-88318-376-5
No. 177	Linear Accelerator and Beam Optics Codes (La Jolla, CA 1988)	88-46074	0-88318-377-3
No. 178	Nuclear Arms Technologies in the 1990s (Washington, DC 1988)	88-83262	0-88318-378-1
No. 179	The Michelson Era in American Science: 1870–1930 (Cleveland, OH, 1987)	88-83369	0-88318-379-X